Orbits

Guochang Xu

Orbits

With 26 Figures and 6 Tables

Springer

Dr. Guochang Xu
GeoForschungsZentrum Potsdam
Dept. 1 Geodesy and Remote Sensing
Telegrafenberg A 17
14473 Potsdam
Germany
xu@gfz-potsdam.de

ISBN: 978-3-642-09728-7 e-ISBN: 978-3-540-78522-4

Softcover reprint of the hardcover 1st edition 2008

Cover design: deblik, Berlin

Printed on acid-free paper

9 8 7 6 5 4 3 2 1

springer.com

To
Liping, Jia, Yuxi, Pan and Yan

Preface

The purpose of this reference and handbook is to describe and to derive the analytic solutions of the equations of satellite motion perturbed by extraterrestrial and geopotential disturbances of the second order. The equations of satellite motion perturbed by extraterrestrial disturbances are solved by means of discretization and approximated potential function as well as Gaussian equations. The equations perturbed by geopotential disturbances are solved by symbolic mathematical operations. The traditional problem of singularity in the solutions is solved by so-called singularity-free orbit theory. Simplified disturbed equations of motion are proposed to simplify the solutions. Applications of the theory for analytic orbit determination are also discussed. Indeed, this is the first book since the satellite era, which describes systematically the orbit theory with analytical solutions, with respect to all of extraterrestrial and geopotential disturbances of the second order, and the solutions are free of singularity. Based on such a theory, the algorithms of orbit determination can be renewed; deeper insight into the physics of disturbances becomes possible; the way to a variety of new applications and refinements is opened.

My primary knowledge of the orbit theory came from my education of mathematics while studying physics and theoretical mechanics (1981). My first practical experience with orbit came from the research activity at the Technical University (TU) Berlin on orbit corrections of the satellite altimetry data (1988–1992). The extensive experience on orbit came from the GPS/Galileo software development for orbit determination and geopotential mapping at the GFZ (2001–2004). The traditional adjustment model of the solar radiation used in numerical orbit determination is investigated and considered not reasonable physically; and a new adjustment model is proposed in the user manual of the Multi-Functional GPS/Galileo software (MFGsoft) (Xu, 2004), which is also reported in the 2nd edition of the book *GPS – Theory, Algorithms and Applications* (Xu, 2007). Indeed, one of the ways to obtain the solutions of the extraterrestrial disturbances of the satellite motion is found during that investigation. However, it has not been realised until two scientists, Dr. Xiaochun Lu and Dr. Xiaohui Li of the National Time Service Center (NTSC) in Xi'an, came to visit and to cooperate with me at GFZ. We discussed the virtual navigation system and tried to solve the stability problem of the 3-D positioning of the system. By considering what is significant in theory and, what is more important than our numerical study, the idea of solving the disturbed equations of motion was obtained, and the solutions of the extraterrestrial disturbances of the equation of satellite motion were found. Because of the importance of the geopotential disturbances, great efforts were then made to derive the related solutions. Thereafter, alternative solutions of the extraterrestrial disturbances were found by using different means (besides the discretization, also approximated potential function and Gaussian disturbed equations). To simplify the solutions, the simplified disturbed equations were proposed. To solve the problem of singularity, the singularity-free theory was also developed.

After publishing my book, *GPS – Theory, Algorithms and Applications*, in 2003, I did not want to ever write another scientific book because this process took more than two years extreme hard work. However, I must finish this book because some of the scientists have contributed their lifetime to the theoretical solutions of the geopotential disturbances of the equation of satellite motion and now the results are here. The solutions of the extraterrestrial disturbances of the orbit motion are of extreme importance for practice, but they are rarely investigated because they are highly complex. From the theory, a special confusion related to the solar radiation from the pure numerical orbit determination has been cleared. Many interesting applications will follow soon. To make the process of writing easy, a small portion of the basic contents of my GPS book is partly modified and imported or rearranged and used.

The book includes ten chapters. After a brief introduction, the coordinate and time systems are described in the second chapter. The third chapter is dedicated to the Keplerian satellite orbits – the orbits of the satellite under the attraction of the central force of the Earth.

The fourth chapter deals with perturbations of the orbit. Perturbed equations of satellite motion are derived. Perturbation forces of the satellite motion are discussed in detail, including the perturbations of the Earth's gravitational field, Earth tide and ocean tide, the sun, the moon and planets, solar radiation pressure, and atmospheric drag, as well as coordinate perturbation.

The fifth chapter covers the analytic solution of $\bar{C}_{20}$ perturbation, including the complete formulas of the long term, and long and short periodic terms. The derivation also gives the algorithm and model of the orbit correction. The solutions of other geopotential disturbances of higher order and degree are described in the sixth chapter. As examples, solutions of disturbances of $\overline{C}_{30}$, D_{21} and D_{22} are given. General solutions of disturbance of D_{lm} are derived. Symbolic operation software for deriving solutions of geopotential disturbances of any order and degrees are designed and used.

The seventh chapter covers the solutions of extraterrestrial disturbances such as solar radiation pressure, atmospheric drag and the disturbances of the sun, the moon and planets. The principle and strategy that lead to the solution are described. The solutions are derived via discretization and approximated potential function as well as Gaussian perturbed equations of motion. Simplified disturbed equations are proposed and used partly. The ephemeris of the sun, the moon and planets are given for practical use.

The eighth chapter is dedicated to numerical orbit determination, including its principle, the algebraic solutions of the variation equations, and the numerical integration and interpolation algorithms, as well as the related derivatives.

The ninth chapter describes the principle of analytical orbit determination based on the proposed new solutions. Real time ability and properties of the analytic orbit solutions are discussed.

The final chapter includes algorithms that lead to singularity-free orbit theory and the equations of motion in non-inertial frame as well as discussions concerning the further development of the orbit theory and its applications as well as comments on some remaining problems.

The book has been subjected to an individual review of chapters and sections and a general review. I am grateful to reviewers Prof. Markus Rothacher of GFZ, Prof. Dieter Lelgemann of TU Berlin, Prof. Yuanxi Yang of the Institute of Surveying and Mapping (ISM) in Xi'an, Dr. Jianfeng Guo of Information Engineering University (IEU) in Zhengzhou, Prof. Xuhai Yang of NTSC in Xi'an, Dr. Junping Chen of GFZ. A grammatical check of technical English writing has been performed by Springer Heidelberg.

I wish to sincerely thank Prof. Markus Rothacher for his support and trust during my research activities at GFZ. Dr. Jürgen Kusche is thanked for his encouragement and help. Dr. Christoph Reigber is thanked for granting me special freedom of research. My grateful thanks go to Dr. Xiaochun Lu and Dr. Xiaohui Li of NTSC in Xi'an. Their visit to and cooperation at the GFZ have led to the derivations of the key contents of this book. Dr. Jiangfeng Guo of IEU in Zhengzhou followed a part of my derivation and checked for the correctness. Volker Grund of GFZ helped me greatly by assisting in the application of software tools, which is another key to the solution of geopotential disturbances. Qianxin Wang of GFZ helped to check a part of the formula typing. Dr. Jinghui Liu of the educational department of the Chinese Embassy in Berlin, Prof. Yuanxi Yang of ISM in Xi'an, Prof. Heping Sun of the Institute of Geodesy and Geophysics (IGG) in Wuhan and Prof. Qin Zhang of ChangAn University in Xi'an are thanked for their friendly support during my scientific activities in China. The Chinese Academy of Sciences is thanked for the Outstanding Overseas Chinese Scholars Fund, which supported greatly many valuable scientific activities even outside China.

During this work, many valuable discussions have been held with many scientists and friends. My special thanks go to Dr. Luisa Bastos of the Astronomical Observatory of University Porto, Dr. Rene Forsberg of Danish National Space Center, Prof. Jörg Reinking of Oldenburg University of Applied Sciences, Prof. Jikun Ou and Prof. Yunbin Yuan of IGG in Wuhan, Prof. Wu Chen of Hong Kong Polytechnic University, Prof. Yunzhong Shen of Tongji University in Shanghai, Dr. Yanxiong Liu of the First Oceanic Institute in Qingdao, Prof. Jiancheng Li of Wuhan University, Prof. Ta-Kang Yeh of the ChingYun University of Taiwan, Dr. Jürgen Neumeyer, Dr. Franz Barthelmes, and Dr. Svetozar Petrovic of GFZ, Dr. Uwe Meyer of GeoZentrum Hannover, Dr. Ludger Timmen of University Hannover, Dr. Xiong

Li of Hugro Inc. Houston, Dr. Daniela Morujao of Lisbon University, Prof. Klaus Hehl of Technical University of Applied Sciences Berlin, etc.

I also wish to sincerely thank Angelika Svarovsky and Hartmut Pflug of GFZ for their kind help. I am also grateful to Dr. Chris Bendall of Springer Heidelberg for his valuable advice.

My wife Liping, son Jia and daughters Yuxi, Pan and Yan are thanked for their constant support and understanding, as well as for their help.

October 2007 Guochang Xu

Contents

Abbreviations

AU	Astronomical Units
BDT	Barycentric Dynamic Time
CAS	Chinese Academy of Sciences
CIO	Conventional International Origin
CHAMP	Challenging Mini-satellite Payload
CRF	Conventional Reference Frame
CTS	Conventional Terrestrial System
DGK	Deutsche Geodätische Kommission
DGPS	Differential GPS
ECEF	Earth-Centred-Earth-Fixed (system)
ECI	Earth-Centred Inertial (system)
ECSF	Earth-Centred-Space-Fixed (system)
ESA	European Space Agency
EU	European Union
Galileo	Global Navigation Satellite System of EU
GAST	Greenwich Apparent Sidereal Time
GEO	Geostationary Earth Orbit (satellite)
GFZ	GeoForschungsZentrum Potsdam
GIS	Geographic Information System
GLONASS	Global Navigation Satellite System of Russia
GLOT	GLONASS time
GMST	Greenwich Mean Sidereal Time
GNSS	Global Navigation Satellite System
GPS	Global Positioning System
GPST	GPS Time
GRACE	Gravity Recovery and Climate Experiment
GRS	Geodetic Reference System
GST	Galileo System Time
IAG	International Association of Geodesy
IAT	International Atomic Time
IAU	International Astronomical Union

IERS	International Earth Rotation Service
IGG	Institute of Geodesy and Geophysics (CAS)
IGS	International GPS Geodynamics Service
INS	Inertial Navigation System
ION	Institute of Navigation
ITRF	IERS Terrestrial Reference Frame
IUGG	International Union for Geodesy and Geophysics
JD	Julian Date
JPL	Jet Propulsion Laboratory
KMS	National Survey and Cadastre (Denmark)
KSGsoft	Kinematic/Static GPS Software
LEO	Low Earth Orbit (satellite)
LS	Least Squares (adjustment)
mas	micro arc second
MEO	Medium Earth Orbit (satellite)
MFGsoft	Multi-Functional GPS/Galileo Software
MIT	Massachusetts Institute of Technology
MJD	Modified Julian Date
NASA	National Aeronautics and Space Administration
NAVSTAR	Navigation System with Time and Ranging
NGS	National Geodetic Survey
NTSC	National Time Service Center (CAS)
OD	Orbit Determination
PC	Personal Computer
PZ-90	Parameters of the Earth Year 1990
RMS	Root Mean Square
SLR	Satellite Laser Ranging
SNR	Signal-to-Noise Ratio
SST	Satellite-Satellite Tracking
SV	Space Vehicle
TAI	International Atomic Time
TDB	Barycentric Dynamic Time
TDT	Terrestrial Dynamic Time
TJD	Time of Julian Date
TOPEX	(Ocean) Topography Experiment
TOW	Time of Week
TRANSIT	Time Ranging and Sequential
TT	Terrestrial Time
UT	Universal Time
UTC	Universal Time Coordinated
UTC_{SU}	Moscow Time UTC
WGS	World Geodetic System
ZfV	Zeitschrift für Vermessungswesen

Constants

Symbol	Value	Unit	Explanation
a_e	6378137	m	Semi-major axis of WGS-84
f_e	1/298.2572236		Flattening factor of WGS-84
a_p	6378136	m	Semi-major axis of PZ-90
f_p	1/298.2578393		Flattening factor of PZ-90
a_{eI}	6378136.54	m	Semi-major axis of ITRF-96
f_{eI}	1/298.25645		Flattening factor of ITRF-96
ε	84381."412		Obliquity of the ecliptic at J2000.0
JDGPS	2444244.5	JD	Julian Date of GPS standard epoch (1980 Jan. 6, 0h)
JD2000.0	2451545.0	JD	Julian Date of 2000 January 1, 12h
G	6.67259e–11	$m^3s^{-2}kg^{-1}$	Constant of gravitation
μ_e	3.986004418e14	m^3s^{-2}	Geocentric gravitational constant
ω_e	7.292115e–5	$rads^{-1}$	Nominal mean angular velocity of the Earth
C	299792458	ms^{-1}	Speed of light
μ_s	1.327124e20	m^3s^{-2}	Heliocentric gravitational constant
μ_m	$\mu_e(M_m/M_e)$	m^3s^{-2}	Gravitational constant of the moon
M_m/M_e	0.0123000345		Moon-Earth mass ratio
h_2, h_3	0.6078, 0.292		Love number
l_2	0.0847		Shida number
P_s	4.5605e–6	Nm^{-1}	Luminosity of the sun
a_s	1.0000002AU	m	Semi-major axis of the orbit of the sun
AU	149597870691	m	Astronomical units
a_m	384401000	m	Semi-major axis of the orbit of the moon

Chapter 1
Introduction

The desire to understand the orbits of the planets has a history as long as that of mankind. How and why the planets orbit around the sun are questions in two categories. One focuses on geometry and the other on physics. However, without knowing the answer to why the how may not be answered theoretically; with the exception made by astronomical genius Kepler. After the Newton's second law, all the three Kepler's laws may be derived theoretically.

Without any doubt, the milestones of the orbit theory were crossed by Nicolaus Copernicus (1473–1543) with his heliocentric cosmology in "*De revolutionibus orbium coelestium*" (1543), Johannes Kepler (1571–1630) with his laws of planetary motion in "*Astronomia nova*" (1609), Isaac Newton (1643–1727) with his universal gravitation and laws of motion in "*Principia mathematica*" (1687). The Keplerian orbit describes the satellite (or planet) motion under the attracting of the central force of the Earth (or the sun). After the first satellite was launched in 1957, William Kaula (1926–2000) crossed the milestones with the first order solution of the equation of satellite motion disturbed by geopotential perturbations in "*Theory of Satellite Geodesy*" (1966). Thereafter, many scientists devoted themselves to the second order orbit solution of geopotential disturbances. The complexity of the theory is such that only a few people understand the theory, and the theory, in turn, is rarely applied in practice. Numerical orbit determination is developed directly to meet the needs of the satellite missions and to overcome the problem caused by the missing of analysis solutions of the equations of satellite motion.

Apparently most studies of the orbit theory are focused on the solution of the geopotential disturbances. Therefore, there exists a blank in literature on the solution of extraterrestrial disturbances. Meanwhile, it appears that the numerical algorithms are very robust and are not affected much by the obvious unphysical models and by the singularity caused by the parameterisation of the problem.

The descriptions of the Keplerian motion of the satellite under the influence of the central force of the Earth are perfect and exact, and have mathematical beauty (see Chap. 3). As soon as it is found that a satellite is moving in an orbital plane, the equations of motion are re-represented in the orbital plane and the Keplerian motion is then derived completely. Even in the case of central force field, without the coordinate transformation step, it would be nearly impossible to derive the solution. This indicates the extreme importance of the selection of the coordinate system.

G. Xu, *Orbits*,

Recall the Kaula's solution to satellite motion under the influence of the geopotential field. The equations of satellite motion are represented in inertial coordinate system according to Newton's law. However, the geopotential function is represented in the Earth-fixed system. To transform the geopotential function from the Earth-fixed system to the inertial one, a so-called Kaula's function is created, which is extremely complicated and leads to an extremely complicated solution. Some expressions of the solution are implicit. It is very difficult even to try to get the explicit expressions of the $\overline{C}_{20}$ solutions from the Kaula's solution. After Kaula's theory, studies on orbit theory are partly based on alternative variables, i.e., alternative coordinate systems.

The use of alternative coordinate systems is the first key to the solution of the extraterrestrial disturbances of the equation of motion. The approximation of the models of the disturbance forces is the second. Two adjustment models of the solar radiation and atmospheric drag used in numerical orbit determination are proposed by Xu (2004), in which the alternative coordinate systems are suggested and approximation methods to simplify the force (adjustment) models are given. The way to the solutions of the extraterrestrial disturbances is then open; however, this was realised first at the beginning of 2007. The solutions are then derived and given via discretization, approximated potential function and Gaussian equations.

The extraterrestrial disturbances are the second order ones. To derive complete solutions of the second order, the solutions of the geopotential disturbances are searched intensively. A method to derive the solutions of geopotential disturbance of l order and m degree is described generally and used to derive several examples. In GNSS orbit determination, only a few lower order and degrees of the geopotential disturbances need to be considered. Therefore, the examples given will be enough for the analytic orbit determination of satellites at higher altitudes. The higher the order and degrees of the geopotential disturbances are, the more complex the analytic solutions will be. Therefore, the application of the complete solution of geopotential disturbances will be a challenge for the future.

To describe a complete theory of the satellite orbit, coordinate and time systems, the Keplerian orbit of the satellite, have to be discussed (Chaps. 2 and 3). The perturbation forces such as gravitational field, tide, the sun, the moon, planets, solar radiation and atmospheric drag, etc. and the disturbed equations of the satellite motion also have to be discussed (Chap. 4). Then the solutions of $\overline{C}_{20}$ disturbance and other higher order and degree geopotential disturbances can be derived (Chaps. 5 and 6). The solution of extraterrestrial disturbances such as solar radiation pressure, atmospheric drag and the disturbance of the sun, the moon and planets, are then given (Chap. 7). Numerical orbit determination is dealt with (Chap. 8) before discussing the analytic orbit determination and application of the orbit theory (Chap. 9). Singularity-free orbit theory and discussions are given in the last chapter.

This book covers satellite orbit theory in theoretical and numerical aspects, with an emphasis on analytic solutions and applications. The analytic solutions of the extraterrestrial disturbances and the geopoptential disturbances, the singularity-free theory and simplified disturbed equations, are newly derived. The theory has opened and will further open very interesting research areas concerning satellite orbits. A part of the contents are refined theory, obtained from extensive research on individual problems. Because of the strong research and application background, the theories are conformably described with complexity. A brief summary of the contents is given in the Preface.

Some literature (books) is recommended for further reading: Kaula, 1966/2001; Chobotov, 1991; Cui, 1997; Montenbruck and Gill, 2000; Xu, 2003/2007.

Chapter 2
Coordinate and Time Systems

Satellites orbit around the Earth or travel in the planet system of the sun. They are generally observed from the Earth. To describe the orbits of the satellites (positions and velocities), suitable coordinate and time systems have to be defined.

2.1 Geocentric Earth-Fixed Coordinate Systems

It is convenient to use the Earth-Centred Earth-Fixed (ECEF) coordinate system to describe the location of a station on the Earth's surface. The ECEF coordinate system is a right-handed Cartesian system (x, y, z). Its origin and the Earth's centre of mass coincide, while its z-axis and the mean rotational axis of the Earth coincide; the x-axis points to the mean Greenwich meridian, while the y-axis is directed to complete a right-handed system (Fig. 2.1). In other words, the z-axis points to a mean pole of the Earth's rotation. Such a mean pole, defined by international convention, is called the Conventional International Origin (CIO). The xy-plane is called the mean equatorial plane, and the xz-plane is called the mean zero-meridian.

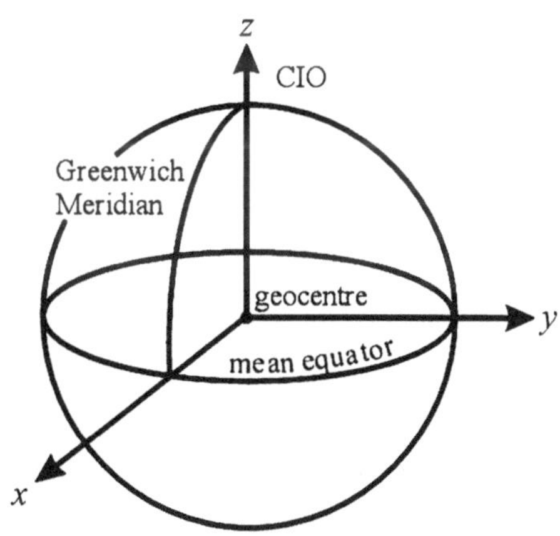

Fig. 2.1 Earth-Centred Earth-Fixed coordinates

The ECEF coordinate system is also known as the Conventional Terrestrial System (CTS). The mean rotational axis and mean zero-meridian used here are necessary. The true rotational axis of the Earth changes its direction all the time with respect to the Earth's body. If such a pole is used to define a coordinate system, then the coordinates of the station would also change all the time. Because the survey is made in our true world, it is obvious that the polar motion has to be taken into account and will be discussed later.

The ECEF coordinate system can, of course, be represented by a spherical coordinate system (r,ϕ,λ), where r is the radius of the point (x,y,z), and ϕ and λ are the geocentric latitude and longitude, respectively (Fig. 2.2). λ is counted eastward from the zero-meridian. The relationship between (x,y,z) and (r,ϕ,λ) is obvious:

$$\begin{pmatrix} x \\ y \\ z \end{pmatrix} = \begin{pmatrix} r\cos\phi\cos\lambda \\ r\cos\phi\sin\lambda \\ r\sin\phi \end{pmatrix} \quad \text{or} \quad \begin{cases} r = \sqrt{x^2+y^2+z^2}, \\ \tan\lambda = y/x, \\ \tan\phi = z/\sqrt{x^2+y^2}. \end{cases} \tag{2.1}$$

An ellipsoidal coordinate system (φ, λ, h) may also be defined on the basis of the ECEF coordinates; however, geometrically, two additional parameters are needed to define the shape of the ellipsoid (Fig. 2.3). φ, λ and h are geodetic latitude, longitude and height, respectively. The ellipsoidal surface is a rotational ellipse. The ellipsoidal system is also called the geodetic coordinate system. Geocentric longitude and geodetic longitude are identical. The two geometric parameters could be the semi-major radius (denoted by a) and the semi-minor radius (denoted by b) of the rotating ellipse, or the semi-major radius and the flattening (denoted by f) of the ellipsoid. They are equivalent sets of parameters. The relationship between (x,y,z) and (φ, λ, h) is (see, e.g., Torge, 1991):

$$\begin{pmatrix} x \\ y \\ z \end{pmatrix} = \begin{pmatrix} (N+h)\cos\varphi\cos\lambda \\ (N+h)\cos\varphi\sin\lambda \\ (N(1-e^2)+h)\sin\varphi \end{pmatrix} \tag{2.2}$$

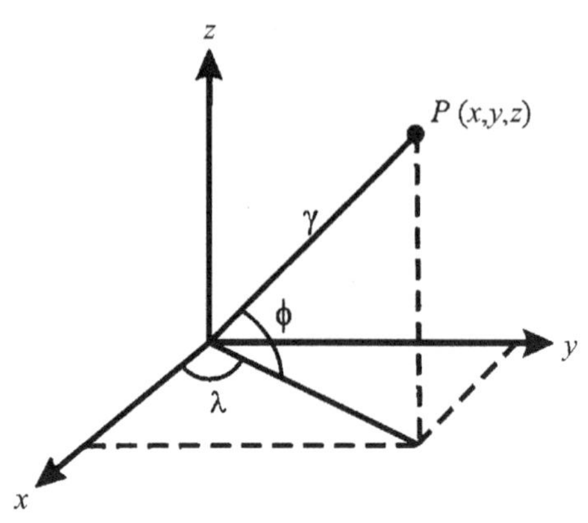

Fig. 2.2 Cartesian and spherical coordinates

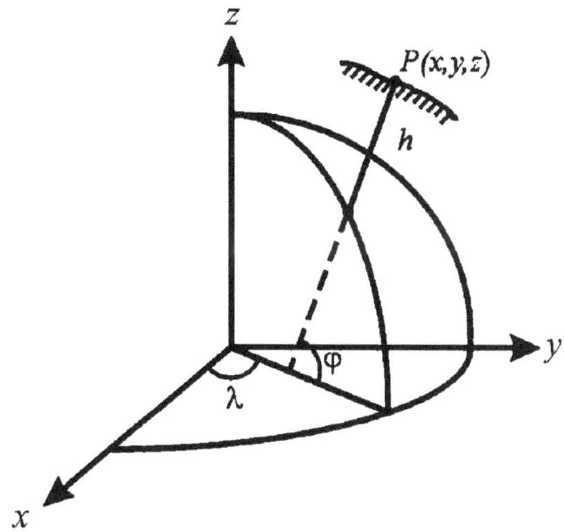

Fig. 2.3 Ellipsoidal coordinate system

or

$$\begin{cases} \tan\varphi = \dfrac{z}{\sqrt{x^2+y^2}}\left(1-e^2\frac{N}{N+h}\right)^{-1}, \\ \tan\lambda = \dfrac{y}{x}, \\ h = \dfrac{\sqrt{x^2+y^2}}{\cos\varphi} - N, \end{cases} \quad (2.3)$$

where

$$N = \frac{a}{\sqrt{1-e^2\sin^2\varphi}}. \quad (2.4)$$

N is the radius of curvature in the prime vertical, and e is the first eccentricity. The geometric meaning of N is shown in Fig. 2.4. In (2.3), the φ and h have to be solved by iteration; however, the iteration process converges quickly, since $h << N$. The flattening and the first eccentricity are defined as

$$f = \frac{a-b}{a} \quad \text{and} \quad e = \frac{\sqrt{a^2-b^2}}{a}. \quad (2.5)$$

In cases where $\varphi = \pm 90°$ or h is very large, the iteration formulas of (2.3) could be instable. Alternatively, using

$$\mathrm{c}\tan\varphi = \frac{\sqrt{x^2+y^2}}{z+\Delta z} \quad \text{and} \quad \Delta z = e^2 N\sin\varphi = \frac{ae^2\sin\varphi}{\sqrt{1-e^2\sin^2\varphi}},$$

may lead to a stably iterated result of φ (see Lelgemann, 2002). Δz and e^2N are the lengths of $\overline{OB}$ and $\overline{AB}$ (see Fig. 2.4), respectively. The geodetic height h can be obtained using Δz, i.e.,

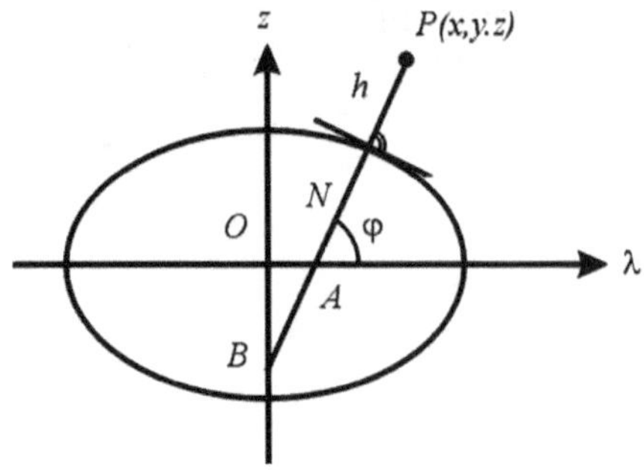

Fig. 2.4 Radius of curvature in the prime vertical

$$h = \sqrt{x^2 + y^2 + (z + \Delta z)^2} - N.$$

The two geometric parameters used in the World Geodetic System 1984 (WGS-84) are ($a = 6378137\,\mathrm{m}$, $f = 1/298.2572236$). In International Terrestrial Reference Frame 1996 (ITRF-96), the two parameters are ($a = 6378136.49\,\mathrm{m}$, $f = 1/298.25645$). ITRF uses the International Earth Rotation Service (IERS) Conventions (see McCarthy, 1996). In the PZ-90 (Parameters of the Earth Year 1990) coordinate system of GLONASS, the two parameters are ($a = 6378136\,\mathrm{m}$, $f = 1/298.2578393$).

The relation between the geocentric and geodetic latitude ϕ and φ (see (2.1) and (2.3)) may be given by

$$\tan\phi = \left(1 - e^2 \frac{N}{N+h}\right) \tan\varphi. \tag{2.6}$$

2.2 Coordinate System Transformations

Any Cartesian coordinate system can be transformed to another Cartesian coordinate system through three successive rotations if their origins are the same and if they are both right-handed or left-handed coordinate systems. These three rotational matrices are

$$R_1(\alpha) = \begin{pmatrix} 1 & 0 & 0 \\ 0 & \cos\alpha & \sin\alpha \\ 0 & -\sin\alpha & \cos\alpha \end{pmatrix},$$

$$R_2(\alpha) = \begin{pmatrix} \cos\alpha & 0 & -\sin\alpha \\ 0 & 1 & 0 \\ \sin\alpha & 0 & \cos\alpha \end{pmatrix}, \tag{2.7}$$

$$R_3(\alpha) = \begin{pmatrix} \cos\alpha & \sin\alpha & 0 \\ -\sin\alpha & \cos\alpha & 0 \\ 0 & 0 & 1 \end{pmatrix},$$

where α is the rotating angle, which has a positive sign for a counter-clockwise rotation as viewed from the positive axis to the origin. R_1, R_2, and R_3 are called the rotating matrix around the x, y, and z-axis, respectively. For any rotational matrix R, there are properties of $R^{-1}(\alpha) = R^T(\alpha)$ and $R^{-1}(\alpha) = R(-\alpha)$; that is, the rotational matrix is an orthogonal one, where R^{-1} and R^T are the inverse and transpose of the matrix R.

For two Cartesian coordinate systems with different origins and different length units, the general transformation can be given in vector (matrix) form as

$$X_n = X_0 + \mu R X_{\mathrm{old}} \tag{2.8}$$

or

$$\begin{pmatrix} x_n \\ y_n \\ z_n \end{pmatrix} = \begin{pmatrix} x_0 \\ y_0 \\ z_0 \end{pmatrix} + \mu R \begin{pmatrix} x_{\mathrm{old}} \\ y_{\mathrm{old}} \\ z_{\mathrm{old}} \end{pmatrix},$$

where μ is the scale factor (or the ratio of the two length units), and R is a transformation matrix that can be formed by three suitably successive rotations. x_n and x_{old} denote the new and old coordinates, respectively; x_0 denotes the translation vector and is the coordinate vector of the origin of the old coordinate system in the new one.

If rotational angle α is very small, then one has $\sin\alpha \approx \alpha$ and $\cos\alpha \approx 1$. In such a case, the rotational matrix can be simplified. If the three rotational angles α_1, α_2, α_3 in R of (2.8) are very small, then R can be written as

$$R = \begin{pmatrix} 1 & \alpha_3 & -\alpha_2 \\ -\alpha_3 & 1 & \alpha_1 \\ \alpha_2 & -\alpha_1 & 1 \end{pmatrix}, \tag{2.9}$$

where α_1, α_2, α_3 are small rotating angles around the x, y and z-axis, respectively (see, e.g., Lelgemann and Xu, 1991). Using the simplified R, the transformation (2.8) is called the Helmert transformation.

As an example, the transformation from WGS-84 to ITRF-90 (McCarthy, 1996) is given by:

$$\begin{pmatrix} x_{\mathrm{ITRF\text{-}90}} \\ y_{\mathrm{ITRF\text{-}90}} \\ z_{\mathrm{ITRF\text{-}90}} \end{pmatrix} = \begin{pmatrix} 0.060 \\ -0.517 \\ -0.223 \end{pmatrix} + \mu \begin{pmatrix} 1 & -0.0070'' & -0.0003'' \\ 0.0070'' & 1 & -0.0183'' \\ 0.0003'' & 0.0183'' & 1 \end{pmatrix} \begin{pmatrix} x_{\mathrm{WGS\text{-}84}} \\ y_{\mathrm{WGS\text{-}84}} \\ z_{\mathrm{WGS\text{-}84}} \end{pmatrix},$$

where $\mu = 0.999999989$, and the translation vector has the unit of meter.

The transformation between two coordinate systems can be generally represented by (2.8), where the scale factor $\mu = 1$ (i.e., the units of length used nowadays are the same). A formula of velocity transformation between different coordinate systems can be obtained by differentiating (2.8) with respect to the time.

2.3 Local Coordinate System

The local left-handed Cartesian coordinate system (x',y',z') can be defined by placing the origin to the local point $P_1(x_1,y_1,z_1)$, whose z'-axis is pointed to the vertical, x'-axis is directed to the north, and y' is pointed to the east (see Fig. 2.5). The $x'y'$-plane is called the horizontal plane; the vertical is defined perpendicular to the ellipsoid. Such a coordinate system is also called a local horizontal coordinate system. For any point P_2, whose coordinates in the global and local coordinate system are (x_2,y_2,z_2) and (x',y',z'), respectively, one has relations of

$$\begin{pmatrix} x' \\ y' \\ z' \end{pmatrix} = d \begin{pmatrix} \cos A \sin Z \\ \sin A \sin Z \\ \cos Z \end{pmatrix}, \quad \text{and} \quad \begin{pmatrix} d = \sqrt{x'^2 + y'^2 + z'^2} \\ \tan A = y'/x' \\ \cos Z = z'/d \end{pmatrix}, \tag{2.10}$$

where A is the azimuth, Z is the zenith distance and d is the radius of the P_2 in the local system. A is measured from the north clockwise; Z is the angle between the vertical and the radius d.

The local coordinate system (x',y',z') can indeed be obtained by two successive rotations of the global coordinate system (x,y,z) by $R_2(90° - \varphi)R_3(\lambda)$ and then by changing the x-axis to a right-handed system. In other words, the global system has to be rotated around the z-axis with angle λ, then around the y-axis with angle $90° - \varphi$, and then change the sign of the x-axis. The total transformation matrix R is then

$$R = \begin{pmatrix} -\sin\varphi\cos\lambda & -\sin\varphi\sin\lambda & \cos\varphi \\ -\sin\lambda & \cos\lambda & 0 \\ \cos\varphi\cos\lambda & \cos\varphi\sin\lambda & \sin\varphi \end{pmatrix}, \tag{2.11}$$

and there are

$$X_{\text{local}} = R X_{\text{global}} \quad \text{and} \quad X_{\text{global}} = R^T X_{\text{local}}, \tag{2.12}$$

where X_{local} and X_{global} are the same vector represented in local and global coordinate systems. (φ,λ) are the geodetic latitude and longitude of the local point.

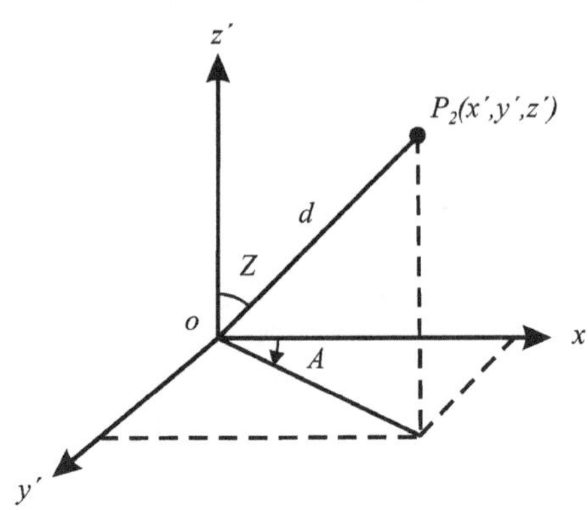

Fig. 2.5 Astronomical coordinate system

If the vertical direction is defined as the plumb line of the gravitational field at the local point, then such a local coordinate system is called an astronomic horizontal system (its x'-axis is pointed to the north, left-handed system). The plumb line of gravity g and the vertical line of the ellipsoid at the point p are generally not coinciding with each other; however, the difference is very small. The difference is omitted in GPS practice.

Combining (2.10) and (2.12), the zenith angle and azimuth of a point P_2 (satellite) related to the station P_1 can be directly computed by using the global coordinates of the two points by

$$\cos Z = \frac{z'}{d} \quad \text{and} \quad \tan A = \frac{y'}{x'}, \tag{2.13}$$

where

$$\begin{aligned} d &= \sqrt{(x_2-x_1)^2+(y_2-y_1)^2+(z_2-z_1)^2}, \\ x' &= -(x_2-x_1)\sin\varphi\cos\lambda-(y_2-y_1)\sin\varphi\sin\lambda+(z_2-z_1)\cos\varphi, \\ y' &= -(x_2-x_1)\sin\lambda+(y_2-y_1)\cos\lambda \end{aligned}$$

and

$$z' = (x_2-x_1)\cos\varphi\cos\lambda+(y_2-y_1)\cos\varphi\sin\lambda+(z_2-z_1)\sin\varphi.$$

2.4 Earth-Centred Inertial Coordinate System

To describe the motion of the GPS satellites, an inertial coordinate system has to be defined. The motion of the satellites follows the Newtonian mechanics, and the Newtonian mechanics is valid and expressed in an inertial coordinate system. For various reasons, the Conventional Celestial Reference Frame (CRF) is suitable for our purpose. The xy-plane of the CRF is the plane of the Earth's equator; the coordinates are celestial longitude, measured eastward along the equator from the vernal equinox, and celestial latitude. The vernal equinox is a crossover point of the ecliptic and the equator. So the right-handed Earth-centred inertial (ECI) system uses the Earth centre as the origin, CIO (Conventional International Origin) as the z-axis, and its x-axis is directed to the equinox of J2000.0 (Julian Date of 12h 1st January 2000). Such a coordinate system is also called equatorial coordinates of date. Because of the motion (acceleration) of the Earth's centre, ECI is indeed a quasi-inertial system, and the general relativistic effects have to be taken into account in this system. The system moves around the sun, however, without rotating with respect to the CIO. This system is also called the Earth-centred space-fixed (ECSF) coordinate system.

An excellent figure has been given by Torge (1991) to illustrate the motion of the Earth's pole with respect to the ecliptic pole (see Fig. 2.6). The Earth's flattening, combined with the obliquity of the ecliptic, results in a slow turning of the equator on the ecliptic due to the differential gravitational effect of the moon and the sun. The slow circular motion with a period of about 26000 years is called precession, and the other quicker motion with periods ranging from 14 days to 18.6 years is

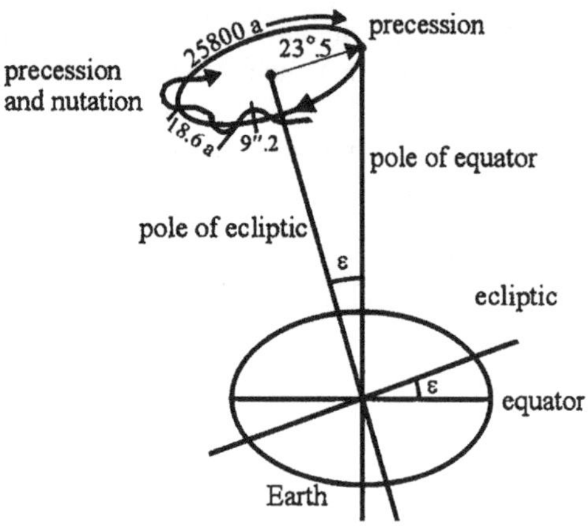

Fig. 2.6 Precession and nutation

called nutation. Taking the precession and nutation into account, the Earth's mean pole (related to the mean equator) is transformed to the Earth's true pole (related to the true equator). The x-axis of the ECI is pointed to the vernal equinox of date.

The angle of the Earth's rotation from the equinox of date to the Greenwich meridian is called Greenwich Apparent Sidereal Time (GAST). Taking GAST into account (called the Earth's rotation), the ECI of date is transformed to the true equatorial coordinate system. The difference between the true equatorial system and the ECEF system is the polar motion. So we have transformed the ECI system in a geometric way to the ECEF system. Such a transformation process can be written as

$$X_{\mathrm{ECEF}} = R_{\mathrm{M}} R_{\mathrm{S}} R_{\mathrm{N}} R_{\mathrm{P}} X_{\mathrm{ECI}}, \tag{2.14}$$

where R_{P} is the precession matrix, R_{N} is the nutation matrix, R_{S} is the Earth rotation matrix, R_{M} is the polar motion matrix, X is the coordinate vector, and indices ECEF and ECI denote the related coordinate systems.

Precession

The precession matrix consists of three successive rotational matrices, i.e. (see, e.g., Hofmann-Wellenhof et al., 1997/2001; Leick, 1995/2004; McCarthy, 1996),

$$\begin{aligned} R_{\mathrm{P}} &= R_3(-z)R_2(\theta)R_3(-\zeta) \\ &= \begin{pmatrix} \cos z\cos\theta\cos\zeta - \sin z\sin\zeta & -\cos z\cos\theta\sin\zeta - \sin z\cos\zeta & -\cos z\sin\theta \\ \sin z\cos\theta\cos\zeta + \cos z\sin\zeta & -\sin z\cos\theta\sin\zeta + \cos z\cos\zeta & -\sin z\sin\theta \\ \sin\theta\cos\zeta & -\sin\theta\sin\zeta & \cos\theta \end{pmatrix}, \end{aligned} \tag{2.15}$$

where z, θ, ζ are precession parameters and

$$\begin{aligned} z &= 2306.''2181T + 1.''09468T^2 + 0.''018203T^3, \\ \theta &= 2004.''3109T - 0.''42665T^2 - 0.''041833T^3 \end{aligned} \tag{2.16}$$

and

$$\zeta = 2306.''2181T + 0.''30188T^2 + 0.''017998T^3,$$

where T is the measuring time in Julian centuries (36525 days) counted from J2000.0 (see Sect. 2.8 time systems).

Nutation

The nutation matrix consists of three successive rotational matrices, i.e. (see, e.g., Hofmann-Wellenhof et al., 1997/2001; Leick, 1995/2004; McCarthy, 1996)

$$\begin{aligned} R_N &= R_1(-\varepsilon - \Delta\varepsilon)R_3(-\Delta\psi)R_1(\varepsilon) \\ &= \begin{pmatrix} \cos\Delta\psi & -\sin\Delta\psi\cos\varepsilon & -\sin\Delta\psi\sin\varepsilon \\ \sin\Delta\psi\cos\varepsilon_t & \cos\Delta\psi\cos\varepsilon_t\cos\varepsilon + \sin\varepsilon_t\sin\varepsilon & \cos\Delta\psi\cos\varepsilon_t\sin\varepsilon - \sin\varepsilon_t\cos\varepsilon \\ \sin\Delta\psi\sin\varepsilon_t & \cos\Delta\psi\sin\varepsilon_t\cos\varepsilon - \cos\varepsilon_t\sin\varepsilon & \cos\Delta\psi\sin\varepsilon_t\sin\varepsilon + \cos\varepsilon_t\cos\varepsilon \end{pmatrix} \\ &\approx \begin{pmatrix} 1 & -\Delta\psi\cos\varepsilon & -\Delta\psi\sin\varepsilon \\ \Delta\psi\cos\varepsilon_t & 1 & -\Delta\varepsilon \\ \Delta\psi\sin\varepsilon_t & \Delta\varepsilon & 1 \end{pmatrix}, \end{aligned} \tag{2.17}$$

where ε is the mean obliquity of the ecliptic angle of date, $\Delta\psi$ and $\Delta\varepsilon$ are nutation angles in longitude and obliquity, $\varepsilon_t = \varepsilon + \Delta\varepsilon$, and

$$\varepsilon = 84381.''448 - 46.''8150T - 0.''00059T^2 + 0.''001813T^3. \tag{2.18}$$

The approximation is made by letting $\cos\Delta\psi = 1$ and $\sin\Delta\psi = \Delta\psi$ for very small $\Delta\psi$. For precise purposes, the exact rotation matrix shall be used. The nutation parameters $\Delta\psi$ and $\Delta\varepsilon$ can be computed using the International Astronomical Union (IAU) theory or IERS theory:

$$\Delta\Psi = \sum_{i=1}^{106} (A_i + A_i'T)\sin\beta,$$

$$\Delta\varepsilon = \sum_{i=1}^{106} (B_i + B_i'T)\cos\beta,$$

or

$$\Delta\Psi = \sum_{i=1}^{263} (A_i + A_i'T)\sin\beta + A_i''\cos\beta,$$

$$\Delta\varepsilon = \sum_{i=1}^{263} (B_i + B_i'T)\cos\beta + B_i''\cos\beta,$$

where argument

$$\beta = N_{1i}l + N_{2i}l' + N_{3i}F + N_{4i}D + N_{5i}\Omega,$$

where l is the mean anomaly of the moon, l' is the mean anomaly of the sun, $F = L - \Omega, D$ is the mean elongation of the moon from the sun, Ω is the mean longitude of the ascending node of the moon, and L is the mean longitude of the moon. The formulas of l, l', F, D, and Ω, are given in Sect. 7.8. The coefficient values of N_{1i}, N_{2i}, N_{3i}, N_{4i}, N_{5i}, A_i, B_i, A'_i, B'_i, A''_i, and B''_i can be found in, e.g., McCarthy (1996). The updated formulas and tables can be found in updated IERS conventions. For convenience, the coefficients of the IAU 1980 nutation model are given in Appendix 1.

Earth Rotation

The Earth rotation matrix can be represented as

$$R_S = R_3(\text{GAST}), \tag{2.19}$$

where GAST is Greenwich Apparent Sidereal Time and

$$\text{GAST} = \text{GMST} + \Delta\Psi\cos\varepsilon + 0.''00264\sin\Omega + 0.''000063\sin 2\Omega, \tag{2.20}$$

where GMST is Greenwich Mean Sidereal Time. Ω is the mean longitude of the ascending node of the moon; the second term on the right-hand side is the nutation of the equinox. Furthermore,

$$\begin{aligned} \text{GMST} &= \text{GMST}_0 + \alpha\text{UT1}, \\ \text{GMST}_0 &= 6\times 3600.''0 + 41\times 60.''0 + 50.''54841 \\ &\quad + 8640184.''812866T_0 + 0.''093104T_0^2 - 6.''2\times 10^{-6}T_0^3, \\ \alpha &= 1.002737909350795 + 5.9006\times 10^{-11}T_0 - 5.9\times 10^{-15}T_0^2, \end{aligned} \tag{2.21}$$

where GMST_0 is Greenwich Mean Sidereal Time at midnight on the day of interest. α is the rate of change. UT1 is the polar motion corrected Universal Time (see Sect. 2.8). T_0 is the measuring time in Julian centuries (36525 days) counted from J2000.0 to 0h UT1 of the measuring day. By computing GMST, UT1 is used (see Sect. 2.8).

Polar Motion

As shown in Fig. 2.7, the polar motion is defined as the angles between the pole of date and the CIO pole. The polar motion coordinate system is defined by xy-plane coordinates, whose x-axis is pointed to the south and is coincided to the mean Greenwich meridian, and whose y-axis is pointed to the west. x_p and y_p are the angles of the pole of date, so the rotation matrix of polar motion can be represented as

Fig. 2.7 Polar motion

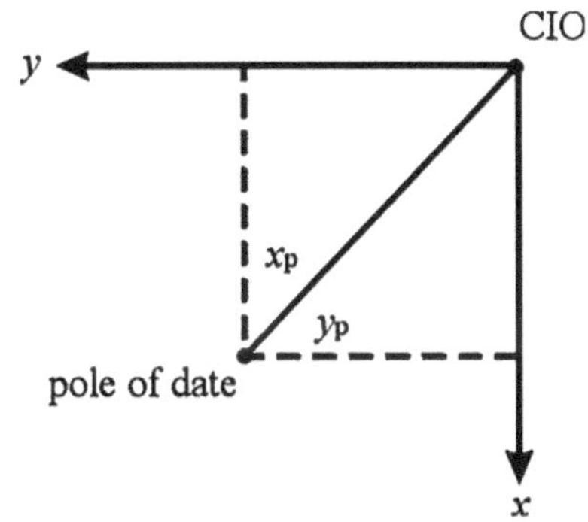

$$R_M = R_2(-x_p)R_1(-y_p) = \begin{pmatrix} \cos x_p & \sin x_p \sin y_p & \sin x_p \cos y_p \\ 0 & \cos y_p & -\sin y_p \\ -\sin x_p & \cos x_p \sin y_p & \cos x_p \cos y_p \end{pmatrix}$$
$$\approx \begin{pmatrix} 1 & 0 & x_p \\ 0 & 1 & -y_p \\ -x_p & y_p & 1 \end{pmatrix}. \tag{2.22}$$

The IERS determined x_p and y_p can be obtained from the home pages of IERS.

2.5 IAU 2000 Framework

At its 2000 General Assembly, the International Astronomical Union (IAU) adopted a set of resolutions that provide a consistent framework for defining the barycentric and geocentric celestial reference systems (Petit, 2002). The consequence of the resolution is that the coordinate transformation from celestial reference system (CRS, i.e., the ECI system) to the terrestrial reference system (TRS, i.e., the ECEF system) has the form

$$X_{\text{ECEF}} = R_M R_S R_{NP} X_{\text{ECI}}, \tag{2.23}$$

where R_{NP} is the precession-nutation matrix, R_S is the Earth rotation matrix, R_M is the polar motion matrix, X is the coordinate vector, and indices ECEF and ECI denote the related coordinate systems. The rotation matrices are functions of time T which is defined (see McCarthy and Petit, 2003) by

$$T = (\text{TT} - 2000\text{January 1d 12h TT}) \text{ in days/36525}, \tag{2.24}$$

where TT is the Terrestrial Time (for details see Sect. 2.8) and

$$\begin{aligned} R_M &= R_2(-x_p)R_1(-y_p)R_3(s'), \\ R_S &= R_3(\vartheta) \end{aligned} \tag{2.25}$$

and

$$R_{\rm NP} = R_3(-s)R_3(-E)R_2(d)R_3(E),$$

where $x_{\rm p}$ and $y_{\rm p}$ are the angles of the pole of date (or polar coordinates of the Celestial Intermediate Pole (CIP) in TRS), and s' is a function of $x_{\rm p}$ and $y_{\rm p}$:

$$s' = \frac{1}{2}\int_{T0}^{T}(x_{\rm p}\dot{y}_{\rm p} - \dot{x}_{\rm p}y_{\rm p})\mathrm{d}t$$

or approximately (see McCarthy and Capitaine, 2002)

$$s' = (-47\mu as)T, \tag{2.26}$$

where T is time in Julian Century counted from J2000.0 and

$$\vartheta = 2\pi(0.7790572732640 + 1.00273781191135448\,T_u), \tag{2.27}$$

where T_u = (Julian UT1 date − 2451545.0) and $\mathrm{UT1} = \mathrm{UTC} + (\mathrm{UT1} - \mathrm{UTC})\cdot$ $(\mathrm{UT1} - \mathrm{UTC})$ is published by the IERS.

E and d being such that the coordinates of the CIP in the CRS are

$$\begin{aligned} X &= \sin d \cos E, \\ Y &= \sin d \sin E, \\ Z &= \cos d. \end{aligned} \tag{2.28}$$

Equivalently $R_{\rm NP}$ can be given by

$$R_{\rm NP} = R_3(-s)\cdot\begin{pmatrix} 1-aX^2 & -aXY & X \\ -aXY & 1-aY^2 & Y \\ -X & -Y & 1-a(X^2+Y^2) \end{pmatrix}^{-1}, \tag{2.29}$$

where

$$a = \frac{1}{1+\cos d} \approx \frac{1}{2} + \frac{1}{8}(X^2+Y^2). \tag{2.30}$$

The developments of X and Y can be found on the website of the IERS Conventions and have the following form (in mas: microarcsecond) (Capitaine, 2002)

$$\begin{aligned} X = &-16616.99'' + 2004191742.88''T - 427219.05''T^2 \\ &- 198620.54''T^3 - 46.05''T^4 + 5.98''T^5 \\ &+ \textstyle\sum_i [(a_{s,0})_i \sin\beta + (a_{c,0})_i \cos\beta] \\ &+ \textstyle\sum_i [(a_{s,1})_i T \sin\beta + (a_{c,1})_i T \cos\beta] \\ &+ \textstyle\sum_i [(a_{s,2})_i T^2 \sin\beta + (a_{c,2})_i T^2 \cos\beta] + \cdots, \end{aligned} \tag{2.31}$$

$$
\begin{aligned}
Y = & -6950.78'' - 25381.99''T - 22407250.99''T^2 \\
& + 1842.28''T^3 - 1113.06''T^4 + 0.99''T^5 \\
& + \sum_i [(b_{s,0})_i \sin\beta + (b_{c,0})_i \cos\beta] \\
& + \sum_i [(b_{s,1})_i T \sin\beta + (b_{c,1})_i T \cos\beta] \\
& + \sum_i [(b_{s,2})_i T^2 \sin\beta + (b_{c,2})_i T^2 \cos\beta] + \cdots
\end{aligned}
\tag{2.32}
$$

s in (2.29) is the accumulated rotation, between the reference epoch and the date T, of CEO on the true equator due to the celestial motion of CIP, and can be expressed as

$$
s(T) = -\frac{1}{2}[X(T)Y(T) - X(T_0)Y(T_0)] + \int_{T_0}^{T} \dot{X}Y\mathrm{d}t - (\sigma_0 N_0 - \Sigma_0 N_0),
$$

where σ_0 and Σ_0 are the positions of CEO at J2000.0 and the x-origin of CRS, respectively and N_0 is the ascending node at J2000.0 in the equator of CRS. In above equation, terms$(T) + \frac{1}{2}[X(T)Y(T)]$ can be expressed as (in mas):

$$
\begin{aligned}
s + XY/2 = & \, 94.0 + 3808.35T - 119.94T^2 \\
& - 72574.09T^3 + 27.70T^4 + 15.61T^5 \\
& + \sum_i [(c_{s,0})_i \sin\beta + (c_{c,0})_i \cos\beta] \\
& + \sum_i [(c_{s,1})_i T \sin\beta + (c_{c,1})_i T \cos\beta] \\
& + \sum_i [(c_{s,2})_i T^2 \sin\beta + (c_{c,2})_i T^2 \cos\beta] + \cdots
\end{aligned}
\tag{2.33}
$$

In (2.31), (2.32) and (2.33), coefficients $(a_{s,j})_i, (a_{c,j})_i, (b_{s,j})_i, (b_{c,j})_i$ and $(c_{s,j})_i, (c_{c,j})_i$ can be extracted from table5.2a, table5.2b and table5.2c (available at ftp://tai.bipm.org/iers/conv2003/chapter5/). β is the combination of the fundamental arguments of nutation theory

$$
\beta = \sum_{j=1}^{14} N_j F_j. \tag{2.34}
$$

The first five F_j are the Delaunary variables l, l', F, D, Ω (given in Sect. 7.8); the amplitudes of sines and cosines β can be derived from the amplitudes of the precession and nutation series (see McCarthy and Petit, 2003); F_6 to F_{13} are the mean longitudes of the planets (Mercury to Neptune), including the Earth; F_{14} is the general precession in longitude. They are given in radians and T in Julian Centuries of TDB (see Sect. 2.8). The coefficients N_j are functions of index i and can be found in IERS website.

$$
\begin{aligned}
F_6 &= l_{Me} = 4.402608842 + 2608.7903141574T, \\
F_7 &= l_{Ve} = 3.176146697 + 1021.3285546211T, \\
F_8 &= l_E = 1.753470314 + 628.3075849991T,
\end{aligned}
$$

$$\begin{aligned}
F_9 &= l_{Ma} = 6.203480913 + 334.0612426700T, \\
F_{10} &= l_{Ju} = 0.599546497 + 52.9690962641T, \\
F_{11} &= l_{Sa} = 0.874016757 + 21.3299104960T, \\
F_{12} &= l_{Ur} = 5.481293872 + 7.4781598567T, \\
F_{13} &= l_{Ne} = 5.311886287 + 3.8133035638T, \\
F_{14} &= P_a = 0.024381750T + 0.00000538691T^2.
\end{aligned} \tag{2.35}$$

Using the new paradigm, the complete procedure of transforming the GCRS to the ITRS, which is compatible with the IAU2000 precession-nutation, is based on the expressions of (2.31), (2.32) and (2.33).

An equivalent way to realise the transformation between TRS and CRS under the definition of IAU 2000 can be implemented in a classical way by adding IAU2000 corrections to the corresponding rotating angles. Using the transformation formula (2.14), where the three precession rotating angles (see McCarthy and Petit, 2003) are

$$\begin{aligned}
z &= -2.5976176'' + 2306.0803226''T + 1.0947790''T^2 \\
&\quad + 0.0182273''T^3 + 0.0000470''T^4 - 0.0000003''T^5, \\
\theta &= 2004.1917476''T - 0.4269353''T^2 - 0.0418251''T^3 \\
&\quad - 0.0000601''T^4 - 0.0000001''T^5
\end{aligned} \tag{2.36}$$

and

$$\begin{aligned}
\zeta &= 2.5976176'' + 2306.0809506''T + 0.3019015''T^2 \\
&\quad + 0.0179663''T^3 - 0.0000327''T^4 - 0.0000002''T^5.
\end{aligned}$$

The IAU 2000 nutation model is given by series for nutation in longitude $\Delta\psi$ and obliquity $\Delta\varepsilon$, referred to the mean equator and equinox of date, with T measured in Julian centuries from epoch J2000.0:

$$\begin{aligned}
\Delta\psi &= \sum_{i=1}^{N} (A_i + A_i'T)\cos\beta + (A_i'' + A_i'''T)\cos\beta, \\
\Delta\varepsilon &= \sum_{i=1}^{N} (B_i + B_i'T)\cos\beta + (B_i'' + B_i'''T)\cos\beta,
\end{aligned} \tag{2.37}$$

where argument β can be found on the IERS website. For these two formulas, rate and bias corrections are necessary because of the new definition of the Celestial Intermediate Pole and the Celestial and Terrestrial ephemeris Origin:

$$\begin{aligned}
d\Delta\psi &= (-0.0166170 \pm 0.0000100)'' + (-0.29965 \pm 0.00040)''T, \\
d\Delta\varepsilon &= (-0.0068192 \pm 0.0000100)'' + (-0.02524 \pm 0.00010)''T.
\end{aligned} \tag{2.38}$$

The Earth rotation angle (i.e. the apparent Greenwich Sidereal Time GST or GAST) can be computed by adding a correction *EO* to the GMST in (2.27) (in mas)

$$\begin{aligned} EO = {} & 14506 + 4612157399.66T + 1396677.21T^2 - 93.44T^3 + 18.82T^4 \\ & + \Delta\psi \cos\varepsilon + \sum_i [(d_{s,0})_i \sin\beta + (d_{c,0})_i \cos\beta] \\ & + \sum_i [(d_{s,1})_i T \sin\beta + (d_{c,1})_i T \cos\beta] + \cdots, \end{aligned} \quad (2.39)$$

where coefficients $(d_{s,j})_i, (d_{c,j})_i$ can be extracted from table5.4 (available at ftp://tai.bipm.org/iers/conv2003/chapter5/). $\Delta\psi$ is defined in (2.37) and ε is defined in (2.18).

Similarly, the rotation matrix of polar motion shall be represented as the first formula of (2.25) and (2.26).

2.6 Geocentric Ecliptic Inertial Coordinate System

As discussed above, ECI uses the CIO pole in the space as the *z*-axis (through consideration of the polar motion, nutation and precession). If the ecliptic pole is used as the *z*-axis, then an ecliptic coordinate system is defined, and it may be called the Earth Centred Ecliptic Inertial (ECEI) coordinate system. ECEI places the origin at the mass centre of the Earth, its *z*-axis is directed to the ecliptic pole (or, the *xy*-plane is the mean ecliptic), and its *x*-axis is pointed to the vernal equinox of date. The coordinate transformation between the ECI and ECEI systems can be represented as

$$X_{\text{ECEI}} = R_1(-\varepsilon) X_{\text{ECI}}, \quad (2.40)$$

where ε is the ecliptic angle (mean obliquity) of the ecliptic plane related to the equatorial plane. The formula for ε is given in Sect. 2.4. Usually, coordinates of the sun and the moon, as well as planets, are given in the ECEI system.

2.7 Satellite Fixed Coordinate System

The orbit data, which describes the position of the satellite, is usually referred to the mass centre of the satellite. However, the orbit determination is usually measured through an instrument which is not exactly at the mass centre of the satellite. Therefore, a satellite fixed coordinate system is necessary to be defined for describing the position of the instrument (e.g., antenna or reflector). Such antenna centre correction (also called mass centre correction) has to be applied to the satellite coordinates in precise applications.

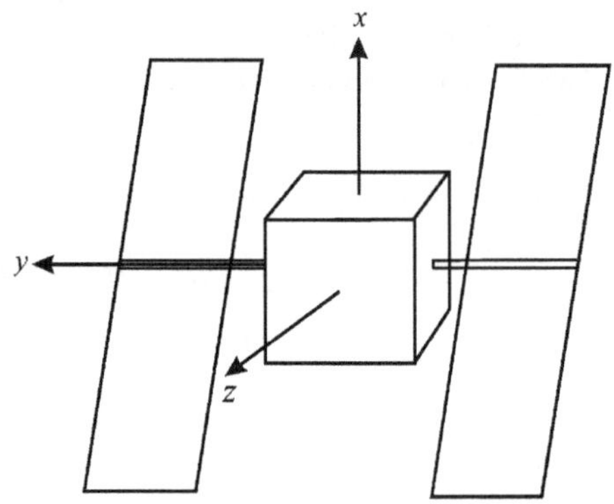

Fig. 2.8 Satellite fixed coordinate system

A satellite fixed coordinate system shall be set up for describing the antenna phase centre offset to the mass centre of the satellite. As shown in Fig. 2.8, the origin of the frame coincides with the mass centre of the satellite, the z-axis is parallel to the antenna pointing direction, the y-axis is parallel to the solar-panel axis, and the x-axis is selected to complete the right-handed frame. A solar vector is a vector from the satellite mass centre pointed to the sun. During the motion of the satellite, the z-axis is always pointing to the Earth, and the y-axis (solar-panel axis) shall be kept perpendicular to the solar vector. In other words, the y-axis is always perpendicular to the plane, which is formed by the sun, the Earth and satellite. The solar-panel can be rotated around its axis to keep the solar-panel perpendicular to the ray of the sun for optimally collecting the solar energy. The solar angle β is defined as the angle between the z-axis and the solar identity vector $\vec{n}_{\text{sun}}$ (see Fig. 2.9). Denoting the identity vector of the satellite fixed frame as $(\vec{e}_x, \vec{e}_y, \vec{e}_z)$, then the solar identity vector can be represented as

$$\vec{n}_{\text{sun}} = (\sin\beta, 0, \cos\beta). \tag{2.41}$$

β is needed for computation of the solar radiation pressure in orbit determination.

Denoting $\vec{r}$ as the geocentric satellite vector and $\vec{r}_{\text{s}}$ as the geocentric solar vector (Fig. 2.10),

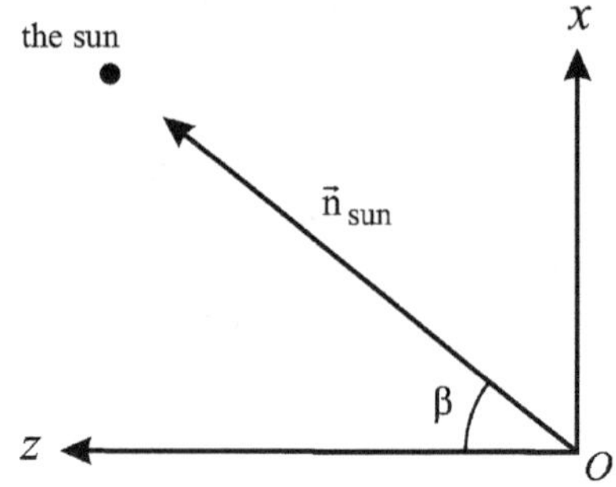

Fig. 2.9 The sun vector in satellite fixed frame

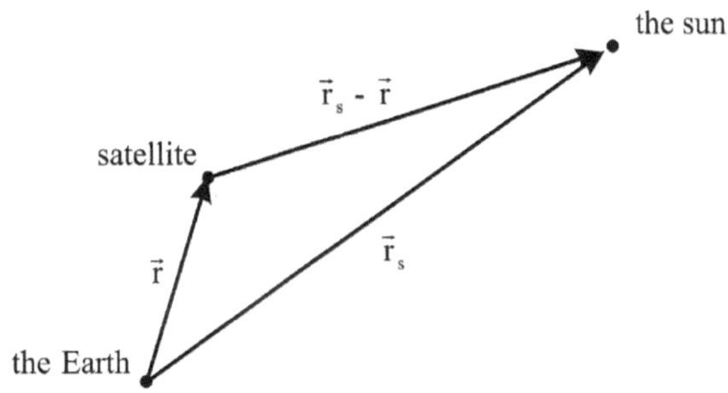

Fig. 2.10 The Earth-sun-satellite vectors

$$\vec{r} = \begin{pmatrix} X \\ Y \\ Z \end{pmatrix}, \qquad \vec{r}_S = \begin{pmatrix} X_{\text{sun}} \\ Y_{\text{sun}} \\ Z_{\text{sun}} \end{pmatrix}, \tag{2.42}$$

then in a geocentric coordinate system one has

$$\vec{e}_z = -\frac{\vec{r}}{|\vec{r}|}, \tag{2.43}$$

$$\vec{e}_y = \frac{\vec{e}_z \times \vec{n}_{\text{sun}}}{|\vec{e}_z \times \vec{n}_{\text{sun}}|},$$

$$\vec{e}_x = \vec{e}_y \times \vec{e}_z, \tag{2.44}$$

$$\vec{n}_{\text{sun}} = \frac{\vec{r}_s - \vec{r}}{|\vec{r}_s - \vec{r}|} \tag{2.45}$$

and

$$\cos\beta = \vec{n}_{\text{sun}} \cdot \vec{e}_z, \tag{2.46}$$

or

$$\vec{e}_z = \frac{-1}{r} \begin{pmatrix} X \\ Y \\ Z \end{pmatrix}, \quad r = \sqrt{X^2 + Y^2 + Z^2}, \tag{2.47}$$

$$\vec{n}_{\text{sun}} = \frac{1}{R} \begin{pmatrix} X_{\text{sun}} - X \\ Y_{\text{sun}} - Y \\ Z_{\text{sun}} - Z \end{pmatrix}, \tag{2.48}$$

$$\vec{e}_y = \frac{-1}{S} \begin{pmatrix} YZ_{\text{sun}} - Y_{\text{sun}}Z \\ ZX_{\text{sun}} - Z_{\text{sun}}X \\ XY_{\text{sun}} - X_{\text{sun}}Y \end{pmatrix} \tag{2.49}$$

Table 2.1 GPS satellite antenna phase centre offset

Satellite	x	y	z
Block I	0.2100	0.0	0.8540
Block II/IIA	0.2794	0.0	1.0259
Block IIR	0.0000	0.0	1.2053

and

$$\vec{e}_x = \frac{1}{S \cdot r} \begin{pmatrix} (ZX_{\text{sun}} - Z_{\text{sun}}X)Z - (XY_{\text{sun}} - X_{\text{sun}}Y)Y \\ (XY_{\text{sun}} - X_{\text{sun}}Y)X - (YZ_{\text{sun}} - Y_{\text{sun}}Z)Z \\ (YZ_{\text{sun}} - Y_{\text{sun}}Z)Y - (ZX_{\text{sun}} - Z_{\text{sun}}X)X \end{pmatrix}, \tag{2.50}$$

where

$$R = \sqrt{(X_{\text{sun}} - X)^2 + (Y_{\text{sun}} - Y)^2 + (Z_{\text{sun}} - Z)^2} \tag{2.51}$$

and

$$S = \sqrt{(YZ_{\text{sun}} - Y_{\text{sun}}Z)^2 + (ZX_{\text{sun}} - Z_{\text{sun}}X)^2 + (XY_{\text{sun}} - X_{\text{sun}}Y)^2}. \tag{2.52}$$

Suppose the satellite antenna phase centre in the satellite fixed frame is (x,y,z), then the offset vector in the geocentric frame can be obtained by substituting (2.47), (2.49) and (2.50) into the following formula:

$$\vec{d} = x\vec{e}_x + y\vec{e}_y + z\vec{e}_z, \tag{2.53}$$

which may be added to the vector $\vec{r}$.

GPS satellite antenna phase centre offsets in the satellite fixed frame are given in Table 2.1.

The dependence of the phase centre on the signal direction and frequencies is not considered for the satellite here. A mis-orientation of the $\vec{e}_y$ ($\vec{e}_x$ too) of the satellite with respect to the sun may cause errors in the geometrical phase centre correction. In the Earth's shadow (for up to 55 min), the mis-orientation becomes worse. The geometrical mis-orientation may be modelled and estimated.

2.8 Time Systems

The three time systems used in satellite surveying are sidereal time, dynamic time and atomic time (see, e.g., Hofmann-Wellenhof et al., 1997/2001; Leick, 1995/2004; McCarthy, 1996; King et al., 1987).

Sidereal time is a measure of the Earth's rotation and is defined as the hour angle of the vernal equinox. If the measure is counted from the Greenwich meridian, the sidereal time is called Greenwich Sidereal Time. Universal Time (UT) is the Greenwich hour angle of the apparent sun, which is orbiting uniformly in the equatorial plane. Because the angular velocity of the Earth's rotation is not a constant, sidereal time is not a uniformly-scaled time. The oscillation of UT is also partly caused by the polar motion of the Earth. The universal time corrected for the polar motion is denoted by UT1.

Dynamical time is a uniformly-scaled time used to describe the motion of bodies in a gravitational field. Barycentric Dynamic Time (TDB) is applied in an inertial coordinate system (its origin is located at the centre-of-mass (Barycentre)). Terrestrial Dynamic Time (TDT) is used in a quasi-inertial coordinate system (such as ECI). Because of the motion of the Earth around the sun (or say, in the sun's gravitational field), TDT will have a variation with respect to TDB. However, both the satellite and the Earth are subject to almost the same gravitational perturbations. TDT may be used for describing the satellite motion without taking into account the influence of the gravitational field of the sun. TDT is also called Terrestrial Time (TT).

Atomic Time is a time system kept by atomic clocks such as International Atomic Time (TAI). It is a uniformly-scaled time used in the ECEF coordinate system. TDT is realised by TAI in practice with a constant offset (32.184 s). Because of the slowing down of the Earth's rotation with respect to the sun, Coordinated Universal Time (UTC) is introduced to keep the synchronisation of TAI to the solar day (by inserting the leap seconds). GPS Time (GPST) is also atomic time.

The relationships between different time systems are given as follows:

$$\begin{aligned} \text{TAI} &= \text{GPST} + 19.0\,\text{sec}, \\ \text{TAI} &= \text{TDT} - 32.184\,\text{sec}, \\ \text{TAI} &= \text{UTC} + n\,\text{sec} \\ \text{UT1} &= \text{UTC} + \text{dUT1}, \end{aligned} \tag{2.54}$$

where dUT1 can be obtained by IERS, (dUT1 < 0.7 s, see Zhu et al., 1996), (dUT1 is also broadcasted with the navigation data), n is the number of leap seconds of date and is inserted into UTC on the 1st of January and 1st of July of the years. The actual n can be found in the IERS report.

Time argument T (Julian centuries) is used in the formulas given in Sect. 2.4. For convenience, T is denoted by TJD, and TJD can be computed from the civil date (Year, Month, Day, and Hour) as follows:

$$\text{JD} = \text{INT}(365.25Y) + \text{INT}(30.6001\,(M+1)) + \text{Day} + \text{Hour}/24 + 1720981.5$$

and

$$\text{TJD} = \text{JD}/36525, \tag{2.55}$$

where

$$\begin{aligned} &Y = \text{Year} - 1, \quad M = \text{Month} + 12, \quad \text{if Month} \le 2, \\ &Y = \text{Year}, \qquad\;\; M = \text{Month}, \qquad\quad\; \text{if Month} > 2, \end{aligned}$$

where JD is the Julian Date, Hour is the time of UT and INT denotes the integer part of a real number. The Julian Date counted from JD2000.0 is then JD2000 = JD–JD2000.0, where JD2000.0 is the Julian Date of 2000 January 1st 12h and has the value of 2451 545.0 days. One Julian century is 36 525 days.

Inversely, the civil date (Year, Month, Day and Hour) can be computed from the Julian Date (JD) as follows:

$$\begin{aligned}
b &= \mathrm{INT}(\mathrm{JD}+0.5)+1537, \\
c &= \mathrm{INT}\left(\frac{b-122.1}{365.25}\right), \\
d &= \mathrm{INT}(365.25c), \\
e &= \mathrm{INT}\left(\frac{b-d}{30.6001}\right), \\
\mathrm{Hour} &= \mathrm{JD}+0.5-\mathrm{INT}(\mathrm{JD}+0.5), \\
\mathrm{Day} &= b-d-\mathrm{INT}(30.6001e), \\
\mathrm{Month} &= e-1-12\mathrm{INT}\left(\frac{e}{14}\right)
\end{aligned}$$

and

$$\mathrm{Year} = c-4715-\mathrm{INT}\left(\frac{7+\mathrm{Month}}{10}\right), \tag{2.56}$$

where b, c, d, and e are auxiliary numbers.

Because the GPS standard epoch is defined as JD = 2444244.5 (1980 January 6, 0h), GPS week and the day of week (denoted by Week and N) can be computed by

$$N = \mathrm{modulo}(\mathrm{INT}(\mathrm{JD}+1.5), 7)$$

and

$$\mathrm{Week} = \mathrm{INT}\left(\frac{\mathrm{JD}-2444244.5}{7}\right), \tag{2.57}$$

where N is the day of week ($N = 0$ for Monday, $N = 1$ for Tuesday, and so on).

For saving digits and counting the date from midnight instead of noon, the Modified Julian Date (MJD) is defined as

$$\mathrm{MJD} = (\mathrm{JD}-2400000.5). \tag{2.58}$$

GLONASS time (GLOT) is defined by Moscow time $\mathrm{UTC_{SU}}$, which equals UTC plus three hours (corresponding to the offset of Moscow time to Greenwich time), theoretically. GLOT is permanently monitored and adjusted by the GLONASS Central Synchroniser (see Roßbach, 2006). UTC and GLOT then have a simple relation

$$\mathrm{UTC} = \mathrm{GLOT} + \tau_c - 3\mathrm{h},$$

where τ_c is the system time correction with respect to UTC_{SU}, which is broadcasted by the GLONASS ephemeris and is less than one microsecond. Therefore there is approximately

$$\mathrm{GPST} = \mathrm{GLOT} + m - 3\mathrm{h},$$

where m is the number of "leap seconds" between GPS and GLONASS (UTC) time and is given in the GLONASS ephemeris. m is indeed the leap seconds since GPS standard epoch (1980 January 6, 0h).

Galileo system time (GST) will be maintained by a number of UTC laboratory clocks. GST and GPST are time systems of various UTC laboratories. After the offset of GST and GPST is made available to the user, the interoperability will be ensured.

Chapter 3
Keplerian Orbits

Satellite motion can be considered a motion of the satellite under the central force field of the Earth and the disturbed motion caused by other perturbation forces. Therefore, the Keplerian orbits are important in orbit theory and will be discussed in this chapter.

3.1 Keplerian Motion

The simplified satellite orbiting is called Keplerian motion, and the problem is called the two-bodies problem. The satellite is supposed to move in a central force field. The equation of satellite motion is described by Newton's second law of motion by

$$\vec{f} = m \cdot a = m \cdot \ddot{\vec{r}}, \tag{3.1}$$

where $\vec{f}$ is the attracting force, m is the mass of the satellite, a, or alternatively, $\ddot{\vec{r}}$ is the acceleration of the motion (second order differentiation of vector $\vec{r}$ with respect to the time), and according to Newton's law

$$\vec{f} = -\frac{GMm}{r^2}\frac{\vec{r}}{r}, \tag{3.2}$$

where G is the universal gravitational constant, M is the mass of the Earth, r is the distance between the mass centre of the Earth and the mass centre of the satellite. The equation of satellite motion is then

$$\ddot{\vec{r}} = -\frac{\mu}{r^2}\frac{\vec{r}}{r}, \tag{3.3}$$

where $\mu(= GM)$ is called Earth's gravitational constant.

Equation (3.3) of satellite motion is valid only in an inertial coordinate system, so the ECSF coordinate system discussed in Sect. 2.4 will be used for describing the orbit of the satellite. The vector form of the equation of motion can be rewritten through the three components x, y and $z(\vec{r} = (x, y, z))$ as

G. Xu, *Orbits*,

$$\begin{aligned}
\ddot{x} &= -\frac{\mu}{r^3}x, \\
\ddot{y} &= -\frac{\mu}{r^3}y, \\
\ddot{z} &= -\frac{\mu}{r^3}z.
\end{aligned} \tag{3.4}$$

Multiplying y, z to the first equation of (3.4), and x, z to the second, x, y to the third, and then forming the differences of them, one gets

$$\begin{aligned}
y\ddot{z} - z\ddot{y} &= 0, \\
z\ddot{x} - x\ddot{z} &= 0, \\
x\ddot{y} - y\ddot{x} &= 0,
\end{aligned} \tag{3.5}$$

or in vector form:

$$\vec{r} \times \ddot{\vec{r}} = 0. \tag{3.6}$$

Equations (3.5) and (3.6) are equivalent to

$$\begin{aligned}
\frac{\mathrm{d}(y\dot{z} - z\dot{y})}{\mathrm{d}t} &= 0, \\
\frac{\mathrm{d}(z\dot{x} - x\dot{z})}{\mathrm{d}t} &= 0, \\
\frac{\mathrm{d}(x\dot{y} - y\dot{x})}{\mathrm{d}t} &= 0,
\end{aligned} \tag{3.7}$$

$$\frac{\mathrm{d}(\vec{r} \times \dot{\vec{r}})}{\mathrm{d}t} = 0. \tag{3.8}$$

Integrating (3.7) and (3.8) lead to

$$\begin{aligned}
y\dot{z} - z\dot{y} &= A, \\
z\dot{x} - x\dot{z} &= B, \\
x\dot{y} - y\dot{x} &= C,
\end{aligned} \tag{3.9}$$

$$\vec{r} \times \dot{\vec{r}} = \vec{h} = \begin{pmatrix} A \\ B \\ C \end{pmatrix}, \tag{3.10}$$

where A, B, C are integration constants; they form the integration constant vector $\vec{h}$. That is

$$h = \sqrt{A^2 + B^2 + C^2} = \left|\vec{r} \times \dot{\vec{r}}\right|. \tag{3.11}$$

The constant h is two times of the area that the radius vector sweeps during a unit time. This is indeed the Kepler's second law. Then $h/2$ is called the area velocity of the radius of the satellite.

Multiplying x, y and z to the three equations of (3.9) and adding them together, one has

$$Ax + By + Cz = 0. \tag{3.12}$$

That is, the satellite motion fulfils the equation of a plane, and the origin of the coordinate system is in the plane. In other words, the satellite moves in a plane in the central force field of the Earth. The plane is called the orbital plane of the satellite.

The angle between the orbital plane and the equatorial plane is called inclination of the satellite (denoted by i, see Fig. 3.1). Alternatively, the inclination i is the angle between the vector $\vec{z} = (0,0,1)$ and $\vec{h} = (A,B,C)$, i.e.,

$$\cos i = \frac{\vec{z} \cdot \vec{h}}{|\vec{z}| \cdot |\vec{h}|} = \frac{C}{h}. \tag{3.13}$$

The orbital plane cuts the equator at two points. They are called ascending node N and descending node (see the next section for details). Vector $\vec{s}$ denotes the vector from the Earth centre pointed to the ascending point. The angle between the ascending node and the x-axis (vernal equinox) is called the right ascension of the ascending node (denoted by Ω). Thus

$$\vec{s} = \vec{z} \times \vec{h},$$

and

$$\begin{aligned} \cos\Omega &= \frac{\vec{s} \cdot \vec{x}}{|\vec{s}| \cdot |\vec{x}|} = \frac{-B}{\sqrt{A^2 + B^2}}, \\ \sin\Omega &= \frac{\vec{s} \cdot \vec{y}}{|\vec{s}| \cdot |\vec{y}|} = \frac{A}{\sqrt{A^2 + B^2}}. \end{aligned} \tag{3.14}$$

Parameters i and Ω uniquely define the place of the orbital plane and are therefore called orbital plane parameters. Ω, i and h are then selected as integration constants, which have significant geometric meanings in the satellite orbits.

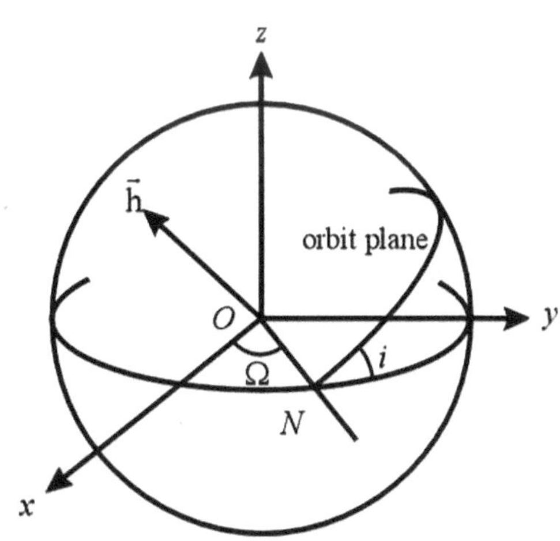

Fig. 3.1 Orbital plane

3.2 Satellite Motion in the Orbital Plane

In the orbital plane, a two-dimensional rectangular coordinate system is given in Fig. 3.2. The coordinates can be represented in polar coordinate r and ϑ as

$$\begin{aligned} p &= r\cos\vartheta, \\ q &= r\sin\vartheta. \end{aligned} \tag{3.15}$$

The equation of motion in pq-coordinates is similar to (3.4) as

$$\begin{aligned} \ddot{p} &= -\frac{\mu}{r^3}p, \\ \ddot{q} &= -\frac{\mu}{r^3}q. \end{aligned} \tag{3.16}$$

From (3.15), one has

$$\begin{aligned} \dot{p} &= \dot{r}\cos\vartheta - r\dot{\vartheta}\sin\vartheta, \\ \dot{q} &= \dot{r}\sin\vartheta + r\dot{\vartheta}\cos\vartheta, \\ \ddot{p} &= (\ddot{r} - r\dot{\vartheta}^2)\cos\vartheta - (r\ddot{\vartheta} + 2\dot{r}\dot{\vartheta})\sin\vartheta, \\ \ddot{q} &= (\ddot{r} - r\dot{\vartheta}^2)\sin\vartheta + (r\ddot{\vartheta} + 2\dot{r}\dot{\vartheta})\cos\vartheta. \end{aligned} \tag{3.17}$$

Substituting (3.17) and (3.15) into (3.16), one gets

$$\begin{aligned} (\ddot{r} - r\dot{\vartheta}^2)\cos\vartheta - (r\ddot{\vartheta} + 2\dot{r}\dot{\vartheta})\sin\vartheta &= -\frac{\mu}{r^2}\cos\vartheta, \\ (\ddot{r} - r\dot{\vartheta}^2)\sin\vartheta + (r\ddot{\vartheta} + 2\dot{r}\dot{\vartheta})\cos\vartheta &= -\frac{\mu}{r^2}\sin\vartheta. \end{aligned} \tag{3.18}$$

The point from which the polar angle ϑ is measured is arbitrary. So setting ϑ as zero, the equation of motion is then

$$\begin{aligned} \ddot{r} - r\dot{\vartheta}^2 &= -\frac{\mu}{r^2}, \\ r\ddot{\vartheta} + 2\dot{r}\dot{\vartheta} &= 0. \end{aligned} \tag{3.19}$$

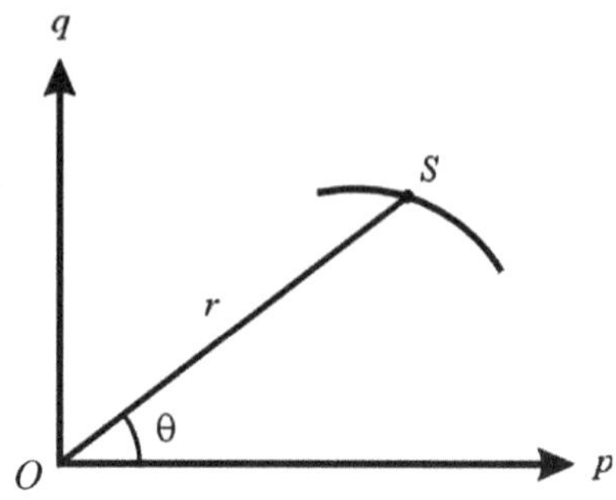

Fig. 3.2 Polar coordinates in the orbital plane

Multiplying r to the second equation of (3.19), it turns out to be

$$\frac{\mathrm{d}(r^2\dot{\vartheta})}{\mathrm{d}t} = 0. \tag{3.20}$$

Because $r\dot{\vartheta}$ is the tangential velocity, $r^2\dot{\vartheta}$ is the two times of the area velocity of the radius of the satellite. Integrating (3.20) and comparing it with the discussion in Sect. 3.1, one has

$$r^2\dot{\vartheta} = h. \tag{3.21}$$

$h/2$ is the area velocity of the radius of the satellite.

For solving the first differential equation (3.19), the equation has to be transformed into a differential equation of r with respect to variable f. Let

$$u = \frac{1}{r}, \tag{3.22}$$

then from (3.21), one gets

$$\frac{\mathrm{d}\vartheta}{\mathrm{d}t} = hu^2 \tag{3.23}$$

and

$$\begin{aligned}\frac{\mathrm{d}r}{\mathrm{d}t} &= \frac{\mathrm{d}r}{\mathrm{d}\vartheta}\frac{\mathrm{d}\vartheta}{\mathrm{d}t} = \frac{\mathrm{d}}{\mathrm{d}\vartheta}\left(\frac{1}{u}\right)hu^2 = -h\frac{\mathrm{d}u}{\mathrm{d}\vartheta},\\ \frac{\mathrm{d}^2r}{\mathrm{d}t^2} &= -h\frac{\mathrm{d}^2u}{\mathrm{d}\vartheta^2}\frac{\mathrm{d}\vartheta}{\mathrm{d}t} = -h^2u^2\frac{\mathrm{d}^2u}{\mathrm{d}\vartheta^2}.\end{aligned} \tag{3.24}$$

Substituting (3.22) and (3.24) into the first equation of (3.19), the equation of motion is then

$$\frac{\mathrm{d}^2u}{\mathrm{d}\vartheta^2} + u = \frac{\mu}{h^2}, \tag{3.25}$$

and its solution is

$$u = d_1\cos\vartheta + d_2\sin\vartheta + \frac{\mu}{h^2},$$

where d_1 and d_2 are constants of integration. The above equation may be simplified as

$$u = \frac{\mu}{h^2}(1 + e\cos(\vartheta - \omega)), \tag{3.26}$$

where

$$d_1 = \frac{\mu}{h^2}e\cos\omega, \quad d_2 = \frac{\mu}{h^2}e\sin\omega.$$

Thus the moving equation of satellite in the orbital plane is

$$r = \frac{h^2/\mu}{1 + e\cos(\vartheta - \omega)}. \tag{3.27}$$

Comparing (3.27) with a standard polar equation of conic:

$$r = \frac{a(1-e^2)}{1-e\cos\varphi}, \tag{3.28}$$

orbit (3.27) is obviously a polar equation of conic section with the origin at one of the foci. Where parameter e is the eccentricity, for $e = 0, e < 1, e = 1, e > 1$, the conic is a circle, an ellipse, a parabola, and a hyperbola, respectively. For the satellite orbiting around the Earth, generally, $e < 1$. Thus the satellite orbit is an ellipse, and this is indeed the Kepler's first law. Parameter a is the semi-major axis of the ellipse, and

$$\frac{h^2}{\mu} = a(1-e^2). \tag{3.29}$$

It is obvious that parameter a has more significant geometric sense than that of h, so a is preferred to be used. Parameters a and e define the size and shape of the ellipse and are called ellipse parameters. The ellipse cuts the equator at the ascending and descending nodes. Polar angle φ is counted from the apogee of the ellipse. This can be seen by let $\varphi = 0$, thus $r = a(1+e)$. φ has a 180 degree difference with the angle $\vartheta - \omega$. Letting $f = \vartheta - \omega$, where f is called the true anomaly of the satellite counted from the perigee, then the orbit (3.27) can be written as

$$r = \frac{a(1-e^2)}{1+e\cos f}. \tag{3.30}$$

In the case of $f = 0$, i.e., the satellite is in the point of perigee, $\omega = \vartheta$, ϑ is the polar angle of the perigee counted from the p-axis. Supposing the p-axis is an axis in the equatorial plane and is pointed to the ascending node N, then ω is the angle of perigee counted from the ascending node (see Fig. 3.3) and is called the argument of perigee. The argument of perigee defines the axis direction of the ellipse related to the equatorial plane.

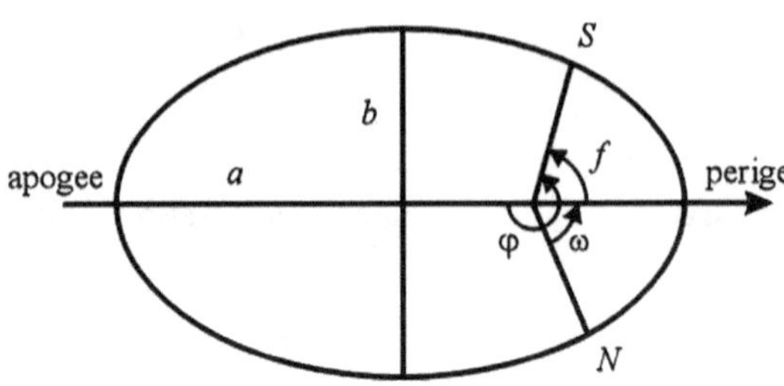

Fig. 3.3 Ellipse of the satellite motion

3.3 Keplerian Equation

Up to now, five integration constants have been derived. They are inclination angle i, right ascension of ascending node Ω, semi-major axis a, eccentricity e of the ellipse, and argument of perigee ω. Parameters i and Ω decide the place of the orbital plane,

Fig. 3.4 Orbital geometry

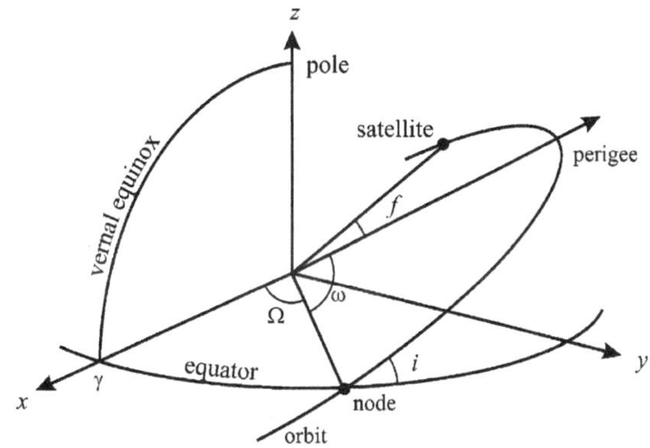

a and e decide the size and shape of the ellipse and ω decides the direction of the ellipse (see Fig. 3.4). To describe the satellite position in the ellipse, velocity of the motion has to be discussed.

The period T of the satellite motion is the area of ellipse divided by area velocity:

$$T = \frac{\pi ab}{\frac{1}{2}h} = \frac{2\pi ab}{\sqrt{\mu a(1-e^2)}} = 2\pi a^{3/2}\mu^{-1/2}. \tag{3.31}$$

The average angular velocity n is then

$$n = \frac{2\pi}{T} = a^{-3/2}\mu^{1/2}. \tag{3.32}$$

Equation (3.32) is the Kepler's third law. It is obvious that it is easier to describe the angular motion of the satellite under the average angular velocity n in the geometric centre of the ellipse (than in the geocentre). For simplifying the problem, an angle called the eccentric anomaly is defined (denoted by E, see Fig. 3.5). S' is the vertical

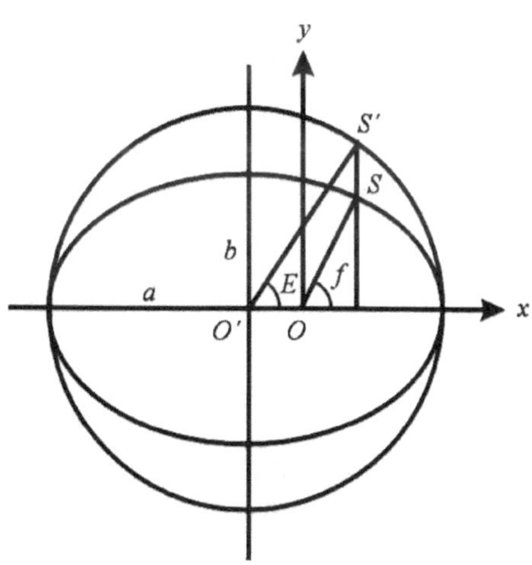

Fig. 3.5 Mean anomaly of satellite

projection of the satellite S on the circle with a radius of a (semi-major axis of the ellipse). The distance between the geometric centre O' of the ellipse and the geocentre O is ae. Thus,

$$\begin{aligned} x &= r\cos f = a\cos E - ae, \\ y &= r\sin f = b\sin E = a\sqrt{1-e^2}\sin E, \end{aligned} \tag{3.33}$$

where the second equation can be obtained by substituting the first into the standard ellipse equation ($x^2/a^2 + y^2/b^2 = 1$) and omitting the small terms that contain e (for the satellite, generally, $e << 1$), where b is the semi-minor axis of the ellipse. The orbit equation can then be represented by variable E as

$$r = a(1 - e\cos E). \tag{3.34}$$

The relation between true and eccentric anomalies can be derived by using (3.33) and (3.34):

$$\tan\frac{f}{2} = \frac{\sin f}{1+\cos f} = \frac{\sin E}{1+\cos E}\frac{\sqrt{1-e^2}}{1-e} = \frac{\sqrt{1+e}}{\sqrt{1-e}}\tan\frac{E}{2}. \tag{3.35}$$

If the xyz-coordinates are rotated so that the xy-plane coincides with the orbital plane, then the area velocity formulas of (3.9) and (3.10) have only one component along the z-axis, i.e.,

$$x\dot{y} - y\dot{x} = h = \sqrt{\mu a(1-e^2)}. \tag{3.36}$$

From (3.33), one has

$$\begin{aligned} \dot{x} &= -a\sin E\frac{\mathrm{d}E}{\mathrm{d}t}, \\ \dot{y} &= a\sqrt{1-e^2}\cos E\frac{\mathrm{d}E}{\mathrm{d}t}. \end{aligned} \tag{3.37}$$

Substituting (3.33) and (3.37) into (3.36) and taking (3.32) into account, a relation between E and t is obtained

$$(1 - e\cos E)\mathrm{d}E = \sqrt{\mu}a^{-3/2}\mathrm{d}t = n\mathrm{d}t. \tag{3.38}$$

Suppose at the time t_p satellite is at the point perigee, i.e. $E(t_\mathrm{p}) = 0$, and at any time t, $E(t) = E$, then integration of (3.38) from 0 to E, namely from t_p to t is

$$E - e\sin E = M, \tag{3.39}$$

where

$$M = n(t - t_p). \tag{3.40}$$

Equation (3.39) is the Keplerian equation. E is given as a function of M, namely t. Because of (3.34), the Keplerian equation indirectly assigns r as a function of t.

M is called the mean anomaly. M describes the satellite as orbiting the Earth with a mean angular velocity n. t_p is called the perigee passage and is the sixth integration constant of the equation of satellite motion in a centre-force field.

Knowing M to compute E, the Keplerian equation (3.39) may be solved iteratively. Because of the small e, the convergence can be achieved very quickly.

Three anomalies (true anomaly f, eccentric anomaly E and mean anomaly M) are equivalent through the relations of (3.35) and (3.39). They are functions of time t (including the perigee passage t_p), and they describe the position changes of the satellite with the time in the ECSF coordinates.

3.4 State Vector of the Satellite

Consider the orbital right-handed coordinate system: if the xy-plane is the orbital plane, the x-axis is pointing to the perigee, the z-axis is in the direction of vector $\vec{h}$, and the origin is in the geocentre, the position vector $\vec{q}$ of the satellite is then (see (3.33))

$$\vec{q} = \begin{pmatrix} a(\cos E - e) \\ a\sqrt{1-e^2}\sin E \\ 0 \end{pmatrix} = \begin{pmatrix} r\cos f \\ r\sin f \\ 0 \end{pmatrix}. \tag{3.41}$$

Differentiating (3.41) with respect to time t and taking (3.38) into account, the velocity vector of the satellite is then

$$\dot{\vec{q}} = \begin{pmatrix} -\sin E \\ \sqrt{1-e^2}\cos E \\ 0 \end{pmatrix} \frac{na}{1-e\cos E} = \begin{pmatrix} -\sin f \\ e+\cos f \\ 0 \end{pmatrix} \frac{na}{\sqrt{1-e^2}}. \tag{3.42}$$

The second part of above equation can be derived from the relation between E and f. The state vector of the satellite in the orbital coordinate system can be rotated to the ECSF coordinate system by three successive rotations. First, a clockwise rotation around the 3rd-axis from the perigee to the node is given by (see Fig. 3.4)

$$R_3(-\omega).$$

Next, a clockwise rotation around the 1st-axis with the angle of inclination i is given by

$$R_1(-i).$$

Finally, a clockwise rotation around the 3rd-axis from the node to the vernal equinox is given by

$$R_3(-\Omega).$$

So the state vector of the satellite in the ECSF coordinate system is

$$\begin{pmatrix} \vec{r} \\ \dot{\vec{r}} \end{pmatrix} = R_3(-\Omega)R_1(-i)R_3(-\omega)\begin{pmatrix} \vec{q} \\ \dot{\vec{q}} \end{pmatrix}, \tag{3.43}$$

where

$$\vec{r} = \begin{pmatrix} x \\ y \\ z \end{pmatrix}, \quad \dot{\vec{r}} = \begin{pmatrix} \dot{x} \\ \dot{y} \\ \dot{z} \end{pmatrix}.$$

For given six Keplerian elements (Ω, i, ω, a, e, M_0) of t_0, where $M_0 = n(t_0 - t_\mathrm{p})$, the satellite state vector of time t can be computed, e.g., as follows:

1. Using (3.32) to compute the mean angular velocity n;
2. Using (3.40), (3.39), (3.33) and (3.30) to compute the three anomalies M, E, f and r;
3. Using (3.41) and (3.42) to compute the state vector $\vec{q}$ and $\dot{\vec{q}}$ in orbital coordinates;
4. Using (3.43) to rotate state vector $\vec{q}$ and $\dot{\vec{q}}$ to the ECSF coordinates.

Keplerian elements can be given in practice at any time. For example, with t_0, where only f is a function of t_0, other parameters are constants. In this case, the related E and M can be computed by (3.35) and (3.39), thus t_p can be computed by (3.40).

From (3.42), one has

$$v^2 = \frac{a^2n^2}{(1-e\cos E)^2}[\sin^2 E + (1-e^2)\cos^2 E] = \frac{a^2n^2(1+e\cos E)}{1-e\cos E}. \tag{3.44}$$

Taking (3.32) and (3.34) into account leads to

$$v^2 = \frac{\mu(1+e\cos E)}{r} = \frac{\mu(2-r/a)}{r} = \mu\left(\frac{2}{r} - \frac{1}{a}\right), \tag{3.45}$$

where $v^2/2$ is the kinetic energy scaled by mass, μ/r is the potential energy, and a is the semi-major axis of the ellipse. This is the total energy conservative law of mechanics.

Rotate the vector $\vec{q}$ and $\dot{\vec{q}}$ in (3.41) and (3.42) by $R_3(-\omega)$ and denote by $\vec{p}$ and $\dot{\vec{p}}$, i.e.

$$\vec{p} = \begin{pmatrix} p_1 \\ p_2 \\ p_3 \end{pmatrix} = R_3(-\omega)\begin{pmatrix} r\cos f \\ r\sin f \\ 0 \end{pmatrix} = \begin{pmatrix} r\cos(\omega+f) \\ r\sin(\omega+f) \\ 0 \end{pmatrix} \tag{3.46}$$

and

$$\dot{\vec{p}} = \begin{pmatrix} \dot{p}_1 \\ \dot{p}_2 \\ \dot{p}_3 \end{pmatrix} = R_3(-\omega)\begin{pmatrix} -\sin f \\ e+\cos f \\ 0 \end{pmatrix}\frac{na}{\sqrt{1-e^2}} = \begin{pmatrix} -\sin(\omega+f) - e\sin\omega \\ \cos(\omega+f) + e\cos\omega \\ 0 \end{pmatrix}\frac{na}{\sqrt{1-e^2}}. \tag{3.47}$$

The reverse problem of (3.43), i.e., for given rectangular satellite state vector $(\vec{r},\dot{\vec{r}})^T$ to compute the Keplerian elements, can be carried out as follows. $\omega+f$ is called argument of latitude and denoted by u.

1. Using the given state vector to compute the modulus r and v ($r=|\vec{r}|$, $v=|\dot{\vec{r}}|$);
2. Using (3.10) and (3.11) to compute vector $\vec{h}$ and its modulus h;
3. Using (3.13) and (3.14) to compute inclination i and the right ascension of ascending node Ω;
4. Using (3.45), (3.29) and (3.32) to compute semi-major axis a, eccentricity e and average angular velocity n;
5. Rotating $\vec{r}$ by $\vec{p}=R_1(i)R_3(\Omega)\vec{r}$ and then using (3.46) to compute $\omega+f$;
6. Rotating $\dot{\vec{r}}$ by $\dot{\vec{p}}=R_1(i)R_3(\Omega)\dot{\vec{r}}$ and then using (3.47) to compute ω and f;
7. Using (3.33), (3.39) and (3.40) to compute E, M and t_p.

To transform the GPS state vector from the ECSF coordinate system to other coordinate systems, the formulas discussed in Chap. 2 can be used.

Chapter 4
Perturbations on the Orbits

Satellites are attracted not only by the central force of the Earth, but also by the non-central force of the Earth, the attracting forces of the sun, the moon and planets, and the drag force of the atmosphere. They are also affected by solar radiation pressure, Earth and ocean tides, general relativity effects and coordinate perturbations. Equations of satellite motion have to be represented by perturbed equations. In this chapter, after discussions of the perturbed equations of motion, emphasis is given to the attracting forces and the order estimation of the disturbances.

4.1 Perturbed Equation of Satellite Motion

The perturbed equation of motion of the satellite is described by Newton's second law in an inertial Cartesian coordinate system as

$$m\ddot{\vec{r}} = \vec{f}, \tag{4.1}$$

where $\vec{f}$ is the summated force vector acting on the satellite, and $\vec{r}$ is the radius vector of the satellite with mass m. $\ddot{\vec{r}}$ is the acceleration. Equation (4.1) is a second-order differential equation. For convenience, it can be written as two first-order differential equations as follows

$$\begin{aligned} \frac{\mathrm{d}\vec{r}}{\mathrm{d}t} &= \dot{\vec{r}}, \\ \frac{\mathrm{d}\dot{\vec{r}}}{\mathrm{d}t} &= \frac{1}{m}\vec{f}. \end{aligned} \tag{4.2}$$

Denoting the state vector of the satellite as

$$\vec{X} = \begin{pmatrix} \vec{r} \\ \dot{\vec{r}} \end{pmatrix}, \tag{4.3}$$

equation (4.2) can be written as

$$\dot{\vec{X}} = \vec{F}, \tag{4.4}$$

where

$$\vec{F} = \begin{pmatrix} \dot{\vec{r}} \\ \vec{f}/m \end{pmatrix}. \tag{4.5}$$

Equation (4.4) is called the state equation of the satellite motion. Integrating (4.4) from t_0 to t, one has

$$\vec{X}(t) = \vec{X}(t_0) + \int_{t_0}^{t} \vec{F} \mathrm{d}t, \tag{4.6}$$

where $\vec{X}(t)$ is the instantaneous state vector of the satellite, $\vec{X}(t_0)$ is the initial state vector at time t_0, and $\vec{F}$ is a function of the state vector $\vec{X}(t)$ and time t. Denoting the initial state vector as $\vec{X}_0$, then the perturbed satellite orbit problem turns out to be a problem of solving the differential state equation under the initial condition as

$$\begin{cases} \dot{\vec{X}}(t) = \vec{F}, \\ \vec{X}(t_0) = \vec{X}_0. \end{cases} \tag{4.7}$$

4.1.1 Lagrangian Perturbed Equation of Satellite Motion

If the force $\vec{f}$ includes only the conservative forces, then there is a potential function V so that

$$\frac{\vec{f}}{m} = \mathrm{grad}V = \left(\frac{\partial V}{\partial x}\ \frac{\partial V}{\partial y}\ \frac{\partial V}{\partial z} \right) = \left(\frac{\partial V}{\partial r}\ \frac{\partial V}{\partial \varphi}\ \frac{\partial V}{\partial \lambda} \right), \tag{4.8}$$

where (x,y,z) and (r,φ,λ) are Cartesian coordinates and spherical coordinates, respectively. Denoting R as the disturbance potential and V_0 as the potential of the centred force $\vec{f}_0$, then

$$R = V - V_0, \quad \frac{\vec{f} - \vec{f}_0}{m} = \mathrm{grad}R. \tag{4.9}$$

The perturbed equation (4.2) of satellite motion in Cartesian coordinates is then

$$\begin{aligned} \frac{\mathrm{d}x}{\mathrm{d}t} &= \dot{x}, \\ \frac{\mathrm{d}y}{\mathrm{d}t} &= \dot{y}, \\ \frac{\mathrm{d}z}{\mathrm{d}t} &= \dot{z}, \\ \frac{\mathrm{d}\dot{x}}{\mathrm{d}t} &= -\frac{\mu}{r^3}x + \frac{\partial R}{\partial x}, \\ \frac{\mathrm{d}\dot{y}}{\mathrm{d}t} &= -\frac{\mu}{r^3}y + \frac{\partial R}{\partial y}, \\ \frac{\mathrm{d}\dot{z}}{\mathrm{d}t} &= -\frac{\mu}{r^3}z + \frac{\partial R}{\partial z}, \end{aligned} \tag{4.10}$$

where μ is the gravitational constant of the Earth. The state vector $(\vec{r}, \dot{\vec{r}})$ of the satellite corresponds to an instantaneous Keplerian ellipse $(a, e, \omega, i, \Omega, M)$. Using the relationships between the two sets of parameters (see Chap. 3), the perturbed equation of motion (4.10) can be transformed into a so-called Lagrangian perturbed equation system (see, e.g., Kaula, 1966/2001)

$$\begin{aligned}
\frac{\mathrm{d}a}{\mathrm{d}t} &= \frac{2}{na}\frac{\partial R}{\partial M},\\
\frac{\mathrm{d}e}{\mathrm{d}t} &= \frac{1-e^2}{na^2e}\frac{\partial R}{\partial M} - \frac{\sqrt{1-e^2}}{na^2e}\frac{\partial R}{\partial \omega},\\
\frac{\mathrm{d}\omega}{\mathrm{d}t} &= \frac{\sqrt{1-e^2}}{na^2e}\frac{\partial R}{\partial e} - \frac{\cos i}{na^2\sqrt{1-e^2}\sin i}\frac{\partial R}{\partial i},\\
\frac{\mathrm{d}i}{\mathrm{d}t} &= \frac{1}{na^2\sqrt{1-e^2}\sin i}\left(\cos i\frac{\partial R}{\partial \omega} - \frac{\partial R}{\partial \Omega}\right),\\
\frac{\mathrm{d}\Omega}{\mathrm{d}t} &= \frac{1}{na^2\sqrt{1-e^2}\sin i}\frac{\partial R}{\partial i},\\
\frac{\mathrm{d}M}{\mathrm{d}t} &= n - \frac{2}{na}\frac{\partial R}{\partial a} - \frac{1-e^2}{na^2e}\frac{\partial R}{\partial e}.
\end{aligned} \tag{4.11}$$

On the basis of this equation system, Kaula derived the first order perturbed analysis solution (see Kaula, 1966/2001). In the case of a small e (e $\ll 1$), the orbit is nearly circular, so that the perigee and the related Keplerian elements f and ω are not defined (this is not to be confused with the force vector $\vec{f}$ and true anomaly f). To overcome this problem, let $u = f + \omega$, and a parameter set of (a, i, Ω, ξ, η, λ) is used to describe the motion of the satellite, where

$$\begin{aligned}
\xi &= e\cos\omega,\\
\eta &= -e\sin\omega,\\
\lambda &= M + \omega.
\end{aligned} \tag{4.12}$$

Thus, one has

$$\begin{aligned}
\frac{\mathrm{d}\xi}{\mathrm{d}t} &= \frac{\xi}{e}\frac{\mathrm{d}e}{\mathrm{d}t} + \eta\frac{\mathrm{d}\omega}{\mathrm{d}t},\\
\frac{\mathrm{d}\eta}{\mathrm{d}t} &= \frac{\eta}{e}\frac{\mathrm{d}e}{\mathrm{d}t} - \xi\frac{\mathrm{d}\omega}{\mathrm{d}t},\\
\frac{\mathrm{d}\lambda}{\mathrm{d}t} &= \frac{\mathrm{d}M}{\mathrm{d}t} + \frac{\mathrm{d}\omega}{\mathrm{d}t}
\end{aligned} \tag{4.13}$$

and

$$\begin{aligned}
\frac{\partial R}{\partial \omega} &= \frac{\partial R}{\partial(\xi,\eta,\lambda)}\frac{\partial(\xi,\eta,\lambda)}{\partial \omega} = \frac{\partial R}{\partial(\xi,\eta,\lambda)}(\eta,-\xi,1)^T = \eta\frac{\partial R}{\partial \xi} - \xi\frac{\partial R}{\partial \eta} + \frac{\partial R}{\partial \lambda},\\
\frac{\partial R}{\partial e} &= \frac{\partial R}{\partial(\xi,\eta,\lambda)}\frac{\partial(\xi,\eta,\lambda)}{\partial e} = \frac{\partial R}{\partial(\xi,\eta,\lambda)}\left(\frac{\xi}{e},\frac{\eta}{e},0\right)^T = \frac{\xi}{e}\frac{\partial R}{\partial \xi} + \frac{\eta}{e}\frac{\partial R}{\partial \eta},\\
\frac{\partial R}{\partial M} &= \frac{\partial R}{\partial(\xi,\eta,\lambda)}\frac{\partial(\xi,\eta,\lambda)}{\partial M} = \frac{\partial R}{\partial(\xi,\eta,\lambda)}(0,0,1)^T = \frac{\partial R}{\partial \lambda}.
\end{aligned} \tag{4.14}$$

Substituting (4.14) into (4.11) and then substituting the 2nd, 3rd and 6th equations of (4.11) into (4.13), one has

$$\begin{aligned}
\frac{\mathrm{d}a}{\mathrm{d}t} &= \frac{2}{na}\frac{\partial R}{\partial \lambda},\\
\frac{\mathrm{d}i}{\mathrm{d}t} &= \frac{1}{na^2\sqrt{1-e^2}\sin i}\left[\cos i\left(\eta\frac{\partial R}{\partial \xi} - \xi\frac{\partial R}{\partial \eta} + \frac{\partial R}{\partial \lambda}\right) - \frac{\partial R}{\partial \Omega}\right],\\
\frac{\mathrm{d}\Omega}{\mathrm{d}t} &= \frac{1}{na^2\sqrt{1-e^2}\sin i}\frac{\partial R}{\partial i},\\
\frac{\mathrm{d}\xi}{\mathrm{d}t} &= \frac{\sqrt{1-e^2}}{na^2}\frac{\partial R}{\partial \eta} - \eta\frac{\cos i}{na^2\sqrt{1-e^2}\sin i}\frac{\partial R}{\partial i} + \xi\frac{1-e^2-\sqrt{1-e^2}}{na^2e^2}\frac{\partial R}{\partial \lambda},\\
\frac{\mathrm{d}\eta}{\mathrm{d}t} &= -\frac{\sqrt{1-e^2}}{na^2}\frac{\partial R}{\partial \xi} + \xi\frac{\cos i}{na^2\sqrt{1-e^2}\sin i}\frac{\partial R}{\partial i} + \eta\frac{1-e^2-\sqrt{1-e^2}}{na^2e^2}\frac{\partial R}{\partial \lambda},\\
\frac{\mathrm{d}\lambda}{\mathrm{d}t} &= n - \frac{2}{na}\frac{\partial R}{\partial a} - \frac{\cos i}{na^2\sqrt{1-e^2}\sin i}\frac{\partial R}{\partial i} - \frac{1-e^2-\sqrt{1-e^2}}{na^2e^2}\left(\xi\frac{\partial R}{\partial \xi} + \eta\frac{\partial R}{\partial \eta}\right).
\end{aligned} \tag{4.15}$$

The new variables of (4.12) do not have clear geometric meanings. An alternative is to use the Hill variables (see, e.g., Cui, 1990).

4.1.2 Gaussian Perturbed Equation of Satellite Motion

Considering the non-conservative disturbance forces such as solar radiation and air drag, no potential functions exist for use; therefore, the Lagrangian perturbed equation of motion cannot be directly used in such a case. The equation of motion perturbed by non-conservative disturbance force has to be derived.

Considering any force vector $\vec{f} = (f_x,\ f_y,\ f_z)^T$ in ECSF coordinate system, one has

$$\begin{pmatrix} f_x \\ f_y \\ f_z \end{pmatrix} = R_3(-\Omega)R_1(-i)R_3(-u)\begin{pmatrix} f_r \\ f_\alpha \\ f_h \end{pmatrix}, \tag{4.16}$$

where $(f_r, f_\alpha, f_h)^T$ is a force vector with three orthogonal components in an orbital plane coordinate system, the first two components are in the orbital plane, f_r is the radial force component, f_α is the force component perpendicular to f_r and pointed in the direction of satellite motion, and f_h completes a right-handed system. For convenience, the force vector may also be represented by tangential, central components in the orbital plane (f_t, f_c) as well as f_h (see Fig. 4.1). It is obvious that

$$\begin{pmatrix} f_r \\ f_\alpha \\ f_h \end{pmatrix} = R_3(-\beta) \begin{pmatrix} f_t \\ f_c \\ f_h \end{pmatrix}, \tag{4.17}$$

where

$$\tan\beta = r\frac{\mathrm{d}f}{\mathrm{d}r} = \frac{a(1-e^2)}{1+e\cos f}\frac{\mathrm{d}f}{\frac{a(1-e^2)}{(1+e\cos f)^2}e\sin f\mathrm{d}f} = \frac{1+e\cos f}{e\sin f}, \tag{4.18}$$

or

$$\begin{aligned} \sin\beta &= \frac{1+e\cos f}{\sqrt{1+2e\cos f+e^2}}, \\ \cos\beta &= \frac{e\sin f}{\sqrt{1+2e\cos f+e^2}}. \end{aligned} \tag{4.19}$$

To replace the partial derivatives $\partial R/\partial\sigma$ by force components, the relationships between them have to be derived, where σ is a symbol for all Keplerian elements. Using the regulation of partial derivatives, one has

$$\begin{aligned} \frac{\partial R}{\partial \sigma} &= \frac{\partial R}{\partial \vec{r}}\cdot\frac{\partial \vec{r}}{\partial \sigma} = \vec{f}\cdot\left(\frac{\partial r}{\partial \sigma}\vec{e}_r + r\frac{\partial \vec{e}_r}{\partial \sigma}\right), \\ &= R_3(-\Omega)R_1(-i)R_3(-u)\begin{pmatrix} f_r \\ f_\alpha \\ f_h \end{pmatrix}\cdot\left(\frac{\partial r}{\partial \sigma}\vec{e}_r + r\frac{\partial \vec{e}_r}{\partial \sigma}\right), \end{aligned} \tag{4.20}$$

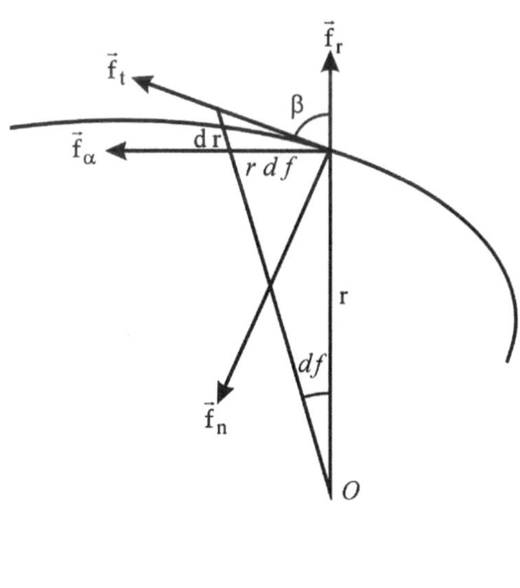

Fig. 4.1 Relation of radial and tangential forces

where $\vec{e}_r$ is the radial identity vector of the satellite, the dot is the vector dot product, and

$$\vec{e}_r = \begin{pmatrix} \varepsilon_1 \\ \varepsilon_2 \\ \varepsilon_3 \end{pmatrix} = R_3(-\Omega)R_1(-i)R_3(-u)\begin{pmatrix} 1 \\ 0 \\ 0 \end{pmatrix} = \begin{pmatrix} \cos\Omega\cos u - \sin\Omega\cos i\sin u \\ \sin\Omega\cos u + \cos\Omega\cos i\sin u \\ \sin i\sin u \end{pmatrix},$$

$$\frac{\partial\vec{e}_r}{\partial\sigma} = \begin{pmatrix} \sin\Omega\sin i\sin u\frac{\partial i}{\partial\sigma} - \varepsilon_2\frac{\partial\Omega}{\partial\sigma} - (\cos\Omega\sin u + \sin\Omega\cos i\cos u)\frac{\partial u}{\partial\sigma} \\ -\cos\Omega\sin i\sin u\frac{\partial i}{\partial\sigma} + \varepsilon_1\frac{\partial\Omega}{\partial\sigma} - (\sin\Omega\sin u - \cos\Omega\cos i\cos u)\frac{\partial u}{\partial\sigma} \\ \cos i\sin u\frac{\partial i}{\partial\sigma} + \sin i\cos u\frac{\partial u}{\partial\sigma} \end{pmatrix}. \tag{4.21}$$

Substituting (4.21) into (4.20) and simplifying it, one has

$$\frac{\partial R}{\partial\sigma} = \frac{\partial r}{\partial\sigma}f_r + r\left(\cos i\frac{\partial\Omega}{\partial\sigma} + \frac{\partial u}{\partial\sigma}\right)f_\alpha + r\left(\sin u\frac{\partial i}{\partial\sigma} - \sin i\cos u\frac{\partial\Omega}{\partial\sigma}\right)f_h. \tag{4.22}$$

For deriving the partial derivatives of r and $u(=f+\omega)$ with respect to the six Keplerian elements, the following basic relations (see Chap. 3) are used:

$$\begin{aligned} r &= \frac{a(1-e^2)}{1+e\cos f} = a(1-e\cos E), \\ r\cos f &= a(\cos E - e), \\ r\sin f &= a\sqrt{1-e^2}\sin E, \\ \tan\frac{f}{2} &= \sqrt{\frac{1+e}{1-e}}\tan\frac{E}{2}, \\ E - e\sin E &= M, \end{aligned} \tag{4.23}$$

where E is a function of (e,M), f is a function of (e,E), i.e., (e,M), r is a function of (a,e,M), and u is a function of (ω,f), i.e., (ω,e,M). Thus

$$\begin{aligned} \frac{\partial E}{\partial(e,M)} &= \left(\frac{a}{r}\sin E,\ \frac{a}{r}\right), \\ \frac{\partial f}{\partial(e,M)} &= \left(\frac{2+e\cos f}{1-e^2}\sin f,\ \left(\frac{a}{r}\right)^2\sqrt{1-e^2}\right), \\ \frac{\partial r}{\partial M} &= ae\sin E\frac{\partial E}{\partial M} = \frac{a^2e}{r}\sin E = \frac{ae}{\sqrt{1-e^2}}\sin f, \\ \frac{\partial r}{\partial(a,e,i,\Omega,\omega)} &= \left(\frac{r}{a},\ -a\cos f,\ 0,\ 0,\ 0\right), \end{aligned} \tag{4.24}$$

$$\frac{\partial u}{\partial e} = \frac{\partial u}{\partial f}\frac{\partial f}{\partial e} = \frac{2+e\cos f}{1-e^2}\sin f,$$
$$\frac{\partial u}{\partial M} = \frac{\partial u}{\partial f}\frac{\partial f}{\partial M} = \left(\frac{a}{r}\right)^2\sqrt{1-e^2},$$
$$\frac{\partial u}{\partial(a,i,\Omega,\omega)} = (0,\ 0,\ 0,\ 1).$$

Substituting (4.24) into (4.22), one has

$$\begin{aligned}
\frac{\partial R}{\partial a} &= \frac{r}{a}f_r,\\
\frac{\partial R}{\partial e} &= -a\cos f\cdot f_r + \frac{r\sin f}{1-e^2}(2+e\cos f)\cdot f_\alpha,\\
\frac{\partial R}{\partial i} &= r\sin u\cdot f_h,\\
\frac{\partial R}{\partial \Omega} &= i\cos i\cdot f_\alpha - r\sin i\cos u\cdot f_h,\\
\frac{\partial R}{\partial \omega} &= r\cdot f_\alpha,\\
\frac{\partial R}{\partial M} &= \frac{ae}{\sqrt{1-e^2}}\sin f\cdot f_r + \frac{a(1+e\cos f)}{\sqrt{1-e^2}}\cdot f_\alpha.
\end{aligned} \tag{4.25}$$

Putting (4.25) into Lagrangian perturbed equations of motion (4.11), the so-called Gaussian perturbed equations of motion are then

$$\begin{aligned}
\frac{\mathrm{d}a}{\mathrm{d}t} &= \frac{2}{n\sqrt{1-e^2}}\left[e\cos f\cdot f_r + (1+e\cos f)\cdot f_\alpha\right],\\
\frac{\mathrm{d}e}{\mathrm{d}t} &= \frac{\sqrt{1-e^2}}{na}\left[\sin f\cdot f_r + (\cos E+\cos f)\cdot f_\alpha\right],\\
\frac{\mathrm{d}i}{\mathrm{d}t} &= \frac{(1-e\cos E)\cos u}{na\sqrt{1-e^2}}\cdot f_h,\\
\frac{\mathrm{d}\Omega}{\mathrm{d}t} &= \frac{(1-e\cos E)\sin u}{na\sqrt{1-e^2}\sin i}\cdot f_h,\\
\frac{\mathrm{d}\omega}{\mathrm{d}t} &= \frac{\sqrt{1-e^2}}{nae}\left[-\cos f\cdot f_r + \frac{2+e\cos f}{1+e\cos f}\sin f\cdot f_\alpha\right] - \cos i\frac{\mathrm{d}\Omega}{\mathrm{d}t},\\
\frac{\mathrm{d}M}{\mathrm{d}t} &= n - \frac{1-e^2}{nae}\left[-\left(\cos f - \frac{2e}{1+e\cos f}\right)\cdot f_r + \frac{2+e\cos f}{1+e\cos f}\sin f\cdot f_\alpha\right].
\end{aligned} \tag{4.26}$$

The force components of $(f_r,\ f_\alpha,\ f_h)$ are used. Using (4.17), the Gaussian perturbed equations of motion can be represented by a disturbed force vector of (f_t, f_c, f_h).

4.2 Perturbation Forces of Satellite Motion

Perturbation forces of satellite motion will be discussed in this section. They are the gravitational forces of the Earth, the attracting forces of the sun, the moon and the planets, the drag force of the atmosphere, solar radiation pressure, Earth and ocean tides, and coordinate perturbations.

4.2.1 Perturbation of the Earth's Gravitational Field

After a brief review of the Earth's gravitational field, the perturbation force of the Earth will be outlined here.

The Earth's Gravitational Field

The complete real solution of the Laplace equation is called potential function V of the Earth. In spherical coordinates, V can be expressed by (Moritz, 1980; Sigl, 1989):

$$V = \sum_{lmi} \frac{1}{r^{l+1}} V_{lmi} = \sum_{l=0}^{\infty} \sum_{m=0}^{l} \frac{1}{r^{l+1}} P_{lm}(\sin\varphi)[C_{lm}\cos m\lambda + S_{lm}\sin m\lambda], \tag{4.27}$$

where r is the radius, φ is the latitude, and λ is the longitude measured eastward (counter-clockwise looking toward the origin from the positive end of the z-axis). One can, of course, use the co-latitude ϑ (or polar distance) instead of the latitude $\varphi(\sin\varphi = \cos\vartheta)$. The subscript i in the first term denotes the $\cos m\lambda$ or $\sin m\lambda$ term. $P_{lm}(\sin\varphi)$ is the so-called associated Legendre function, V_{lmi} denotes surface spherical harmonics, C_{lm}, S_{lm} are coefficients of the spherical functions, and

$$P_{lm}(\sin\varphi) = \cos^m\varphi \sum_{t=0}^{k} T_{lmt} \sin^{l-m-2t}\varphi, \tag{4.28}$$

where k is the integer part of $(l\text{-}m)/2$, and

$$T_{lmt} = \frac{(-1)^t(2l-2t)!}{2^l t!(l-t)!(l-m-2t)!}. \tag{4.29}$$

An important property of surface spherical harmonics V_{lmi} is that they are orthogonal ones. For the integration over the surface of a sphere there is (Heiskanen and Moritz, 1967; Kaula, 1966/2001):

$$\int_{\text{sphere}} V_{LMI}V_{lmi}\mathrm{d}\sigma = 0, \qquad \text{if} \quad L \neq l \quad \text{or} \quad M \neq m \quad \text{or} \quad I \neq i. \tag{4.30}$$

The integral of the square of V_{lmi} for $C_{lm} = 1$ or $S_{lm} = 1$ is

$$\int_{\text{sphere}} V_{lmi}^2 \mathrm{d}\sigma = \left[\frac{(l+m)!}{(l-m)!(2l+1)(2-\delta_{0m})}\right] 4\pi, \tag{4.31}$$

where the Kronecker delta δ_{0m} is equal to 1 for $m = 0$ and 0 for $m \neq 0$.

The normalised Legendre functions can be defined and denoted by

$$\bar{P}_{lm}(x) = P_{lm}(x) \left[\frac{(l-m)!(2l+1)(2-\delta_{0m})}{(l+m)!}\right]^{1/2}, \tag{4.32}$$

where $x = \sin\varphi = \cos\vartheta$. Recurrence formulae can be easily derived (Wenzel, 1985):

$$\begin{aligned}
\bar{P}_{(l+1)(l+1)}(x) &= \bar{P}_{ll}(x) \left[\frac{(2l+3)}{(l+1)(2-\delta_{0l})}\right]^{1/2} (1-x^2)^{1/2}, \\
\bar{P}_{(l+1)l}(x) &= \bar{P}_{ll}(x)\,[2l+3]^{1/2} x, \; l \geq 1, \\
\bar{P}_{(l+1)m}(x) &= \bar{P}_{lm}(x) \left[\frac{(2l+1)(2l+3)}{(l+m+1)(l-m+1)}\right]^{1/2} x \\
&\quad - \bar{P}_{(l-1)m}(x) \left[\frac{(l+m)(l-m)(2l+3)}{(l+m+1)(l-m+1)(2l-1)}\right]^{1/2}
\end{aligned}$$

and

$$\bar{P}_{00}(x) = 1, \quad \bar{P}_{10}(x) = \sqrt{3}x, \quad \bar{P}_{11}(x) = \sqrt{3(1-x^2)}. \tag{4.33}$$

Since the first term of V (i.e., $l = 0$) is represented by GM/r, the fully normalised geopotential function is taken as follows (Torge, 1989; Rapp, 1986):

$$V(r,\varphi,\lambda) = \frac{GM}{r}\left[1 + \sum_{l=2}^{\infty}\sum_{m=0}^{l}\left(\frac{a}{r}\right)^l \bar{P}_{lm}(\sin\varphi)[\bar{C}_{lm}\cos m\lambda + \bar{S}_{lm}\sin m\lambda]\right], \tag{4.34}$$

where GM is the geocentric gravitational constant, $\bar{C}_{lm}$, $\bar{S}_{lm}$ are normalised coefficients and a is the mean equatorial radius of the Earth. The first term of V is the potential of the central force of the Earth. The perturbation potential of the Earth is then (denoting $GM = \mu$)

$$R_{\text{geo}}(r,\varphi,\lambda) = \frac{\mu}{r}\sum_{l=2}^{\infty}\sum_{m=0}^{l}\left(\frac{a}{r}\right)^l \bar{P}_{lm}(\sin\varphi)\left[\bar{C}_{lm}\cos m\lambda + \bar{S}_{lm}\sin m\lambda\right]. \tag{4.35}$$

For any initial external potential of the Earth

$$U(r,\varphi,\lambda) = \frac{\mu}{r}\left[1 + \sum_{l=2}^{L}\sum_{m=0}^{l}\left(\frac{a}{r}\right)^l \bar{P}_{lm}(\sin\varphi)\left[\bar{C}_{lm}^N\cos m\lambda + \bar{S}_{lm}^N\sin m\lambda\right]\right], \tag{4.36}$$

the disturbing potential T is then

$$T = V - U = \frac{\mu}{r}\left[\sum_{l=2}^{\infty}\sum_{m=0}^{l}\left(\frac{a}{r}\right)^{l}\overline{P}_{lm}(\sin\varphi)[\Delta\bar{C}_{lm}\cos m\lambda + \Delta\bar{S}_{lm}\sin m\lambda]\right], \quad (4.37)$$

where $\bar{C}^N_{lm}$, $\bar{S}^N_{lm}$ are known normalised coefficients of the disturbing potential and

$$\bar{C}_{lm} = \Delta\bar{C}_{lm} - \bar{C}^N_{lm}, \quad \bar{S}_{lm} = \Delta\bar{S}_{lm} - \bar{S}^N_{lm} \quad l \le L. \quad (4.38)$$

Perturbation Force of the Earth's Gravitational Field

Denoting (x', y', z') as three orthogonal Cartesian coordinates in the ECEF system, then the force vector is

$$\vec{f}_{\text{ECEF}} = \begin{pmatrix} \dfrac{\partial V}{\partial x'} \\ \dfrac{\partial V}{\partial y'} \\ \dfrac{\partial V}{\partial z'} \end{pmatrix} = \begin{pmatrix} \dfrac{\partial V}{\partial(r,\varphi,\lambda)}\dfrac{\partial(r,\varphi,\lambda)}{\partial x'} \\ \dfrac{\partial V}{\partial(r,\varphi,\lambda)}\dfrac{\partial(r,\varphi,\lambda)}{\partial y'} \\ \dfrac{\partial V}{\partial(r,\varphi,\lambda)}\dfrac{\partial(r,\varphi,\lambda)}{\partial z'} \end{pmatrix} = \left(\frac{\partial V}{\partial(r,\varphi,\lambda)}\frac{\partial(r,\varphi,\lambda)}{\partial(x',y',z')}\right)^T. \quad (4.39)$$

From the relation between the Cartesian and spherical coordinates

$$\begin{pmatrix} x' \\ y' \\ z' \end{pmatrix} = \begin{pmatrix} r\cos\varphi\cos\lambda \\ r\cos\varphi\sin\lambda \\ r\sin\varphi \end{pmatrix}, \quad \begin{pmatrix} r = \sqrt{x'^2 + y'^2 + z'^2} \\ \varphi = \tan^{-1}\dfrac{z'}{\sqrt{x'^2+y'^2}} \\ \lambda = \tan^{-1}\dfrac{y'}{x'} \end{pmatrix}, \quad (4.40)$$

one has

$$\frac{\partial(r,\varphi,\lambda)}{\partial(x',y',z')} = \begin{pmatrix} \cos\varphi\cos\lambda & \cos\varphi\sin\lambda & \sin\varphi \\ -\frac{1}{r}\sin\varphi\cos\lambda & -\frac{1}{r}\sin\varphi\sin\lambda & \frac{1}{r}\cos\varphi \\ -\frac{1}{r\cos\varphi}\sin\lambda & \frac{1}{r\cos\varphi}\cos\lambda & 0 \end{pmatrix}. \quad (4.41)$$

For differentiations of the associated Legendre function, from (4.33) one has similar recurrence formulas:

$$\frac{\mathrm{d}\bar{P}_{00}(\sin\varphi)}{\mathrm{d}\varphi} = 0, \quad (4.42)$$

$$\frac{\mathrm{d}\bar{P}_{10}(\sin\varphi)}{\mathrm{d}\varphi} = \sqrt{3}\cos\varphi,$$

$$\frac{\mathrm{d}\bar{P}_{11}(\sin\varphi)}{\mathrm{d}\varphi} = -\sqrt{3}\sin\varphi,$$

$$\frac{\mathrm{d}\bar{P}_{(l+1)(l+1)}(\sin\varphi)}{\mathrm{d}\varphi} = -q\sin\varphi\bar{P}_{ll}(\sin\varphi) + q\cos\varphi\frac{\mathrm{d}\bar{P}_{ll}(\sin\varphi)}{\mathrm{d}\varphi},$$

$$q = \sqrt{\frac{2l+3}{2l+2}}, \quad l \geq 1,$$

$$\frac{\mathrm{d}\bar{P}_{(l+1)l}(\sin\varphi)}{\mathrm{d}\varphi} = g\cos\varphi\bar{P}_{ll}(\sin\varphi) + g\sin\varphi\frac{\mathrm{d}\bar{P}_{ll}(\sin\varphi)}{\mathrm{d}\varphi}, \quad l \geq 1,$$

$$g = \sqrt{2l+3},$$

$$\frac{\mathrm{d}\bar{P}_{(l+1)m}(\sin\varphi)}{\mathrm{d}\varphi} = h\cos\varphi\bar{P}_{lm}(\sin\varphi) + h\sin\varphi\frac{\mathrm{d}\bar{P}_{lm}(\sin\varphi)}{\mathrm{d}\varphi} - k\frac{\mathrm{d}\bar{P}_{(l-1)m}(\sin\varphi)}{\mathrm{d}\varphi},$$

$$h = \sqrt{\frac{(2l+1)(2l+3)}{(l+m+1)(l-m+1)}},$$

$$k = \sqrt{\frac{(l+m)(l-m)(2l+3)}{(l+m+1)(l-m+1)(2l-1)}}$$

and

$$\frac{\mathrm{d}^2\bar{P}_{00}(\sin\varphi)}{\mathrm{d}\varphi^2} = 0, \tag{4.43}$$

$$\frac{\mathrm{d}^2\bar{P}_{10}(\sin\varphi)}{\mathrm{d}\varphi^2} = -\sqrt{3}\sin\varphi,$$

$$\frac{\mathrm{d}^2\bar{P}_{11}(\sin\varphi)}{\mathrm{d}\varphi^2} = -\sqrt{3}\cos\varphi,$$

$$\frac{\mathrm{d}^2\bar{P}_{(l+1)(l+1)}(\sin\varphi)}{\mathrm{d}\varphi^2} = -q\cos\varphi\bar{P}_{ll}(\sin\varphi) - 2q\sin\varphi\frac{\mathrm{d}\bar{P}_{ll}(\sin\varphi)}{\mathrm{d}\varphi} + q\cos\varphi\frac{\mathrm{d}^2\bar{P}_{ll}(\sin\varphi)}{\mathrm{d}\varphi^2},$$

$$\frac{\mathrm{d}^2\bar{P}_{(l+1)l}(\sin\varphi)}{\mathrm{d}\varphi^2} = -g\sin\varphi\bar{P}_{ll}(\sin\varphi) + 2g\cos\varphi\frac{\mathrm{d}\bar{P}_{ll}(\sin\varphi)}{\mathrm{d}\varphi} + g\sin\varphi\frac{\mathrm{d}^2\bar{P}_{ll}(\sin\varphi)}{\mathrm{d}\varphi^2},$$

$$l \geq 1,$$

$$\begin{aligned}\frac{\mathrm{d}^2\bar{P}_{(l+1)m}(\sin\varphi)}{\mathrm{d}\varphi^2} &= -h\sin\varphi\bar{P}_{lm}(\sin\varphi) + 2h\cos\varphi\frac{\mathrm{d}\bar{P}_{lm}(\sin\varphi)}{\mathrm{d}\varphi} \\ &\quad + h\sin\varphi\frac{\mathrm{d}^2\bar{P}_{lm}(\sin\varphi)}{\mathrm{d}\varphi^2} - k\frac{\mathrm{d}^2\bar{P}_{(l-1)m}(\sin\varphi)}{\mathrm{d}\varphi^2}.\end{aligned}$$

The partial derivatives of the potential function with respect to the spherical coordinates are

$$\frac{\partial V}{\partial r} = -\frac{\mu}{r^2}\left[1 + \sum_{l=2}^{\infty}\sum_{m=0}^{l}(l+1)\left(\frac{a}{r}\right)^l \bar{P}_{lm}(\sin\varphi)\left[\bar{C}_{lm}\cos m\lambda + \bar{S}_{lm}\sin m\lambda\right]\right],$$

$$\frac{\partial V}{\partial \varphi} = \frac{\mu}{r}\sum_{l=2}^{\infty}\sum_{m=0}^{l}\left(\frac{a}{r}\right)^{l}\frac{\mathrm{d}\bar{P}_{lm}(\sin\varphi)}{\mathrm{d}\varphi}\left[\bar{C}_{lm}\cos m\lambda + \bar{S}_{lm}\sin m\lambda\right],$$
$$\frac{\partial V}{\partial \lambda} = \frac{\mu}{r}\sum_{l=2}^{\infty}\sum_{m=0}^{l} m\left(\frac{a}{r}\right)^{l}\bar{P}_{lm}(\sin\varphi)\left[-\bar{C}_{lm}\sin m\lambda + \bar{S}_{lm}\cos m\lambda\right]. \tag{4.44}$$

Using the transformation formula of (2.14), the perturbation force of the Earth's gravitational field in the ECSF system is then

$$\vec{f}_{\mathrm{ECSF}} = R_{\mathrm{P}}^{-1}R_{\mathrm{N}}^{-1}R_{\mathrm{S}}^{-1}R_{\mathrm{M}}^{-1}\vec{f}_{\mathrm{ECEF}}. \tag{4.45}$$

The computation process of disturbance force of the Earth's gravitational field in the ECSF coordinate system may be carried out by

1. using (2.14) to transform the satellite coordinates in the ECSF system to the ECEF system;
2. using (4.40) to compute the spherical coordinates of the satellite in the ECEF system;
3. using (4.39) to compute the force vector in the ECEF system;
4. using (4.45) to transform the force vector to the ECSF system.

4.2.2 Perturbation of the Sun and the Moon as well as Planets

The equations of motion of two point-masses M and m under their mutual action can be given by

$$M\ddot{\vec{r}}_M = GMm\frac{\vec{r}_{Mm}}{r_{Mm}^3} \quad \text{and} \quad m\ddot{\vec{r}}_m = GMm\frac{\vec{r}_{mM}}{r_{mM}^3}, \tag{4.46}$$

where r is the length of the vector $\vec{r}$, index Mm means the vector is pointing from point-mass M to m, and single index M or m means the vector is pointing to point-mass M or m. Introducing additional point-masses $m(j)$, $j = 1,2,\ldots$, the attractions of $m(j)$ on M and m can be given as equations similar to (4.46), and the total attractions may be obtained by summations

$$M\ddot{\vec{r}}_M = GMm\frac{\vec{r}_{Mm}}{r_{Mm}^3} + \sum_j GMm(j)\frac{\vec{r}_{Mm(j)}}{r_{Mm(j)}^3},$$
$$m\ddot{\vec{r}}_m = GMm\frac{\vec{r}_{mM}}{r_{mM}^3} + \sum_j Gmm(j)\frac{\vec{r}_{mm(j)}}{r_{mm(j)}^3}. \tag{4.47}$$

By dividing these two equations with $-M$ and m, respectively, then adding them together, one has

$$\ddot{\vec{r}}_m - \ddot{\vec{r}}_M = -G(M+m)\frac{\vec{r}_{Mm}}{r_{mM}^3} + \sum_j Gm(j)\left[\frac{\vec{r}_{mm(j)}}{r_{mm(j)}^3} - \frac{\vec{r}_{Mm(j)}}{r_{Mm(j)}^3}\right]. \tag{4.48}$$

Letting $\vec{r} = \vec{r}_m - \vec{r}_M$, i.e., using the point-mass M as the origin, substituting $\vec{r}_{mm(j)} = -(\vec{r}_m - \vec{r}_{m(j)})$ in the right side of (4.48) and omitting the mass m (mass of satellite), one has

$$\ddot{\vec{r}} = -GM\frac{\vec{r}}{r^3} - \sum_j Gm(j)\left[\frac{\vec{r}-\vec{r}_{m(j)}}{\left|\vec{r}-\vec{r}_{m(j)}\right|^3} + \frac{\vec{r}_{m(j)}}{r^3_{m(j)}}\right]. \tag{4.49}$$

It is obvious that the first term on the right side is the central force of the Earth; therefore, the disturbance forces of multiple point-masses acting on the satellite are then

$$\vec{f}_m = -m\sum_j Gm(j)\left[\frac{\vec{r}-\vec{r}_{m(j)}}{\left|\vec{r}-\vec{r}_{m(j)}\right|^3} + \frac{\vec{r}_{m(j)}}{r^3_{m(j)}}\right], \tag{4.50}$$

where $Gm(j)$ are the gravitational constants of the sun and the moon as well as the planets.

4.2.3 Earth Tide and Ocean Tide Perturbations

The tidal potential generated by the moon and the sun can be written as

$$W_P = \sum_{j=1}^{2}\mu_j\sum_{n=2}^{\infty}\frac{\rho^n}{r_j^{n+1}}P_n(\cos z_j)$$

or

$$W_P = \sum_{j=1}^{2}\mu_j\sum_{n=2}^{\infty}\frac{\rho^n}{r_j^{n+1}}\left[\begin{array}{l}P_n(\sin\varphi)P_n(\sin\delta_j)\\ +2\sum\limits_{k=1}^{n}\frac{(n-k)!}{(n+k)!}P_{nk}(\sin\varphi)P_{nk}(\sin\delta_j)\cos kh_j\end{array}\right], \tag{4.51}$$

where j is the index of the moon ($j = 1$) and the sun ($j = 2$), μ_j is the gravitational constant of body j, ρ is the geocentric distance of the Earth's surface (set as a_e), r_j is the geocentric distance of the body j, $P_n(x)$ and $P_{nk}(x)$ are the Legendre function and associated Legendre function, z_j is the zenith distance of the body j, δ_j and h_j are the declination and local hour angle of body j, $h_j = H_j - \lambda$, and H_j is the hour angle of j (see Fig. 4.2). The tidal deformation of the Earth caused by the tidal potential can be considered a tidal deformation potential acting on the satellite by Dirichlet's theorem (Melchior, 1978; Dow, 1988):

$$\delta V = \sum_{j=1}^{2}\mu_j\sum_{n=2}^{\infty}k_n\left(\frac{\rho}{r}\right)^{n+1}\frac{\rho^n}{r_j^{n+1}}P_n(\cos z_j)$$

or

$$\delta V = \sum_{j=1}^{2}\mu_j\sum_{n=2}^{N}k_n\frac{a_e^{2n+1}}{r^{n+1}r_j^{n+1}}\left[\begin{array}{l}P_n(\sin\varphi)P_n(\sin\delta_j)\\ +2\sum\limits_{k=1}^{n}\frac{(n-k)!}{(n+k)!}P_{nk}(\sin\varphi)P_{nk}(\sin\delta_j)\cos kh_j\end{array}\right], \tag{4.52}$$

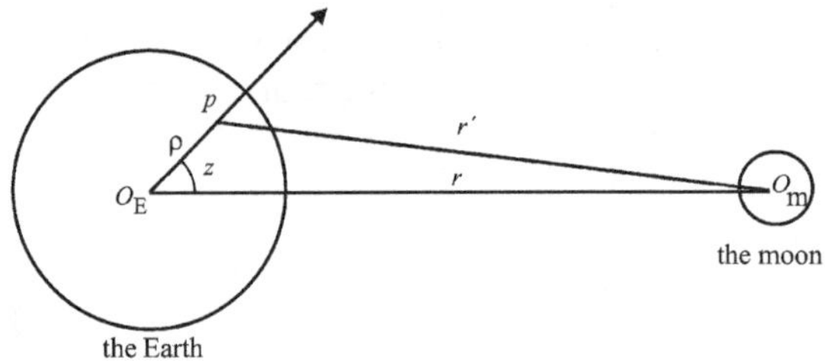

Fig. 4.2 The Earth-moon system

where k_n is the Love number, (r, φ, λ) is the spherical coordinate of the satellite in the ECEF system, and N is the truncating number. The recurrence formulas of the Legendre function are (see, e.g., Xu, 1992)

$$\begin{aligned}
&(n+1)P_{n+1}(x) = (2n+1)xP_n(x) - nP_{n-1}(x),\\
&(1-x^2)\frac{\mathrm{d}P_n(x)}{\mathrm{d}x} = nP_{n-1}(x) - nxP_n(x),\\
&P_0(x) = 1,\, P_1(x) = x.
\end{aligned} \tag{4.53}$$

The disturbing force vector of the tidal potential in the ECEF coordinate system is then

$$\vec{f}_{\mathrm{ECEF}} = \begin{pmatrix} \dfrac{\partial \delta V}{\partial x'} \\ \dfrac{\partial \delta V}{\partial y'} \\ \dfrac{\partial \delta V}{\partial z'} \end{pmatrix} = \begin{pmatrix} \dfrac{\partial \delta V}{\partial (r,\varphi,\lambda)}\dfrac{\partial (r,\varphi,\lambda)}{\partial x'} \\ \dfrac{\partial \delta V}{\partial (r,\varphi,\lambda)}\dfrac{\partial (r,\varphi,\lambda)}{\partial y'} \\ \dfrac{\partial \delta V}{\partial (r,\varphi,\lambda)}\dfrac{\partial (r,\varphi,\lambda)}{\partial z'} \end{pmatrix} = \left(\frac{\partial \delta V}{\partial (r,\varphi,\lambda)}\frac{\partial (r,\varphi,\lambda)}{\partial (x',y',z')}\right)^T, \tag{4.54}$$

where

$$\frac{\partial \delta V}{\partial r} = \sum_{j=1}^{2}\mu_j \sum_{n=2}^{N} -k_n \frac{(n+1)a_\mathrm{e}^{2n+1}}{r^{n+2}r_j^{n+1}} \left[\begin{aligned}&P_n(\sin\varphi)P_n(\sin\delta_j)\\ &+2\sum_{k=1}^{n}\frac{(n-k)!}{(n+k)!}P_{nk}(\sin\varphi)P_{nk}(\sin\delta_j)\cos kh_j\end{aligned}\right],$$

$$\frac{\partial \delta V}{\partial \varphi} = \sum_{j=1}^{2}\mu_j \sum_{n=2}^{N} k_n \frac{a_\mathrm{e}^{2n+1}}{r^{n+1}r_j^{n+1}} \left[\begin{aligned}&\frac{n}{\cos\varphi}(P_{n-1}(\sin\varphi) - \sin\varphi P_n(\sin\varphi))P_n(\sin\delta_j)\\ &+2\sum_{k=1}^{n}\frac{(n-k)!}{(n+k)!}(P_{n(k+1)}(\sin\varphi) - k\tan\varphi P_{nk}(\sin\varphi))\\ &\cdot P_{nk}(\sin\delta_j)\cos kh_j\end{aligned}\right]$$

and

$$\frac{\partial \delta V}{\partial \lambda} = \sum_{j=1}^{2}\mu_j \sum_{n=2}^{N} k_n \frac{a_\mathrm{e}^{2n+1}}{r^{n+1}r_j^{n+1}} \left[2\sum_{k=1}^{n}\frac{(n-k)!}{(n+k)!}kP_{nk}(\sin\varphi)P_{nk}(\sin\delta_j)\sin kh_j\right]. \tag{4.55}$$

Other partial derivatives in (4.54) have been given in Sect. 4.2.1. The transformation of the force vector from the ECEF to the ECSF coordinate system can be made by (4.45).

The ocean tidal potential generated by tide element $\sigma H \mathrm{d}s$ can be written as

$$\frac{G\sigma H \mathrm{d}s}{r'} \quad \text{or} \quad G\sigma H \mathrm{d}s \sum_{n=0}^{\infty} \frac{a_\mathrm{e}^n}{r^{n+1}} P_n(\cos z), \tag{4.56}$$

where H is the ocean tide height of the area $\mathrm{d}s$, G is the gravitational constant, σ is the water density, r' is the distance between the satellite and the water element $\mathrm{d}s$, r is the geocentric distance of the satellite, z is the zenith distance of the $\mathrm{d}s$, and a_e is the radius of the Earth. Using the spherical triangle

$$\cos z = \sin\varphi \sin\varphi_s + \cos\varphi \cos\varphi_s \cos(\lambda_s - \lambda),$$

where (φ_s, λ_s) is the spherical coordinate of $\mathrm{d}s$ and (r, φ, λ) is the spherical coordinate of satellite in the ECEF system, (4.56) turns out to be (denoted by Q)

$$Q = G\sigma H \mathrm{d}s \sum_{n=0}^{\infty} \frac{a_\mathrm{e}^n}{r^{n+1}} \left[\begin{array}{l} P_n(\sin\varphi)P_n(\sin\varphi_s) + (2-\delta_{0n}) \\ \times \sum\limits_{k=0}^{n} \frac{(n-k)!}{(n+k)!} P_{nk}(\sin\varphi) P_{nk}(\sin\varphi_s) \cos k(\lambda_s - \lambda) \end{array} \right]. \tag{4.57}$$

The direct ocean tide potential is then the integration of $Q/\mathrm{d}s$ over the ocean (denoted by O), including the potential of the deformation of the ocean loading. The ocean tide potential is then

$$\delta V_1 = \oint\!\!\oint_O G\sigma H \sum_{n=0}^{\infty} (1+k'_n) \frac{a_\mathrm{e}^n}{r^{n+1}} \left[\begin{array}{l} P_n(\sin\varphi)P_n(\sin\varphi_s) + (2-\delta_{0n}) \\ \times \sum\limits_{k=0}^{n} \frac{(n-k)!}{(n+k)!} P_{nk}(\sin\varphi) P_{nk}(\sin\varphi_s) \cos k(\lambda_s - \lambda) \end{array} \right] \mathrm{d}s, \tag{4.58}$$

where k'_n is the ocean loading Love number. Equation (4.58) does not include the potential changing because of the loading deformation over the continents, which may give a non-negligible contribution to the orbit motion of the satellite (see Knudsen et al., 1999). The loading deformation generated by the ocean tide can be represented as

$$u_r(\varphi, \lambda) = \oint\!\!\oint_{\text{ocean}} \sigma H u(z) \mathrm{d}s$$

and

$$u(z) = \frac{a_\mathrm{e} h'_\infty}{2M \sin(z/2)} + \frac{a_\mathrm{e}}{M} \sum_{n=0}^{N} (h'_n - h'_\infty) P_n(\cos z), \tag{4.59}$$

where a_e is the radius of the Earth, M is the mass of the Earth, z is the geocentric zenith distance of the loading point (related to the computing point, see Fig. 4.3), $P_n(\cos z)$ is the Legendre function, $u(z)$ is the radial loading displacement Green function, h'_n is the loading Love number of order n, and u_r is the radial loading

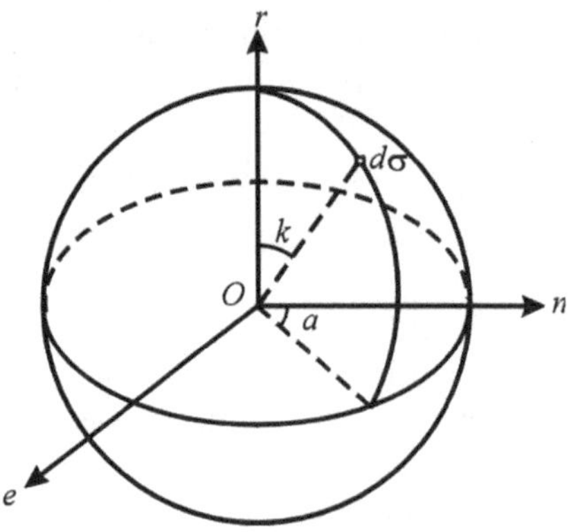

Fig. 4.3 Ocean tide and loading

deformation. Substituting u_r for H in (4.57) and integrating $Q/\mathrm{d}s$ over the continents (denoted by C), the potential of the loading deformation is then

$$\delta V_2 = \oiint\limits_C G\sigma_e u_r \sum_{n=0}^{\infty} \frac{a_e^n}{r^{n+1}} \begin{bmatrix} P_n(\sin\varphi)P_n(\sin\varphi_s) + (2-\delta_{0n}) \\ \times \sum_{k=0}^{n} \frac{(n-k)!}{(n+k)!} P_{nk}(\sin\varphi)P_{nk}(\sin\varphi_s)\cos k(\lambda_s-\lambda) \end{bmatrix} \mathrm{d}s, \tag{4.60}$$

where σ_e is the density of the mass $u_r\mathrm{d}s$ on the Earth's surface. The total ocean tide potential disturbance is the summation of (4.58) and (4.60). Similar to (4.54), the disturbing force can be derived and transformed to the ECSF system. There are

$$\vec{f}_{\text{ECEF}} = \begin{pmatrix} \dfrac{\partial(\delta V_1+\delta V_2)}{\partial x'} \\ \dfrac{\partial(\delta V_1+\delta V_2)}{\partial y'} \\ \dfrac{\partial(\delta V_1+\delta V_2)}{\partial z'} \end{pmatrix} = \left(\frac{\partial(\delta V_1+\delta V_2)}{\partial(r,\varphi,\lambda)} \frac{\partial(r,\varphi,\lambda)}{\partial(x',y',z')} \right)^T, \tag{4.61}$$

where

$$\frac{\partial\delta V_1}{\partial r} = \oiint\limits_O G\sigma H \sum_{n=0}^{\infty} (1+k_n') \frac{-(n+1)a_e^n}{r^{n+2}} \begin{bmatrix} P_n(\sin\varphi)P_n(\sin\varphi_s) + (2-\delta_{0n}) \\ \times \sum_{k=0}^{n} \frac{(n-k)!}{(n+k)!} P_{nk}(\sin\varphi)P_{nk}(\sin\varphi_s)\cos k(\lambda_s-\lambda) \end{bmatrix} \mathrm{d}s,$$

$$\frac{\partial\delta V_1}{\partial \varphi} = \oiint\limits_O G\sigma H \sum_{n=0}^{\infty} (1+k_n') \frac{a_e^n}{r^{n+1}} \begin{bmatrix} \frac{\mathrm{d}P_n(\sin\varphi)}{\mathrm{d}\varphi} P_n(\sin\varphi_s) + (2-\delta_{0n}) \\ \times \sum_{k=0}^{n} \frac{(n-k)!}{(n+k)!} \frac{\mathrm{d}P_{nk}(\sin\varphi)}{\mathrm{d}\varphi} P_{nk}(\sin\varphi_s)\cos k(\lambda_s-\lambda) \end{bmatrix} \mathrm{d}s,$$

$$\frac{\partial\delta V_1}{\partial \lambda} = \oiint\limits_O G\sigma H \sum_{n=0}^{\infty} (1+k_n') \frac{a_e^n}{r^{n+1}} \begin{bmatrix} (2-\delta_{0n}) \\ \times \sum_{k=0}^{n} \frac{(n-k)!}{(n+k)!} P_{nk}(\sin\varphi)P_{nk}(\sin\varphi_s) k \sin k(\lambda_s-\lambda) \end{bmatrix} \mathrm{d}s,$$

$$\frac{\partial \delta V_2}{\partial r} = \oiint\limits_C G\sigma_e u_r \sum_{n=0}^{\infty} \frac{-(n+1)a_e^n}{r^{n+2}} \begin{bmatrix} P_n(\sin\varphi)P_n(\sin\varphi_s) + (2-\delta_{0n}) \\ \times \sum\limits_{k=0}^{n} \dfrac{(n-k)!}{(n+k)!} P_{nk}(\sin\varphi)P_{nk}(\sin\varphi_s)\cos k(\lambda_s - \lambda) \end{bmatrix} \mathrm{d}s,$$

$$\frac{\partial \delta V_2}{\partial \varphi} = \oiint\limits_C G\sigma_e u_r \sum_{n=0}^{\infty} \frac{a_e^n}{r^{n+1}} \begin{bmatrix} \dfrac{\mathrm{d}P_n(\sin\varphi)}{\mathrm{d}\varphi} P_n(\sin\varphi_s) + (2-\delta_{0n}) \\ \times \sum\limits_{k=0}^{n} \dfrac{(n-k)!}{(n+k)!} \dfrac{\mathrm{d}P_{nk}(\sin\varphi)}{\mathrm{d}\varphi} P_{nk}(\sin\varphi_s)\cos k(\lambda_s - \lambda) \end{bmatrix} \mathrm{d}s$$

and

$$\frac{\partial \delta V_2}{\partial \lambda} = \oiint\limits_C G\sigma_e u_r \sum_{n=0}^{\infty} \frac{a_e^n}{r^{n+1}} \begin{bmatrix} (2-\delta_{0n}) \\ \times \sum\limits_{k=0}^{n} \dfrac{(n-k)!}{(n+k)!} P_{nk}(\sin\varphi)P_{nk}(\sin\varphi_s) k \sin k(\lambda_s - \lambda) \end{bmatrix} \mathrm{d}s. \tag{4.62}$$

4.2.4 Solar Radiation Pressure

Solar radiation pressure is sunlight-exerted force acting on the satellite's surface. The radiation force (see, e.g., Seeber, 1993) can be represented as

$$\vec{f}_{\mathrm{solar}} = m\gamma P_{\mathrm{s}} C_{\mathrm{r}} r_{\mathrm{sun}}^2 \frac{S}{m} \frac{\vec{r} - \vec{r}_{\mathrm{sun}}}{\left|\vec{r} - \vec{r}_{\mathrm{sun}}\right|^3}, \tag{4.63}$$

where γ is the shadow factor, P_{s} is the luminosity of the sun, C_{r} is the surface reflectivity, r_{sun} is the geocentric distance of the sun, (S/m) is the surface to mass ratio of the satellite, and $\vec{r}$ and $\vec{r}_{\mathrm{sun}}$ are the geocentric vectors of the satellite and the sun. Usually, P_{s} has the value of $4.5605 \times 10^{-6}\,\mathrm{Nm}^{-1}$ (Newton/meter), C_{r} has values from 1 to 2, 1 is for the complete absorption of the sunlight, and for aluminium, $C_{\mathrm{r}} = 1.95$.

The shadow factor is defined as

$$\gamma = 1 - \frac{A_{\mathrm{ss}}}{A_{\mathrm{s}}}, \tag{4.64}$$

where A_{s} is the sight surface of the sun viewed from the satellite, and A_{ss} is the shadowed sight surface of the sun. The sunlight may be shadowed by the Earth and the moon. For convenience, we will discuss both parameters that are only in the satellite-Earth-sun system (see Fig. 4.4). It is obvious that the half sight angles of the Earth and the moon, as well as the sun, viewed from the satellite are

$$\begin{aligned} \alpha_{\mathrm{e}} &= \sin^{-1}\left(\frac{a_{\mathrm{e}}}{|\vec{r}|}\right), \\ \alpha_{\mathrm{m}} &= \sin^{-1}\left(\frac{a_{\mathrm{m}}}{|\vec{r}_{\mathrm{m}} - \vec{r}|}\right), \\ \alpha_{\mathrm{s}} &= \sin^{-1}\left(\frac{a_{\mathrm{s}}}{|\vec{r}_{\mathrm{s}} - \vec{r}|}\right), \end{aligned} \tag{4.65}$$

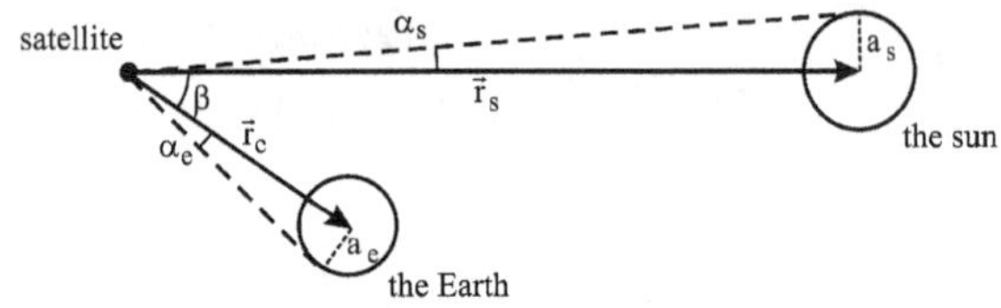

Fig. 4.4 Satellite–Earth–sun system

where a_e, a_s and a_m are semi-major radii of the Earth, sun and moon, respectively; $a_m = 0.272493a_e$, and $a_s = 959.63\pi/(3600 \times 180)$ (AU). For the GPS satellite, $\alpha_s < 0.3°$, $\alpha_e \approx 16.5°$ and $\alpha_m \approx \alpha_s \pm 0.03°$. Furthermore, $A_s = \alpha_s^2\pi$ and $A_m = \alpha_m^2\pi$. The angles between the centre of the Earth and the sun, as well as the centre of the moon and the sun are

$$\beta_{es} = \cos^{-1}\left(\frac{-\vec{r}\cdot(\vec{r}_s - \vec{r})}{r\,|\vec{r}_s - \vec{r}|}\right),$$
$$\beta_{ms} = \cos^{-1}\left(\frac{(\vec{r}_m - \vec{r})\cdot(\vec{r}_s - \vec{r})}{|\vec{r}_m - \vec{r}|\cdot|\vec{r}_s - \vec{r}|}\right), \tag{4.66}$$

where the vectors with indices s and m are the geocentric vectors of the sun and moon, respectively. The vector without an index is the geocentric vector of the satellite, and $r = |\vec{r}|$. If $\beta_{es} \geq \alpha_e + \alpha_s$, then the satellite is not in the shadow of the Earth (i.e., $A_{ss} = 0$). If $\beta_{es} \geq \alpha_e - \alpha_s$, then the sun is not in view of the satellite (i.e., $A_{ss} = A_s$). If $\alpha_e - \alpha_s < \beta_{es} < \alpha_e + \alpha_s$, then the sunlight is partly shadowed by the Earth. The formula of the shadowed surface can be derived as follows (see Fig. 4.5). The two circles with radius α_e and α_s cut each other at points p and q, line $\overline{pq}$ is called a chord (denoted by 2a), the chord-related central angle at origin o_s is denoted by ϕ_1, the surface area between the chord and the arc of the circle α_s on the right side of the chord is denoted by A_1. Line $\overline{pq}$ cuts $\overline{O_sO_e}$ at point g, while $\overline{O_sg}$ and $\overline{gO_e}$ are denoted by b and b_1. Then one has

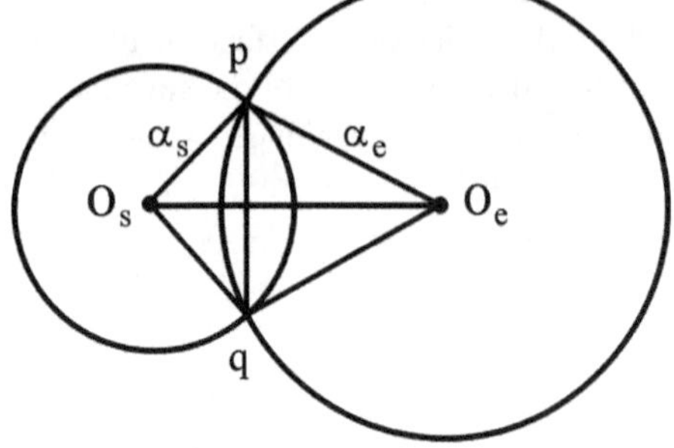

Fig. 4.5 Shadowed surface area

$$a^2 = \alpha_s^2 - b^2, \quad b_1 = \frac{\alpha_e^2 + \beta_{es}^2 - \alpha_s^2}{2\beta_{es}},$$

$$b = \begin{cases} \beta_{es} - b_1 & \text{if} \quad b_1 \le \alpha_e, \\ b_1 - \beta_{es} & \text{if} \quad b_1 > \alpha_e, \end{cases}$$

$$\phi_1 = \begin{cases} 2\cos^{-1}\left(\dfrac{b}{\alpha_s}\right) & \text{if} \quad b_1 \le \alpha_e, \\ 2\pi - 2\cos^{-1}\left(\dfrac{b}{\alpha_s}\right) & \text{if} \quad b_1 > \alpha_e, \end{cases} \tag{4.67}$$

$$A_1 = \begin{cases} \dfrac{1}{2}\phi_1\alpha_s^2 - ab & \text{if} \quad b_1 \le \alpha_e, \\ \dfrac{1}{2}\phi_1\alpha_s^2 + ab & \text{if} \quad b_1 > \alpha_e. \end{cases}$$

Similarly, the chord-related central angle at origin o_e is denoted by ϕ_2, while the surface area between the chord and the arc of the circle α_e on the left side of chord is denoted by A_2. Then one has

$$\phi_2 = 2\cos^{-1}\left(\frac{b_1}{\alpha_e}\right) \quad A_2 = \frac{1}{2}\phi_2\alpha_e^2 - ab_1 \tag{4.68}$$

and

$$\gamma = 1 - \frac{A_1 + A_2}{\alpha_s^2 \pi}. \tag{4.69}$$

A similar discussion can be given for the moon. If $\beta_{ms} \ge \alpha_m + \alpha_s$, then the satellite is not in the shadow of the moon, i.e., $A_{ss} = 0$. If $\beta_{ms} \ge \alpha_m - \alpha_s$, then the full shadow has occurred, i.e., $A_{ss} = \min(A_s, A_m)$. If $|\alpha_m - \alpha_s| < \beta_{ms} < \alpha_m + \alpha_s$, then the sunlight is partially shadowed by the moon. The formula of the shadowed surface can be similarly derived by changing the index e to m in (4.67) and (4.68). Because of the small sight angle of the moon viewed from the satellite, the shadowed time will be very short if it happens. By GPS satellite dynamic orbit determination (e.g., in IGS orbit determination), only the data that have the γ value of 0 or 1 are used.

Because of the complex shape of the satellite and the use of constant reflectivity and homogenous luminosity of the sun, as well as the existence of indirect solar radiation (reflected from the Earth's surface), the model of (4.63) discussed earlier is not accurate enough and will be used as a first order approximation. A further model for the adjustment to fit solar radiation effects is needed.

The force vector is pointed from the sun to the satellite. The satellite fixed coordinate system is introduced in Sect. 2.7 (see Sect. 2.7 for details). The solar radiation force vector in the ECSF system is then

$$\begin{aligned}\vec{f}_{\text{solar}} &= m\gamma P_s C_r \frac{S}{m} \frac{r_{\text{sun}}^2}{|\vec{r}-\vec{r}_{\text{sun}}|^2} \vec{n}_{\text{sun}} \\ &= m\gamma P_s C_r \frac{S}{m} \frac{r_{\text{sun}}^2}{|\vec{r}-\vec{r}_{\text{sun}}|^2} (\sin\beta \cdot \vec{e}_x + \cos\beta \cdot \vec{e}_z),\end{aligned} \tag{4.70}$$

where

$$\vec{e}_z = -\frac{\vec{r}}{|\vec{r}|}, \quad \vec{e}_y = \frac{\vec{e}_z \times \vec{n}_{\text{sun}}}{|\vec{e}_z \times \vec{n}_{\text{sun}}|}, \quad \vec{e}_x = \vec{e}_y \times \vec{e}_z \quad \text{and} \quad \vec{n}_{\text{sun}} = \frac{\vec{r}-\vec{r}_{\text{sun}}}{|\vec{r}-\vec{r}_{\text{sun}}|}. \tag{4.71}$$

Further formulas of (4.71) can be found in Sect.2.7. Taking the remaining error of the radiation pressure into account, the solar radiation force model can be represented as (see Fliegel et al., 1992; Beutler et al., 1994)

$$\vec{f}_{\text{solar-force}} = \vec{f}_{\text{solar}} + \begin{pmatrix} a_{11} & a_{12} & a_{13} \\ a_{21} & a_{22} & a_{23} \\ a_{31} & a_{32} & a_{33} \end{pmatrix} \begin{pmatrix} 1 \\ \cos u \\ \sin u \end{pmatrix}. \tag{4.72}$$

That is, nine parameters are used to model the solar radiation force error for every satellite.

An alternative adjustment model of solar radiation is given by introducing a so-called disturbance coordinate system and will be outlined in the next section (see Xu, 2004).

Disturbance Coordinate System and Radiation Error Model

The solar radiation force vector is pointed from the sun to the satellite. If the shadow factor is computed exactly, the luminosity of the sun is a constant, and the surface reflectivity of the satellite is a constant, then the length of the solar force vector can also be considered a constant, because

$$\frac{r_{\text{sun}}^2}{(r_{\text{sun}}+r)^2} \le \frac{r_{\text{sun}}^2}{|\vec{r}-\vec{r}_{\text{sun}}|^2} \le \frac{r_{\text{sun}}^2}{(r_{\text{sun}}-r)^2} \tag{4.73}$$

and

$$\frac{r_{\text{sun}}^2}{(r_{\text{sun}} \pm r)^2} = \left(\frac{r_{\text{sun}}}{r_{\text{sun}} \pm r}\right)^2 \approx \left(1 \mp \frac{r}{r_{\text{sun}}} \pm \ldots\right)^2 \approx 1 \mp \frac{2r}{r_{\text{sun}}} \approx 1 \mp 3 \times 10^{-5}.$$

Any bias error in P_s, C_r and (S/m) may cause a model error of $\alpha\vec{f}_{\text{solar}}$, where α is a parameter. So the $\alpha\vec{f}_{\text{solar}}$ can be considered a main error model of the solar radiation. Because the ratio of the geocentric distances of the satellite and the sun is so small, the direction and distance changes of the sun-satellite vector are negligible. With the motion of the sun, the solar radiation force vector changes its direction with the time in the ECSF (Earth-Centred-Space-Fixed) coordinate system ca. 1 degree per day. Such an effect can only be considered a small drift, not a periodical change for orbit determination. To model such an effect in the ECSF system one needs three bias parameters in three coordinate axes and three drift terms instead of a few periodical parameters. It is obvious that to model such an effect in the direction of $\vec{n}$, just one parameter α is needed. Therefore, it is very advantageous to define a so-called disturbance coordinate system as follows: the origin is the mass-centre of the satellite, and the three axes are defined by $\vec{r}$ (radial vector of the satellite), $\vec{n}$ (the sun-satellite identity vector) and $\vec{p}$ (the atmospheric drag identity vector). These three axes are always in the main disturbance directions of the indirect solar radiation (reflected from the Earth's surface), direct solar radiation and atmospheric drag, respectively (see Fig. 4.6). This coordinate system is not a Cartesian one and the axes are not orthogonal to each other. The parameters in individual axes are mainly used to model the related disturbance effects, and meanwhile to absorb the remained error of other un-modelled effects.

In the so-called disturbance coordinate system, the solar radiation pressure error model can be represented alternatively by (see Xu, 2004)

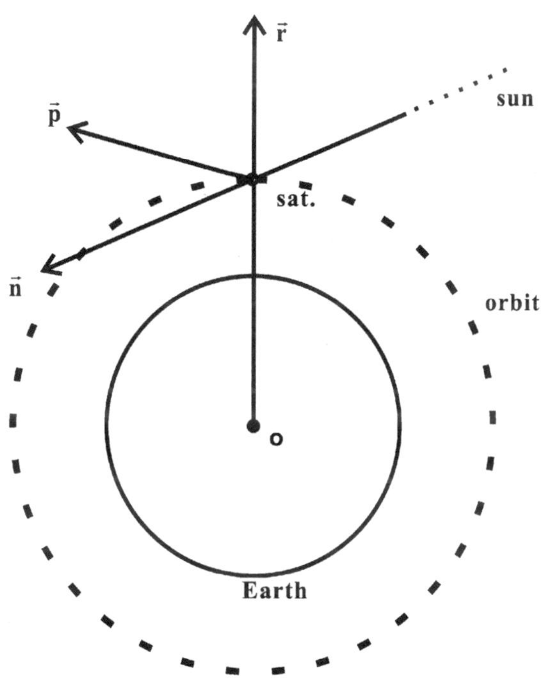

Fig. 4.6 Disturbance coordinate system

$$\alpha \vec{f}_{\text{solar}} = \begin{pmatrix} a_1 & b_1 \\ a_2 & b_2 \\ a_3 & b_3 \end{pmatrix} \begin{pmatrix} 1 \\ t \end{pmatrix}, \tag{4.74}$$

where b–terms are very small.

4.2.5 Atmospheric Drag

Atmospheric drag is the disturbance force acting on the satellite's surface caused by the air. Air drag force can be represented as (see, e.g., Seeber, 1993; Liu and Zhao, 1979)

$$\vec{f}_{\text{drag}} = -m\frac{1}{2}\left(\frac{C_{\text{d}}S}{m}\right)\sigma\left|\dot{\vec{r}} - \dot{\vec{r}}_{\text{air}}\right|\left(\dot{\vec{r}} - \dot{\vec{r}}_{\text{air}}\right), \tag{4.75}$$

where S is the cross section (or effective area) of the satellite, C_{d} is the drag factor, m is the mass of the satellite, $\dot{\vec{r}}$ and $\dot{\vec{r}}_{\text{air}}$ are the geocentric velocity vectors of the satellite and the atmosphere, and σ is the density of the atmosphere. Usually, S has a value of 1/4 of the outer surface area of the satellite, and C_{d} has labour values of 2.2 ± 0.2. The velocity vector of the atmosphere can be modelled by

$$\dot{\vec{r}}_{\text{air}} = k\vec{\omega} \times \vec{r} = k\omega \begin{pmatrix} -y \\ x \\ 0 \end{pmatrix}, \tag{4.76}$$

where $\vec{\omega}$ is the angular velocity vector of the Earth's rotation, and $\omega = |\vec{\omega}|$, k is the atmospheric rotation factor. For the lower layer of the atmosphere, $k = 1$, i.e., the lower layer of the atmosphere is considered to be rotating with the Earth. For the higher layer, $k = 1.2$, because the higher ionosphere is accelerated by the Earth's magnetic field.

The gravity-balanced atmospheric-density model has the exponential form of (see Liu and Zhao, 1979)

$$\sigma = \sigma_0(1+q)\exp\left(-\frac{r-\rho}{H}\right), \tag{4.77}$$

where σ_0 is the atmospheric density at the reference point ρ, q is the daily change factor of the density, r is the geocentric distance of the satellite, and H is the density-height scale factor. For the spherical and rotating ellipsoidal layer atmospheric models, one has

$$\rho = a_{\text{e}} + h_i \tag{4.78}$$

and

$$\rho = (a_{\text{e}} + h_i)\sqrt{1-e^2}\sqrt{\frac{1+\tan^2\varphi}{1+\tan^2\varphi - e^2}}, \tag{4.79}$$

respectively. Where a_{e} is the semi-major radius of the Earth, $h_{\text{i}}\,(i = 1, 2, \ldots)$ is a set of numbers, φ is the geocentric latitude of the satellite, and e is the eccentricity of the ellipsoid. Equations (4.78) and (4.79) represent sphere with radius $a_\mathrm{e}+h_\mathrm{i}$ and rotating ellipsoid with semi-major axis $a_\mathrm{e}+h_\mathrm{i}$. Equation (4.79) can be derived from the relation of $\tan\varphi$ and the ellipsoid equation

$$z^2 = (x^2+y^2)\tan^2\varphi,$$
$$x^2+y^2+z^2\frac{1}{1-e^2} = (a_\mathrm{e}+h_i)^2.$$

A reference of atmospheric densities can be read from Table 4.1 (see Seeber, 1993).

The density–height scale H between every two layers can be then computed from these values. It is notable that the air density may change its value up to a factor of 10 because of the radiation of the sun. The density of the atmosphere at a defined point reaches its maximum value at 14h local time and its minimum at 3.5h. The most significant period of change is the daily change and is represented by the daily changing factor as

$$q = \frac{f-1}{f+1}\cos\psi, \tag{4.80}$$

where f is the ratio of the maximum density and the minimum density, and ψ is the angle between the satellite vector $\vec{r}$ and the daily maximum density direction $\vec{r}_m$. The f may have the value of three and

$$\cos\psi = \frac{\vec{r}\cdot\vec{r}_\mathrm{m}}{|\vec{r}|\cdot|\vec{r}_\mathrm{m}|}, \tag{4.81}$$

where

$$\begin{pmatrix}x\\y\\z\end{pmatrix}_\mathrm{sun} = \begin{pmatrix}r\cos\delta\cos\alpha\\ r\cos\delta\sin\alpha\\ r\sin\delta\end{pmatrix}, \quad \begin{pmatrix} r=\sqrt{x^2+y^2+z^2}\\ \delta=\tan^{-1}\dfrac{z}{\sqrt{x^2+y^2}}\\ \alpha=\tan^{-1}\dfrac{y}{x}\end{pmatrix},$$
$$\vec{r}_\mathrm{m} = \begin{pmatrix}x\\y\\z\end{pmatrix}_\mathrm{m} = \begin{pmatrix}r\cos\delta\cos(\alpha+\pi/6)\\ r\cos\delta\sin(\alpha+\pi/6)\\ r\sin\delta\end{pmatrix}, \tag{4.82}$$

where (α,δ) are the coordinates (right ascension, latitude) of the sun in the ECSF coordinate system.

Table 4.1 Reference of atmospheric densities

h_i (km)	$\sigma_0(i)$ (gkm^{-3})	h_i (km)	$\sigma_0(i)$ (gkm^{-3})
100	497400	600	$0.08-0.64$
200	$255-316$	700	$0.02-0.22$
300	$17-35$	800	$0.07-0.01$
400	$2.2-7.5$	900	$0.003-0.04$
500	$0.4-2.0$	1,000	$0.001-0.02$

Taking the remaining error of the atmospheric drag into account, the air drag force model can be represented as

$$\vec{f}_{\text{air-drag}} = \vec{f}_{\text{drag}} + (1+q)\Delta\vec{f}_{\text{drag}}, \tag{4.83}$$

where the force error vector is denoted by $\Delta\vec{f}_{\text{drag}}$and the time variation part of atmospheric density is considered in parameter q.

Error Model in Disturbance Coordinate System

In the atmospheric drag model (4.75), the velocity vector of the atmosphere is always perpendicular to the z–axis of the ECSF coordinates and the satellite velocity vector is always in the tangential direction of the orbit. The variation of the term $\left|\dot{\vec{r}} - \dot{\vec{r}}_{\text{air}}\right|$ (denoted by g) is dominated by the direction changes of the velocity vectors of the satellite and the atmosphere. Any bias error in S (effective area of the satellite), C_d (drag factor) and σ (density of the atmosphere) may cause a model error of $\mu\vec{f}_{\text{drag}}$, where μ is a parameter. So the $\mu\vec{f}_{\text{drag}}$ can be considered a main error model of the un–modelled atmospheric drag. To simplify our discussion, we consider the velocities of the satellite and atmosphere are constants, and call the satellite positions with max(z) and $-\max(z)$ the highest and lowest points, respectively. With the satellite at the lowest point, the two velocity vectors are in the same direction and therefore the g reaches the minimum. At the ascending node, the two vectors have the maximum angle of inclination i and the g reaches the maximum. Then g reaches the minimum again at the highest point and reaches the maximum again at the descending node, and at the end reaches the minimum at the lowest point. It is obvious that, besides the constant part, g has a dominant periodical component of $\cos 2f$ and $\sin 2f$, where f is the true anomaly of the satellite.

In the so-called disturbance coordinate system the atmospheric drag error model can be represented alternatively by (see Xu, 2004)

$$\mu\vec{f}_{\text{drag}} = [a + b\varphi(2\omega)\cos(2f) + c\varphi(3\omega)\cos(3f) + d\varphi(\omega)\cos f]\,\vec{p}, \tag{4.84}$$

where

$$\varphi(k\omega) = \begin{cases} \sin k\omega, & if\ \cos k\omega = 0 \\ \dfrac{1}{\cos k\omega}, & if\ \cos k\omega \neq 0 \end{cases}, \quad (k = 1,2,3) \tag{4.85}$$

where ω is the angle of perigee and f is the true anomaly of the satellite; a, b, c and d are model parameters to be determined. According to the simulation, a-term and b-term are the most significant terms. The amount of d is just about 1% of the amount of c, and the amount of c is about 1% of that of b.

4.2.6 Additional Perturbations

As mentioned earlier, the disturbed equation of motion of the satellite is valid only in an inertial coordinate system, or ECSF system. Therefore, the state vector and force vectors as well as the disturbing potential function have also to be represented in the ECSF system. As seen earlier, for some reason, the state vector and the force vectors as well as the disturbing potential function R, are sometimes given in the ECEF system and then transformed to the ECSF system by (see Sect. 4.2.4)

$$\begin{aligned} \vec{X}_{\mathrm{ECSF}} &= R_{\mathrm{t}} \cdot \vec{X}_{\mathrm{ECEF}}, \\ \vec{f}_{\mathrm{ECSF}} &= R_{\mathrm{t}} \cdot \vec{f}_{\mathrm{ECEF}}, \\ R_{\mathrm{ECSF}} &= R(R_{\mathrm{t}}^{-1} X_{\mathrm{ECSF}}) \quad \text{for} \quad R(X_{\mathrm{ECEF}}), \end{aligned} \tag{4.86}$$

where R_{t} is the transformation matrix in general. Variable transformation is further denoted by $X_{\mathrm{ECSF}} = R_{\mathrm{t}} X_{\mathrm{ECEF}}$. We have also seen that sometimes the state vectors (of the satellite, the sun, the moon) in the ECSF system have to be transformed to the ECEF system for use, and then the result vectors will be transformed back to the ECSF system again. However, due to the complication of transformation R_{t}^{-1}, quite often a simplified R_{s}^{-1} is used (in later discussions, for example, to represent the disturbing potential function using Keplerian elements, only the Earth rotation is considered). Thus,

$$R_{\mathrm{ECSF}} = \{R(R_{\mathrm{t}}^{-1} X_{\mathrm{ECSF}}) - R(R_{\mathrm{s}}^{-1} X_{\mathrm{ECSF}})\} + R(R_{\mathrm{s}}^{-1} X_{\mathrm{ECSF}}), \tag{4.87}$$

where the first term on the right side is the correction because of the approximation using the second term. The transformations of (4.86) and (4.87) are exact operations, and their differentiation with respect to time t and the partial derivatives with respect to variable X_{ECSF} are then

$$\begin{aligned} \frac{\mathrm{d}\vec{X}_{\mathrm{ECSF}}}{\mathrm{d}t} &= \frac{\mathrm{d}R_{\mathrm{t}}}{\mathrm{d}t} \vec{X}_{\mathrm{ECEF}} + R_{\mathrm{t}} \frac{\mathrm{d}\vec{X}_{\mathrm{ECEF}}}{\mathrm{d}t}, \\ \frac{\mathrm{d}\vec{f}_{\mathrm{ECSF}}}{\mathrm{d}t} &= \frac{\mathrm{d}R_{\mathrm{t}}}{\mathrm{d}t} \vec{f}_{\mathrm{ECEF}} + R_{\mathrm{t}} \frac{\mathrm{d}\vec{f}_{\mathrm{ECEF}}}{\mathrm{d}t}, \\ \frac{\partial R_{\mathrm{ECSF}}}{\partial X_{\mathrm{ECSF}}} &= \frac{\partial \left[R(R_{\mathrm{t}}^{-1} X_{\mathrm{ECSF}}) - R(R_{\mathrm{s}}^{-1} X_{\mathrm{ECSF}})\right]}{\partial X_{\mathrm{ECSF}}} + \frac{\partial R(R_{\mathrm{s}}^{-1} X_{\mathrm{ECSF}})}{\partial X_{\mathrm{ECSF}}}. \end{aligned} \tag{4.88}$$

That is, the time differentiations of the state vector and force vectors cannot be transformed directly as in (4.86). In other words, if the state vector and force vectors are not directly given in the ECSF system, they are not allowed to be differentiated as usual afterward. An approximated and transformed perturbing potential function will introduce an error. The first term on the right-hand side of (4.88) signifies additional perturbations (i.e., coordinate perturbations). The order of such perturbations can be estimated by the first term on the right-hand side. If the relationship between two coordinate systems changes with time or the transformation has not been made exactly, such perturbations will occur. Recalling

$$R = R_{\mathrm{P}}^{-1} R_{\mathrm{N}}^{-1} R_{\mathrm{S}}^{-1} R_{\mathrm{M}}^{-1}$$

and their definitions (see Chap. 2), one has

$$\begin{aligned}\frac{\mathrm{d}R}{\mathrm{d}t} &= R_{\mathrm{P}}^{-1} R_{\mathrm{N}}^{-1} R_{\mathrm{S}}^{-1} \frac{\mathrm{d}R_{\mathrm{M}}^{-1}}{\mathrm{d}t} + R_{\mathrm{P}}^{-1} R_{\mathrm{N}}^{-1} \frac{\mathrm{d}R_{\mathrm{S}}^{-1}}{\mathrm{d}t} R_{\mathrm{M}}^{-1} \\ &+ R_{\mathrm{P}}^{-1} \frac{\mathrm{d}R_{\mathrm{N}}^{-1}}{\mathrm{d}t} R_{\mathrm{S}}^{-1} R_{\mathrm{M}}^{-1} + \frac{\mathrm{d}R_{\mathrm{P}}^{-1}}{\mathrm{d}t} R_{\mathrm{N}}^{-1} R_{\mathrm{S}}^{-1} R_{\mathrm{M}}^{-1},\end{aligned} \tag{4.89}$$

where

$$\begin{aligned}\frac{\mathrm{d}R_{\mathrm{M}}^{-1}}{\mathrm{d}t} &= \begin{pmatrix} 0 & 0 & -\dot{x}_p \\ 0 & 0 & \dot{y}_p \\ \dot{x}_p & -\dot{y}_p & 0 \end{pmatrix}, \qquad \frac{\mathrm{d}R_{\mathrm{S}}^{-1}}{\mathrm{d}t} = \frac{\mathrm{d}R_3(\mathrm{GAST})}{\mathrm{d}t}, \\ \frac{\mathrm{d}R_{\mathrm{N}}^{-1}}{\mathrm{d}t} &= \frac{\mathrm{d}R_1(-\varepsilon)}{\mathrm{d}t} R_3(\Delta\psi) R_1(\varepsilon+\Delta\varepsilon) + R_1(-\varepsilon) \frac{\mathrm{d}R_3(\Delta\psi)}{\mathrm{d}t} R_1(\varepsilon+\Delta\varepsilon) \\ &+ R_1(-\varepsilon) R_3(\Delta\psi) \frac{\mathrm{d}R_1(\varepsilon+\Delta\varepsilon)}{\mathrm{d}t},\end{aligned} \tag{4.90}$$

$$\frac{\mathrm{d}R_{\mathrm{P}}^{-1}}{\mathrm{d}t} = \frac{\mathrm{d}R_3(\zeta)}{\mathrm{d}t} R_2(-\theta) R_3(z) + R_3(\zeta) \frac{\mathrm{d}R_2(-\theta)}{\mathrm{d}t} R_3(z) + R_3(\zeta) R_2(-\theta) \frac{\mathrm{d}R_3(z)}{\mathrm{d}t},$$

where all elements are defined and given in Chap. 2, $(\dot{x}_p, \dot{y}_p)$ is the polar motion rate of time, and

$$\begin{aligned}\frac{\mathrm{d}R_1(\alpha)}{\mathrm{d}t} &= \begin{pmatrix} 0 & 0 & 0 \\ 0 & -\sin\alpha & \cos\alpha \\ 0 & -\cos\alpha & -\sin\alpha \end{pmatrix} \frac{\mathrm{d}\alpha}{\mathrm{d}t}, \\ \frac{\mathrm{d}R_2(\alpha)}{\mathrm{d}t} &= \begin{pmatrix} -\sin\alpha & 0 & -\cos\alpha \\ 0 & 0 & 0 \\ \cos\alpha & 0 & -\sin\alpha \end{pmatrix} \frac{\mathrm{d}\alpha}{\mathrm{d}t}, \\ \frac{\mathrm{d}R_3(\alpha)}{\mathrm{d}t} &= \begin{pmatrix} -\sin\alpha & \cos\alpha & 0 \\ -\cos\alpha & -\sin\alpha & 0 \\ 0 & 0 & 0 \end{pmatrix} \frac{\mathrm{d}\alpha}{\mathrm{d}t}.\end{aligned} \tag{4.91}$$

Further formulas may be easily derived.

4.2.7 Order Estimations of Perturbations

Perturbation forces that are scaled by the mass of the satellite are the accelerations. The accelerations caused by the discussed forces have been estimated for the GPS satellite by several authors and are summarised in Table 4.2.

Table 4.2 Accelerations (ms^{-2}) caused by forces (see Seeber, 1993; Kang, 1998)

Central force acceleration	0.56
Gravitational C_2 acceleration	5×10^{-5}
Other gravitational acceleration	3×10^{-7}
The moon's central force acceleration	5×10^{-6}
The sun's central force acceleration	2×10^{-6}
Planets' central force acceleration	3×10^{-10}
The Earth's tidal acceleration	2×10^{-9}
Ocean's tidal acceleration	5×10^{-10}
Solar pressure acceleration	1×10^{-7}
Atmosphere drag acceleration (Topex)	4×10^{-10}
General relativity acceleration	3×10^{-10}

If the coordinate system is used without taking precession and nutation into account, additional perturbation acceleration can reach up to 3×10^{-10}. Additional acceleration of gravitational potential can reach up to 1×10^{-9} (see Liu and Zhao, 1979).

Chapter 5
Solutions of $\overline{C}_{20}$ Perturbation

Satellites are attracted not only by the central force of the Earth, but also by the non-central force of the Earth, the attracting forces of the sun and the moon as well as planets, and the drag force of the atmosphere, solar radiation pressure, Earth and ocean tides, and coordinate perturbations (see Chap. 4). Equations of satellite motion have to be represented by perturbed equations. In this chapter, Emphasis is given to the analytic solution of the $\overline{C}_{20}$ perturbation. Orbit correction is discussed based on the solution.

5.1 $\overline{C}_{20}$ Perturbed Equations of Motion

The geopotential term $\overline{C}_{20}$ is a zonal term. Compared with other geopotential terms, $\overline{C}_{20}$ has a value that is at least 100 times larger. According to the order estimation discussed in Sect. 4.2.7, $\bar{C}_{20}$ perturbation is one of the most significant disturbing factors and is a perturbation of the first order. The analytic solution of the $\bar{C}_{20}$ perturbation will give a clear insight into orbit disturbance. The related perturbing potential is (see Section *The Earth's Gravitational Field* in Chap. 4)

$$R_2 = \frac{\mu a_e^2}{r^3}\overline{C}_{20}\overline{P}_{20}(\sin\varphi)$$

or

$$R_2 = \frac{b}{r^3}(3\sin^2\varphi - 1), \tag{5.1}$$

where

$$b = \frac{\sqrt{5}\mu a_e^2}{2}\overline{C}_{20}.$$

The variables (r, φ, λ) of the geopotential disturbance function in the ECEF system are transformed into orbital elements in the ECSF system by using the following relations (see Fig. 5.1; see Kaula, 1966/2001):

Fig. 5.1 Orbit-equator-meridian triangle

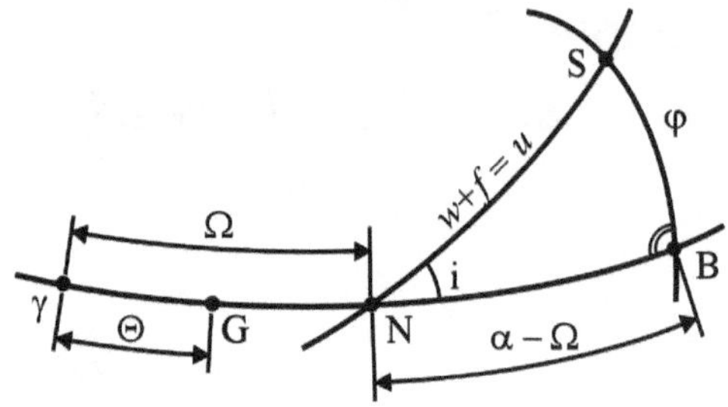

$$\begin{aligned} \sin\varphi &= \sin i \sin u, \\ \lambda &= \alpha - \Theta = \Omega - \Theta + (\alpha - \Omega), \\ \cos(\alpha - \Omega) &= \frac{\cos u}{\cos\varphi}, \\ \sin(\alpha - \Omega) &= \frac{\sin u \cos i}{\cos\varphi}, \end{aligned} \tag{5.2}$$

where α is the right ascension of the satellite, $u = \omega + f$, Θ is the Greenwich Sidereal Time, and other parameters are Keplerian elements. It is obvious that such a coordinate transformation only takes the Earth's rotation into account; this will cause a coordinate perturbation (see Sect. 4.2.6). But such an effect can be neglected by the first order solution. Substituting the first formula of (5.2) into (5.1) and taking the triangle formula (for reducing the order) into account, one has

$$R_2 = \frac{b}{r^3}\left[\frac{3}{2}\sin^2 i\,(1 - \cos 2u) - 1\right], \tag{5.3}$$

where

$$r = \frac{a(1-e^2)}{1 + e\cos f}, \tag{5.4}$$

where Ω does not appear in the zonal disturbance. Taking into account the partial derivatives of f with respect to (M,e) and r with respect to (a,M,e) (see Sect. 4.1), the derivatives of R_2 with respect to Keplerian elements are then

$$\begin{aligned} \frac{\partial R_2}{\partial a} &= \frac{\partial R_2}{\partial r}\frac{\partial r}{\partial a} = \frac{-3}{a}R_2, \quad \frac{\partial R_2}{\partial \Omega} = 0, \\ \frac{\partial R_2}{\partial i} &= \frac{b}{r^3}\left[\frac{3}{2}\sin 2i\,(1 - \cos 2u)\right], \\ \frac{\partial R_2}{\partial \omega} &= \frac{b}{r^3}\left[3\sin^2 i \sin 2u \frac{\partial u}{\partial \omega}\right] = \frac{3b}{r^3}\sin^2 i \sin 2u, \\ \frac{\partial R_2}{\partial e} &= \frac{-3R_2}{r}\frac{\partial r}{\partial e} + \frac{b}{r^3}\left[3\sin^2 i \sin 2u \frac{\partial u}{\partial e}\right] \\ &= \frac{3a\cos f}{r}R_2 + \frac{b}{r^3}\left[3\sin^2 i \sin 2u \frac{2 + e\cos f}{1 - e^2}\sin f\right] \end{aligned}$$

and

$$\begin{aligned}\frac{\partial R_2}{\partial M} &= \frac{-3R_2}{r}\frac{\partial r}{\partial M} + \frac{b}{r^3}\left[3\sin^2 i \sin 2u \frac{\partial u}{\partial M}\right] \\ &= \frac{-3ae\sin f}{r\sqrt{1-e^2}}R_2 + \frac{b}{r^3}\left[3\sin^2 i \sin 2u \left(\frac{a}{r}\right)^2 \sqrt{1-e^2}\right].\end{aligned} \tag{5.5}$$

Substituting these derivatives and R_2 into the equation of motion (4.11), one has the $\overline{C}_{20}$ perturbed equations of motion:

$$\begin{aligned}
\frac{\mathrm{d}a}{\mathrm{d}t} &= \frac{6b\sqrt{1-e^2}}{na^4}\left\{\frac{-e}{(1-e^2)}\frac{a^4}{r^4}\sin f\left[\frac{3}{2}\sin^2 i(1-\cos 2u)-1\right] + \frac{a^5}{r^5}\left[\sin^2 i \sin 2u\right]\right\}, \\
\frac{\mathrm{d}e}{\mathrm{d}t} &= \frac{3b(1-e^2)^{3/2}}{na^5 e}\left\{\frac{-e}{(1-e^2)}\frac{a^4}{r^4}\sin f\left[\frac{3}{2}\sin^2 i(1-\cos 2u)-1\right] + \frac{a^5}{r^5}\left[\sin^2 i \sin 2u\right]\right\} \\
&\quad - \frac{3b\sqrt{1-e^2}}{na^5 e}\frac{a^3}{r^3}\sin^2 i \sin 2u, \\
\frac{\mathrm{d}\omega}{\mathrm{d}t} &= \frac{3b\sqrt{1-e^2}}{na^5 e}\left\{\frac{a^4}{r^4}\cos f\left[\frac{3}{2}\sin^2 i(1-\cos 2u)-1\right] + \frac{a^3}{r^3}\left[\sin^2 i \sin 2u \frac{2+e\cos f}{1-e^2}\sin f\right]\right\} \\
&\quad - \frac{3b}{na^5\sqrt{1-e^2}}\frac{a^3}{r^3}\left[\cos^2 i(1-\cos 2u)\right], \\
\frac{\mathrm{d}i}{\mathrm{d}t} &= \frac{3b}{2na^5\sqrt{1-e^2}}\frac{a^3}{r^3}\sin 2i \sin 2u, \\
\frac{\mathrm{d}\Omega}{\mathrm{d}t} &= \frac{3b}{na^5\sqrt{1-e^2}}\frac{a^3}{r^3}\left[\cos i(1-\cos 2u)\right]
\end{aligned}$$

and

$$\begin{aligned}
\frac{\mathrm{d}M}{\mathrm{d}t} &= n + \frac{6b}{na^5}\frac{a^3}{r^3}\left[\frac{3}{2}\sin^2 i(1-\cos 2u)-1\right] \\
&\quad - \frac{3b(1-e^2)}{na^5 e}\left\{\begin{array}{l}\dfrac{a^4}{r^4}\cos f\left[\dfrac{3}{2}\sin^2 i(1-\cos 2u)-1\right] \\ +\dfrac{a^3}{r^3}\left[\sin^2 i \sin 2u \dfrac{2+e\cos f}{1-e^2}\sin f\right]\end{array}\right\}.
\end{aligned} \tag{5.6}$$

5.2 Solutions of $\overline{C}_{20}$ Perturbed Orbit

For convenience the right-hand side of (5.6) will be separated into three parts:

$$\frac{\mathrm{d}\sigma_j}{\mathrm{d}t} = \left(\frac{\mathrm{d}\sigma_j}{\mathrm{d}t}\right)_0 + \left(\frac{\mathrm{d}\sigma_j}{\mathrm{d}t}\right)_\omega + \left(\frac{\mathrm{d}\sigma_j}{\mathrm{d}t}\right)_f \tag{5.7}$$

or

$$\frac{\mathrm{d}\sigma_j}{\mathrm{d}t} = \dot{\sigma}_{j0} + \left(\frac{\mathrm{d}\sigma_j}{\mathrm{d}t}\right)_{\omega} + \left(\frac{\mathrm{d}\sigma_j}{\mathrm{d}t} - \dot{\sigma}_{j0} - \dot{\sigma}_{j\omega}\right), \tag{5.8}$$

where the first term (denoted by $\dot{\sigma}_{j0}$) on the right-hand side includes all terms that are only functions of (a,i,e), the second term includes all terms of ω (without f) (denoted by $\dot{\sigma}_{j\omega}$), and the third term includes all terms of f. They are denoted by the sub-index of 0, ω and f, respectively. Equation (5.8) is needed for later integral variable transformation. The second terms on the right-hand side of the above two equations are the same. It is notable that r is a function of f. The solution of the R_2 perturbed orbit is the integration of these equations between initial epoch t_0 and any instantaneous epoch t. The three terms on the right side can be integrated with the integral variable of t, ω, and M, respectively.

All terms of ω are represented in the terms of sin2u and cos2u. Omitting the terms of sin2u and cos2u in (5.6), the remaining terms of f are included in the following functions:

$$\left(\frac{a}{r}\right)^3, \quad \left(\frac{a}{r}\right)^4 \sin f \quad \text{and} \quad \left(\frac{a}{r}\right)^4 \cos f, \tag{5.9}$$

where

$$\frac{a}{r} = \frac{1+e\cos f}{1-e^2}, \quad \left(\frac{a}{r}\right)^2 = \frac{1+0.5e^2+2e\cos f+0.5e^2\cos 2f}{(1-e^2)^2},$$

$$\left(\frac{a}{r}\right)^3 = \frac{1+1.5e^2+(3e+0.75e^3)\cos f+1.5e^2\cos 2f+0.25e^3\cos 3f}{(1-e^2)^3},$$

$$\left(\frac{a}{r}\right)^4 = \frac{1}{(1-e^2)^4}\begin{bmatrix} \left(1+3e^2+\frac{3}{8}e^4\right)+(4e+3e^3)\cos f \\ +(3e^2+0.5e^4)\cos 2f+e^3\cos 3f+\frac{1}{8}e^4\cos 4f \end{bmatrix},$$

$$\left(\frac{a}{r}\right)^4 \sin f = \frac{1}{(1-e^2)^4}\begin{bmatrix} \left(1+1.5e^2+\frac{1}{8}e^4\right)\sin f+(2e+e^3)\sin 2f \\ +\left(1.5e^2+\frac{3}{16}e^4\right)\sin 3f+0.5e^3\sin 4f+\frac{1}{16}e^4\sin 5f \end{bmatrix},$$

$$\left(\frac{a}{r}\right)^4 \cos f = \frac{1}{(1-e^2)^4}\begin{bmatrix} (2e+1.5e^3)+\left(1+4.5e^2+\frac{5}{8}e^4\right)\cos f \\ +(2e+2e^3)\cos 2f+\left(1.5e^2+\frac{5}{16}e^4\right)\cos 3f \\ +0.5e^3\cos 4f+\frac{1}{16}e^4\cos 5f \end{bmatrix}, \tag{5.10}$$

and

$$\begin{aligned}\sin jf \sin mf &= -0.5\left[\cos(j+m)f - \cos(j-m)f\right],\\ \cos jf \cos mf &= 0.5\left[\cos(j+m)f + \cos(j-m)f\right],\\ \sin jf \cos mf &= 0.5\left[\sin(j+m)f + \sin(j-m)f\right].\end{aligned} \tag{5.11}$$

Then the first term (equation of long term perturbation) in (5.8) is

$$\begin{aligned}&\left(\frac{\mathrm{d}a}{\mathrm{d}t}\right)_0 = \left(\frac{\mathrm{d}e}{\mathrm{d}t}\right)_0 = \left(\frac{\mathrm{d}i}{\mathrm{d}t}\right)_0 = 0,\\ &\left(\frac{\mathrm{d}\omega}{\mathrm{d}t}\right)_0 = \frac{3b}{na^5(1-e^2)^{3.5}}\left(4\sin^2 i - 3 + \frac{15}{4}e^2\sin^2 i - 3e^2\right),\\ &\left(\frac{\mathrm{d}\Omega}{\mathrm{d}t}\right)_0 = \frac{3b}{2na^5}\cos i\frac{(2+3e^2)}{(1-e^2)^{3.5}},\\ &\left(\frac{\mathrm{d}M}{\mathrm{d}t}\right)_0 = n + \frac{9b}{2na^5}\left(\frac{3}{2}\sin^2 i - 1\right)\frac{e^2}{(1-e^2)^3}.\end{aligned} \tag{5.12}$$

The solutions of the long term perturbation are then the integration of these equations from t_0 to t. t is the instantaneous time of interesting and t_0 is the initial time and is set to 0 for convenience. It is obvious that the long term perturbations are then

$$\Delta\sigma_j(t)_0 = \left(\frac{\mathrm{d}\sigma_j}{\mathrm{d}t}\right)_0 t, \quad j = 1,2,\ldots,6. \tag{5.13}$$

The integral variable transformation between t and ω, Ω and M can be approximated by

$$\mathrm{d}t = \left(\frac{\mathrm{d}\omega}{\mathrm{d}t}\right)_0^{-1}\mathrm{d}\omega, \quad \mathrm{d}t = \left(\frac{\mathrm{d}\Omega}{\mathrm{d}t}\right)_0^{-1}\mathrm{d}\Omega \quad \text{and} \quad \mathrm{d}t = \left(\frac{\mathrm{d}M}{\mathrm{d}t}\right)_0^{-1}\mathrm{d}M. \tag{5.14}$$

The second term (long periodic perturbation) in (5.8) exists only in sin2u and cos2u related terms. All sin2u and cos2u terms are factorised by the following functions:

$$\left(\frac{a}{r}\right)^3,\quad \left(\frac{a}{r}\right)^5,\quad \left(\frac{a}{r}\right)^4\sin f,\quad \left(\frac{a}{r}\right)^4\cos f \quad \text{and} \quad \left(\frac{a}{r}\right)^3\frac{2+e\cos f}{1-e^2}\sin f, \tag{5.15}$$

where

$$\left(\frac{a}{r}\right)^5 = \frac{1}{(1-e^2)^5}\left[\begin{aligned}&\left(1+5e^2+1\frac{7}{8}e^4\right) + \left(5e+7.5e^3+\frac{5}{8}e^5\right)\cos f\\ &+(5e^2+2.5e^4)\cos 2f + \left(2.5e^3+\frac{5}{16}e^5\right)\cos 3f\\ &+\frac{5}{8}e^4\cos 4f + \frac{1}{16}e^5\cos 5f\end{aligned}\right]$$

and

$$\left(\frac{a}{r}\right)^3 \frac{2+e\cos f}{1-e^2}\sin f = \frac{1}{(1-e^2)^4}\begin{bmatrix} \left(2+2.25e^2+\frac{1}{8}e^4\right)\sin f \\ +(3.5e+0.25e^3)\sin 2f \\ +\left(2.5e^2+\frac{3}{16}e^4\right)\sin 3f \\ +\frac{5}{8}e^3\sin 4f+\frac{1}{16}e^4\sin 5f \end{bmatrix}. \tag{5.16}$$

From properties of (5.12) and

$$\begin{aligned} \sin 2u &= \sin 2\omega\cos 2f + \cos 2\omega\sin 2f, \\ \cos 2u &= \cos 2\omega\cos 2f - \sin 2\omega\sin 2f, \end{aligned} \tag{5.17}$$

it is obvious that all ω terms (without f) may be created only by multiplying sin2u and cos2u by sin2f and cos2f in (5.15). In other words, only $\sin^2 2f$ and $\cos^2 2f$ will lead to a constant of 0.5. Therefore, when seeking the ω terms (without f), just sin2f and cos2f related terms in (5.15) have to be taken into account. Thus, the second term (long periodic term perturbation) of (5.8) is:

$$\begin{aligned} \left(\frac{\mathrm{d}a}{\mathrm{d}t}\right)_\omega &= \frac{3be^2(2+e^2)}{na^4(1-e^2)^{4.5}}\sin^2 i\sin 2\omega, \\ \left(\frac{\mathrm{d}e}{\mathrm{d}t}\right)_\omega &= \frac{3be(1+5e^2)}{4na^5(1-e^2)^{3.5}}\sin^2 i\sin 2\omega, \\ \left(\frac{\mathrm{d}\omega}{\mathrm{d}t}\right)_\omega &= \frac{3b}{8na^5(1-e^2)^{3.5}}(6e^2+(2-13e^2)\sin^2 i)\cos 2\omega, \\ \left(\frac{\mathrm{d}i}{\mathrm{d}t}\right)_\omega &= \frac{9be^2}{8na^5(1-e^2)^{3.5}}\sin 2i\sin 2\omega \\ \left(\frac{\mathrm{d}\Omega}{\mathrm{d}t}\right)_\omega &= \frac{-9be^2}{4na^5(1-e^2)^{3.5}}\cos i\cos 2\omega \end{aligned}$$

and

$$\left(\frac{\mathrm{d}M}{\mathrm{d}t}\right)_\omega = -\frac{3b(2+11e^2)}{8na^5(1-e^2)^3}\sin^2 i\cos 2\omega. \tag{5.18}$$

The solutions of the long periodic term perturbation are then the integration of above equations from ω_0 to ω, $\omega_0 = \omega(t_0)$ and $\omega = \omega(t)$. It is obvious that the solutions of the long periodic perturbations are then

$$(\Delta\sigma_j(t))_\omega - (\Delta\sigma_j(t_0))_\omega = (\Delta\sigma_j(\omega))_\omega - (\Delta\sigma_j(\omega_0))_\omega, \tag{5.19}$$

i.e.

$$(\Delta a(\omega))_\omega = -\left(\frac{\mathrm{d}\omega}{\mathrm{dt}}\right)_0^{-1} \frac{3be^2(2+e^2)}{2na^4(1-e^2)^{4.5}} \sin^2 i \cos 2\omega,$$

$$(\Delta e(\omega))_\omega = -\left(\frac{\mathrm{d}\omega}{\mathrm{dt}}\right)_0^{-1} \frac{3be(1+5e^2)}{8na^5(1-e^2)^{3.5}} \sin^2 i \cos 2\omega,$$

$$(\Delta\omega(\omega))_\omega = \left(\frac{\mathrm{d}\omega}{\mathrm{dt}}\right)_0^{-1} \frac{3b}{16na^5(1-e^2)^{3.5}} \left(6e^2 + (2-13e^2)\sin^2 i\right) \sin 2\omega,$$

$$(\Delta i(\omega))_\omega = -\left(\frac{\mathrm{d}\omega}{\mathrm{dt}}\right)_0^{-1} \frac{9be^2}{16na^5(1-e^2)^{3.5}} \sin 2i \cos 2\omega,$$

$$(\Delta\Omega(\omega))_\omega = \left(\frac{\mathrm{d}\omega}{\mathrm{dt}}\right)_0^{-1} \frac{-9be^2}{8na^5(1-e^2)^{3.5}} \cos i \sin 2\omega$$

and

$$(\Delta M(\omega))_\omega = -\left(\frac{\mathrm{d}\omega}{\mathrm{dt}}\right)_0^{-1} \frac{3b(2+11e^2)}{16na^5(1-e^2)^3} \sin^2 i \sin 2\omega. \tag{5.20}$$

The second term on the right side of (5.19) can be obtained by replacing the ω in the first term by ω_0.

The third term of (5.8) includes all terms of f and can be denoted and represented by

$$\left(\frac{\mathrm{d}\sigma_j}{\mathrm{d}t}\right)_f = \left(\frac{\mathrm{d}\sigma_j}{\mathrm{d}t} - \dot{\sigma}_{j0} - \dot{\sigma}_{j\omega}\right). \tag{5.21}$$

This equation can be obtained by withdrawing (5.12) and (5.18) from (5.6). All terms in (5.21) are periodic functions of f. They can be transformed to functions of M by using relations (see Liu and Zhao, 1979):

$$\begin{aligned}\sin f &= \sin M + e\sin 2M + \frac{1}{8}e^2(9\sin 3M - 7\sin M) + \frac{1}{6}e^3(8\sin 4M - 7\sin 2M),\\ \cos f &= \cos M + e(\cos 2M - 1) + \frac{9}{8}e^2(\cos 3M - \cos M) + \frac{4}{3}e^3(\cos 4M - \cos 2M).\end{aligned} \tag{5.22}$$

These two relations have precision of order $O(\mathrm{e}^4)$. After the transformation, the index f in (5.21) can be changed to M

$$\left(\frac{\mathrm{d}\sigma_j}{\mathrm{d}t}\right)_M = \left(\frac{\mathrm{d}\sigma_j}{\mathrm{d}t} - \dot{\sigma}_{j0} - \dot{\sigma}_{j\omega}\right)_M \tag{5.23}$$

and the short periodic disturbances of f can be obtained by

$$(\Delta\sigma_j(t))_M = \int_{M_0}^{M} \left(\frac{\mathrm{d}\sigma_j}{\mathrm{d}t}\right)_M \left(\frac{\mathrm{d}M}{\mathrm{d}t}\right)_0^{-1} \mathrm{d}M. \tag{5.24}$$

For convenience denote these results of integration by

$$(\Delta\sigma_j(t))_M = (\Delta\sigma_j(M))_M - (\Delta\sigma_j(M_0))_M. \tag{5.25}$$

The first term on the right side can be obtained by indefinite integration of (5.24) and will be given below. (The process of an alternative and software based derivation will be outlined in detail in the next chapter by deriving other perturbations of geopotential disturbances). The second term on the right side can be obtained by replacing the M in the first term by M_0. The constant factor in (5.23) is not taken into account in the following solutions, for various reasons. In the application of the following formulas, this factor should be multiplied. Define

$$(\Delta\sigma_j(M))_M = b_j\left(c_j M + d_j(\omega)M + \sum_{k=1}^{10} A_{jk}\cos kM + \sum_{k=1}^{10} B_{jk}\sin kM\right), \tag{5.26}$$

where j is the index of Keplerian elements. Then there are commonly

$$b_0 = \frac{3\sqrt{5}\mu a_e \overline{C}_{20}}{128a^5 e(1-e^2)^{9/2}n}, \quad b_1 = 2aeb_0, \quad b_2 = (1-e^2)b_0,$$
$$b_3 = b_2, \quad b_4 = 8e(1-e^2)b_0, \quad b_5 = b_4, \quad b_6 = -(1-e^2)^{3/2}b_0. \tag{5.27}$$

For $j = 1$, there are

$$c_j = 0,$$
$$d_j = -\left(64 + 40e^2 + 84e^4 + \frac{25}{2}e^6\right)e^2\sin 2\omega\sin^2 i, \tag{5.28}$$
$$A_{j1} = e\left(-64 + 216e^2 - \frac{784}{3}e^4 - \frac{81}{2}e^6 + \left(96 - 324e^2 + 392e^4 + \frac{243}{4}e^6\right.\right.$$
$$\left.\left.+\cos 2\omega\left(16 - \frac{224}{3}e^2 - 252e^4 + \frac{193}{2}e^6\right)\right)\sin^2 i\right),$$
$$A_{j2} = e^2\left(-96 + \frac{1072}{3}e^2 - 286e^4 - 29e^6\right) + \left(e^2\left(144 - 536e^2 + 4292e^4 + \frac{87}{2}e^6\right)\right.$$
$$\left.-\left(32 - 224e^2 + \frac{1592}{3}e^4 + 662e^6 + \frac{343}{12}e^8\right)\cos 2\omega\right)\sin^2 i,$$
$$A_{j3} = e^3\left(-\frac{424}{3} + 536e^2 - \frac{151}{2}e^4\right) + \left(e^3\left(212 - 804e^2 + \frac{453}{4}e^4\right)\right.$$
$$\left.-\left(112 - 750e^2 + 720e^4 + \frac{5023}{12}e^6\right)e\cos 2\omega\right)\sin^2 i,$$
$$A_{j4} = e^4\left(-\frac{616}{3} + 508e^2 + \frac{25}{3}e^4\right) + \left(e^4\left(308 - 762e^2 - \frac{25}{2}e^4\right)\right.$$
$$\left.-\left(272 - \frac{4994}{3}e^2 + \frac{238}{3}e^4 + \frac{725}{12}e^6\right)e^2\cos 2\omega\right)\sin^2 i,$$

$$
\begin{aligned}
A_{j5} &= e^5\left(-\frac{824}{3}+\frac{6413}{30}e^2\right)+\left(e^5\left(412-\frac{6413}{20}e^2\right)\right.\\
&\quad \left.-\left(\frac{1690}{3}-\frac{32552}{15}e^2-\frac{14851}{60}e^4\right)e^3\cos 2\omega\right)\sin^2 i,\\
A_{j6} &= e^6\left(-\frac{742}{3}+29e^2\right)+\left(e^6\left(371-\frac{87}{2}e^2\right)\right.\\
&\quad \left.-\left(\frac{2698}{3}-1414e^2-\frac{627}{8}e^4\right)e^4\cos 2\omega\right)\sin^2 i,\\
A_{j7} &= -\frac{4759}{42}e^7+\left(\frac{4759}{28}e^7-\left(\frac{25468}{21}-\frac{67211}{168}e^2\right)e^5\cos 2\omega\right)\sin^2 i,\\
A_{j8} &= -\frac{235}{12}e^8+\left(\frac{235}{8}e^8-\left(\frac{5209}{6}-\frac{1625}{48}e^2\right)e^6\cos 2\omega\right)\sin^2 i,\\
A_{j9} &= -\frac{22891}{72}e^7\cos 2\omega\sin^2 i,\\
A_{j10} &= -\frac{1099}{24}e^8\cos 2\omega\sin^2 i,
\end{aligned}
\tag{5.29}
$$

$$
\begin{aligned}
B_{j1} &= \left(-16+\frac{220}{3}e^2+\frac{796}{3}e^4-\frac{405}{4}e^6\right)e\sin 2\omega\sin^2 i,\\
B_{j2} &= \left(32-224e^2+\frac{1576}{3}e^4+658e^6+\frac{215}{12}e^8\right)\sin 2\omega\sin^2 i,\\
B_{j3} &= \left(112-750e^2+720e^4+\frac{1330}{3}e^6\right)e\sin 2\omega\sin^2 i,\\
B_{j4} &= \left(272-\frac{4994}{3}e^2+\frac{230}{3}e^4+\frac{425}{6}e^6\right)e^2\sin 2\omega\sin^2 i,\\
B_{j5} &= \left(\frac{1690}{3}-\frac{32552}{15}e^2-\frac{3869}{15}e^4\right)e^3\sin 2\omega\sin^2 i,\\
B_{j6} &= \left(\frac{2968}{3}-1414e^2-\frac{1975}{24}e^4\right)e^4\sin 2\omega\sin^2 i,\\
B_{j7} &= \left(\frac{25468}{21}-\frac{67211}{168}e^2\right)e^5\sin 2\omega\sin^2 i,\\
B_{j8} &= \left(\frac{5029}{6}-\frac{1625}{48}e^2\right)e^6\sin 2\omega\sin^2 i,\\
B_{j9} &= \frac{22891}{72}e^7\sin 2\omega\sin^2 i,\\
B_{j10} &= \frac{1099}{24}e^8\sin 2\omega\sin^2 i.
\end{aligned}
\tag{5.30}
$$

For $j=2$, there are

$$
\begin{aligned}
c_j &= 0,\\
d_j &= -\left(16+80e^2+80e^4+\frac{49}{2}e^6\right)e^2\sin 2\omega\sin^2 i,
\end{aligned}
\tag{5.31}
$$

$$
\begin{aligned}
A_{j1} &= e\left(-64+216e^2-\frac{784}{3}e^4-\frac{81}{2}e^6+\left(96-324e^2+392e^4+\frac{243}{4}e^6\right.\right.\\
&\quad \left.\left.+\cos 2\omega\left(-16+56e^2-\frac{833}{3}e^4+\frac{47}{2}e^6\right)\right)\sin^2 i\right),\\
A_{j2} &= e^2\left(-96+\frac{1072}{3}e^2-286e^4-29e^6+\left(\left(144-536e^2+429e^4+\frac{87}{2}e^6\right)\right.\right.\\
&\quad \left.\left.+\left(16-\frac{308}{3}e^2-856e^4-\frac{1039}{12}e^6\right)\cos 2\omega\right)\sin^2 i\right),\\
A_{j3} &= e^3\left(-\frac{424}{3}+536e^2-\frac{151}{2}e^4\right)+\left(e^3\left(212-804e^2+\frac{453}{4}e^4\right)\right.\\
&\quad \left.-\left(\frac{112}{3}-\frac{862}{3}e^2+229e^4+\frac{6259}{12}e^6\right)e\cos 2\omega\right)\sin^2 i,\\
A_{j4} &= e^4\left(-\frac{616}{3}+508e^2+\frac{25}{3}e^4\right)+\left(e^4\left(308-762e^2-\frac{25}{2}e^4\right)\right.\\
&\quad \left.-\left(136-\frac{2960}{3}e^2-446e^4+\frac{175}{4}e^6\right)e^2\cos 2\omega\right)\sin^2 i,\\
A_{j5} &= e^5\left(-\frac{824}{3}+\frac{6413}{30}e^2\right)+\left(e^5\left(412-\frac{6413}{20}e^2\right)\right.\\
&\quad \left.-\left(338-\frac{8109}{5}e^2-\frac{34231}{60}e^4\right)e^3\cos 2\omega\right)\sin^2 i,\\
A_{j6} &= e^6\left(-\frac{742}{3}+29e^2\right)+\left(e^6\left(371-\frac{87}{2}e^2\right)\right.\\
&\quad \left.-\left(\frac{2132}{3}-\frac{3232}{3}e^2-\frac{1091}{8}e^4\right)e^4\cos 2\omega\right)\sin^2 i,\\
A_{j7} &= -\frac{4759}{42}e^7+\left(\frac{4759}{28}e^7-\left(\frac{3115}{3}-\frac{37907}{168}e^2\right)e^5\cos 2\omega\right)\sin^2 i,\\
A_{j8} &= -\frac{235}{12}e^8+\left(\frac{235}{8}e^8-\left(829+\frac{85}{16}e^2\right)e^6\cos 2\omega\right)\sin^2 i,\\
A_{j9} &= -\frac{22891}{72}e^7\cos 2\omega\sin^2 i,\\
A_{j10} &= -\frac{1099}{24}e^8\cos 2\omega\sin^2 i, \qquad (5.32)\\
B_{j1} &= \left(16-60e^2+\frac{899}{3}e^4-\frac{137}{4}e^6\right)e\sin 2\omega\sin^2 i,\\
B_{j2} &= \left(-16+\frac{292}{3}e^2+864e^4+\frac{767}{12}e^6\right)e^2\sin 2\omega\sin^2 i,
\end{aligned}
$$

$$
\begin{aligned}
B_{j3} &= \left(\frac{112}{3} - \frac{862}{3}e^2 + 223e^4 + \frac{1657}{3}e^6\right) e \sin 2\omega \sin^2 i,\\
B_{j4} &= \left(136 - \frac{2960}{3}e^2 - 454e^4 + \frac{119}{2}e^6\right) e^2 \sin 2\omega \sin^2 i,\\
B_{j5} &= \left(338 - \frac{8109}{5}e^2 - \frac{8714}{15}e^4\right) e^3 \sin 2\omega \sin^2 i,\\
B_{j6} &= \left(\frac{2132}{3} - \frac{3232}{3}e^2 - \frac{3367}{24}e^4\right) e^4 \sin 2\omega \sin^2 i,\\
B_{j7} &= \left(\frac{3115}{3} - \frac{37907}{168}e^2\right) e^5 \sin 2\omega \sin^2 i,\\
B_{j8} &= \left(829 + \frac{85}{16}e^2\right) e^6 \sin 2\omega \sin^2 i,\\
B_{j9} &= \frac{22891}{72}e^7 \sin 2\omega \sin^2 i,\\
B_{j10} &= \frac{1099}{24}e^8 \sin 2\omega \sin^2 i.
\end{aligned} \tag{5.33}
$$

For $j = 3$, there are

$$
c_j = \left(64 + 384e^2 - 32e^4 + 18e^6\right) e + \left(-96 - 480e^2 + 36e^4 - 23e^6\right) e \sin^2 i, \tag{5.34}
$$

$$
\begin{aligned}
d_j &= \Bigg(\left(-48 - 8e^2 - 12e^4\right) e^3\\
&\quad + \left(-16 + 104e^2 + 76e^4 + \frac{59}{2}e^6\right) e \sin^2 i\Bigg) \cos 2\omega,\\
A_{j1} &= e^2 \left(-32 + \frac{296}{3}e^2 + 73e^4\right) \sin 2\omega\\
&\quad + \left(16 - 36e^2 - \frac{589}{3}e^4 + \frac{29}{4}e^6\right) \sin 2\omega \sin^2 i,\\
A_{j2} &= e\left(32 - 176e^2 + 252e^4 + 58e^6\right) \sin 2\omega\\
&\quad + e\left(48 - \frac{580}{3}e^2 - 839e^4 - \frac{1213}{12}e^6\right) \sin 2\omega \sin^2 i,\\
A_{j3} &= e^2 \left(\frac{224}{3} - 388e^2 + 103e^4\right) \sin 2\omega\\
&\quad - \left(\frac{112}{3} - \frac{962}{3}e^2 + 303e^4 + \frac{1606}{3}e^6\right) \sin 2\omega \sin^2 i,\\
A_{j4} &= e^3 \left(136 - 542e^2 - \frac{50}{3}e^4\right) \sin 2\omega\\
&\quad - \left(136 - \frac{3064}{3}e^2 - 463e^4 + \frac{465}{12}e^6\right) e \sin 2\omega \sin^2 i,
\end{aligned}
$$

$$
\begin{aligned}
A_{j5} &= e^4\left(\frac{676}{3}-323e^2\right)\sin 2\omega-\left(338-\frac{24697}{15}e^2-\frac{9377}{15}e^4\right)e^2\sin 2\omega\sin^2 i,\\
A_{j6} &= e^5\left(\frac{836}{3}-58e^2\right)\sin 2\omega-\left(\frac{2132}{3}-\frac{3185}{3}e^2-\frac{1207}{8}e^4\right)e^3\sin 2\omega\sin^2 i,\\
A_{j7} &= \frac{1221}{7}e^6\sin 2\omega-\left(\frac{3115}{3}-\frac{32773}{168}e^2\right)e^4\sin 2\omega\sin^2 i,\\
A_{j8} &= \frac{235}{6}e^7\sin 2\omega-\left(829+\frac{725}{48}e^2\right)e^5\sin 2\omega\sin^2 i,\\
A_{j9} &= -\frac{22891}{72}e^6\sin 2\omega\sin^2 i,\\
A_{j10} &= -\frac{1099}{24}e^7\sin 2\omega\sin^2 i, \qquad (5.35)\\
B_{j1} &= -64-152e^2+\frac{1240}{3}e^4-\frac{631}{2}e^6+\cos 2\omega\left(-32+\frac{304}{3}e^2+67e^4\right)e^2\\
&\quad+\left(96+132e^2-440e^4+\frac{1565}{4}e^6\right.\\
&\quad\left.+\cos 2\omega\left(16-40e^2-197e^4-\frac{11}{2}e^6\right)\right)\sin^2 i,\\
B_{j2} &= \left(-96+\frac{272}{3}e^2+294e^4-74e^6\right)e+\cos 2\omega\left(32-176e^2+252e^4+46e^6\right)e\\
&\quad+\left(\left(144-208e^2-281e^4+94e^6\right)e\right.\\
&\quad\left.+\cos 2\omega\left(48-\frac{596}{3}e^2-817e^4-\frac{1007}{12}e^6\right)e\right)\sin^2 i,\\
B_{j3} &= \left(-\frac{424}{3}+\frac{832}{3}e^2+\frac{653}{2}e^4\right)e^2+\cos 2\omega\left(\frac{224}{3}-388e^2+109e^4\right)e^2\\
&\quad+\left(\left(212-\frac{1460}{3}e^2-\frac{1467}{4}e^4\right)e^2\right.\\
&\quad\left.+\cos 2\omega\left(-\frac{112}{3}+\frac{962}{3}e^2-309e^4-\frac{6289}{12}e^6\right)\right)\sin^2 i,\\
B_{j4} &= \left(-\frac{616}{3}+232e^2+\frac{296}{3}e^4\right)e^3+\cos 2\omega\left(136-542e^2-\frac{34}{3}e^4\right)e^3\\
&\quad+\left((308-425e^2-117e^4)e^3\right.\\
&\quad\left.+\cos 2\omega\left(-136+\frac{3064}{3}e^2+455e^4-\frac{247}{6}e^6\right)e\right)\sin^2 i,\\
B_{j5} &= \left(-\frac{824}{3}+\frac{155}{6}e^2\right)e^4+\cos 2\omega\left(\frac{676}{3}-323e^2\right)e^4\\
&\quad+\left(\left(412-\frac{1923}{20}e^2\right)e^4+\cos 2\omega\left(-338+\frac{24697}{15}e^2+\frac{36883}{60}e^4\right)e^2\right)\sin^2 i,
\end{aligned}
$$

$$
\begin{aligned}
B_{j6} &= \left(-\frac{742}{3} - 18e^2\right)e^5 + \cos 2\omega\left(\frac{836}{3} - 58e^2\right)e^5 \\
&\quad + \left(\left(371 + \frac{34}{3}e^2\right)e^5 + \cos 2\omega\left(-\frac{2132}{3} + \frac{3185}{3}e^2 + \frac{3527}{24}e^4\right)e^3\right)\sin^2 i, \\
B_{j7} &= -\frac{4759}{42}e^6 + \frac{1221}{7}e^6\cos 2\omega \\
&\quad + \left(\frac{4759}{28}e^6 + \cos 2\omega\left(-\frac{3115}{3} + \frac{32773}{168}e^2\right)e^4\right)\sin^2 i, \\
B_{j8} &= -\frac{235}{12}e^7 + \frac{235}{6}e^7\cos 2\omega + \left(\frac{235}{8}e^7 + \cos 2\omega\left(-829 - \frac{725}{48}e^2\right)e^5\right)\sin^2 i, \\
B_{j9} &= -\frac{22891}{72}e^6\cos 2\omega\sin^2 i, \\
B_{j10} &= -\frac{1099}{24}e^7\cos 2\omega\sin^2 i.
\end{aligned}
\tag{5.36}
$$

For $j = 4$, there are

$$
\begin{aligned}
c_j &= 0, \\
d_j &= \left(-3 - \frac{1}{2}e^2 - \frac{3}{4}e^4\right)e^2\sin 2i\sin 2\omega,
\end{aligned}
\tag{5.37}
$$

$$
\begin{aligned}
A_{j1} &= \left(2 - \frac{37}{6}e^2 - \frac{73}{16}e^4\right)e\cos 2\omega\sin 2i, \\
A_{j2} &= \left(-2 + 11e^2 - \frac{63}{4}e^4 - \frac{29}{8}e^6\right)\cos 2\omega\sin 2i, \\
A_{j3} &= \left(-\frac{14}{3} + \frac{97}{4}e^2 - \frac{103}{16}e^4\right)e\cos 2\omega\sin 2i, \\
A_{j4} &= \left(-\frac{17}{2} + \frac{271}{8}e^2 + \frac{25}{24}e^4\right)e^2\cos 2\omega\sin 2i, \\
A_{j5} &= \left(-\frac{169}{12} + \frac{323}{16}e^2\right)e^3\cos 2\omega\sin 2i, \\
A_{j6} &= \left(-\frac{209}{12} + \frac{29}{8}e^2\right)e^4\cos 2\omega\sin 2i, \\
A_{j7} &= -\frac{1221}{112}e^5\cos 2\omega\sin 2i, \\
A_{j8} &= -\frac{235}{96}e^6\cos 2\omega\sin 2i,
\end{aligned}
\tag{5.38}
$$

$$
\begin{aligned}
B_{j1} &= \left(-2 + \frac{19}{3}e^2 + \frac{67}{16}e^4\right)e\sin 2\omega\sin 2i, \\
B_{j2} &= \left(2 - 11e^2 + \frac{63}{4}e^4 + \frac{23}{8}e^6\right)\sin 2\omega\sin 2i,
\end{aligned}
$$

$$
\begin{aligned}
B_{j3} &= \left(\frac{14}{3} - \frac{97}{4}e^2 + \frac{109}{16}e^4\right) e \sin 2\omega \sin 2i, \\
B_{j4} &= \left(\frac{17}{2} - \frac{271}{8}e^2 - \frac{17}{24}e^4\right) e^2 \sin 2\omega \sin 2i, \\
B_{j5} &= \left(\frac{169}{12} - \frac{323}{16}e^2\right) e^3 \sin 2\omega \sin 2i, \\
B_{j6} &= \left(\frac{209}{12} - \frac{29}{8}e^2\right) e^4 \sin 2\omega \sin 2i, \\
B_{j7} &= \frac{1221}{112} e^5 \sin 2\omega \sin 2i, \\
B_{j8} &= \frac{235}{96} e^6 \sin 2\omega \sin 2i.
\end{aligned} \tag{5.39}
$$

For $j = 5$, there are

$$
\begin{aligned}
c_j &= \left(-24 + 3e^2 - e^4\right) e^2 \cos i, \\
d_j &= \left(6 + e^2 + \frac{3}{2}e^4\right) e^2 \cos i \cos 2\omega,
\end{aligned} \tag{5.40}
$$

$$
\begin{aligned}
A_{j1} &= \left(4 - \frac{37}{3}e^2 - \frac{73}{8}e^4\right) e \sin 2\omega \cos i, \\
A_{j2} &= \left(-4 + 22e^2 - \frac{63}{2}e^4 - \frac{29}{4}e^6\right) \sin 2\omega \cos i, \\
A_{j3} &= \left(-\frac{28}{3} + \frac{97}{2}e^2 - \frac{103}{8}e^4\right) e \sin 2\omega \cos i, \\
A_{j4} &= \left(-17 + \frac{271}{4}e^2 + \frac{25}{12}e^4\right) e^2 \sin 2\omega \cos i, \\
A_{j5} &= \left(-\frac{169}{6} + \frac{323}{8}e^2\right) e^3 \sin 2\omega \cos i, \\
A_{j6} &= \left(-\frac{209}{6} + \frac{29}{4}e^2\right) e^4 \sin 2\omega \cos i, \\
A_{j7} &= -\frac{1221}{56} e^5 \sin 2\omega \cos i, \\
A_{j8} &= -\frac{235}{48} e^6 \sin 2\omega \cos i,
\end{aligned} \tag{5.41}
$$

$$
\begin{aligned}
B_{j1} &= \left(\left(24 - 45e^2 + \frac{41}{2}e^4\right) e + \left(4 - \frac{38}{3}e^2 - \frac{67}{8}e^4\right) e \cos 2\omega\right) \cos i, \\
B_{j2} &= \left(\left(18 - 40e^2 + \frac{17}{4}e^4\right) e^2 + \left(-4 + 22e^2 - \frac{63}{2}e^4 - \frac{23}{4}e^6\right) \cos 2\omega\right) \cos i, \\
B_{j3} &= \left(\left(\frac{53}{3} - \frac{123}{4}e^2\right) e^3 + \left(-\frac{28}{3} + \frac{97}{2}e^2 - \frac{109}{8}e^4\right) e \cos 2\omega\right) \cos i,
\end{aligned}
$$

$$
\begin{aligned}
B_{j4} &= \left(\left(\frac{77}{4} - \frac{31}{4}e^2\right)e^4 + \left(-17 + \frac{271}{4}e^2 + \frac{17}{12}e^4\right)e^2\cos 2\omega\right)\cos i,\\
B_{j5} &= \left(\frac{287}{20}e^5 + \left(-\frac{169}{6} + \frac{323}{8}e^2\right)e^2\cos 2\omega\right)\cos i,\\
B_{j6} &= \left(\frac{47}{12}e^6 + \left(-\frac{209}{6} + \frac{29}{4}e^2\right)e^4\cos 2\omega\right)\cos i,\\
B_{j7} &= -\frac{1221}{56}e^5\cos 2\omega\cos i,\\
B_{j8} &= -\frac{235}{48}e^6\cos 2\omega\cos i.
\end{aligned}
\tag{5.42}
$$

For $j = 6$, there are

$$
\begin{aligned}
c_j &= \left(64 - 192e^2 + 40e^4 - 6e^6\right)e + \left(-96 + 288e^2 - 60e^4 + 9e^6\right)e\sin^2 i,\\
d_j &= \left(-16 - 88e^2 + 44e^4 - \frac{37}{2}e^6\right)e\sin^2 i\cos 2\omega,
\end{aligned}
\tag{5.43}
$$

$$
\begin{aligned}
A_{j1} &= \left(16 - 164e^2 + \frac{595}{3}e^4 + \frac{1197}{4}e^6\right)\sin 2\omega\sin^2 i,\\
A_{j2} &= \left(176 - \frac{2692}{3}e^2 + 169e^4 + \frac{1571}{12}e^6\right)e\sin 2\omega\sin^2 i,\\
A_{j3} &= \left(-\frac{112}{3} + \frac{1858}{3}e^2 - 1855e^4 - \frac{370}{3}e^6\right)\sin 2\omega\sin^2 i,\\
A_{j4} &= \left(-136 + \frac{4696}{3}e^2 - 1705e^4 - \frac{425}{4}e^6\right)e\sin 2\omega\sin^2 i,\\
A_{j5} &= \left(-338 + \frac{12739}{5}e^2 - \frac{10003}{15}e^4\right)e^2\sin 2\omega\sin^2 i,\\
A_{j6} &= \left(-\frac{2132}{3} + \frac{6529}{3}e^2 - \frac{649}{8}e^4\right)e^3\sin 2\omega\sin^2 i,\\
A_{j7} &= \left(-\frac{3115}{3} + \frac{21427}{24}e^2\right)e^4\sin 2\omega\sin^2 i,\\
A_{j8} &= \left(-829 + \frac{2265}{16}e^2\right)e^5\sin 2\omega\sin^2 i,\\
A_{j9} &= -\frac{22891}{72}e^6\sin 2\omega\sin^2 i,\\
A_{j10} &= -\frac{1099}{24}e^7\sin 2\omega\sin^2 i,
\end{aligned}
\tag{5.44}
$$

$$
\begin{aligned}
B_{j1} = &\left(-64 + 424e^2 - \frac{2000}{3}e^4 + \frac{353}{2}e^6\right) + \left(96 - 636e^2 + 1000e^4 - \frac{1059}{4}e^6\right.\\
&\left.+\left(16 - 168e^2 + \frac{625}{3}e^4 + \frac{525}{2}e^6\right)\cos 2\omega\right)\sin^2 i,
\end{aligned}
$$

$$
\begin{aligned}
B_{j2} &= \left(-96+\frac{1568}{3}e^2-666e^4+28e^6\right)e+\left(\left(144-784e^2+999e^4-42e^6\right)e\right. \\
&\quad \left.+\left(176-\frac{2708}{3}e^2+191e^4+\frac{1201}{2}e^6\right)e\cos 2\omega\right)\sin^2 i, \\
B_{j3} &= \left(-\frac{424}{3}+\frac{2104}{3}e^2-\frac{823}{2}e^4\right)e^2+\left(\left(212-1052e^2+\frac{2469}{4}e^4\right)e^2\right. \\
&\quad \left.+\left(-\frac{112}{3}+\frac{1858}{3}e^2-1861e^4-\frac{1057}{12}e^6\right)\cos 2\omega\right)\sin^2 i, \\
B_{j4} &= \left(-\frac{616}{3}+694e^2-\frac{262}{3}e^4\right)e^3+\left(\left(308-1041e^2+131e^4\right)e^3\right. \\
&\quad \left.+\left(-136+\frac{4696}{3}e^2-1713e^4-\frac{173}{2}e^6\right)e\cos 2\omega\right)\sin^2 i, \\
B_{j5} &= \left(-\frac{824}{3}+\frac{11107}{30}e^2\right)e^4+\left(\left(412-\frac{11107}{20}e^2\right)e^4\right. \\
&\quad \left.+\left(-338+\frac{12739}{5}e^2-\frac{40637}{60}e^4\right)e^2\cos 2\omega\right)\sin^2 i, \\
B_{j6} &= \left(-\frac{742}{3}+76e^2\right)e^5+\left(\left(371-114e^2\right)e^5\right. \\
&\quad \left.+\left(-\frac{2132}{3}+\frac{6529}{3}e^2-\frac{2041}{24}e^4\right)e^3\cos 2\omega\right)\sin^2 i, \\
B_{j7} &= -\frac{4759}{42}e^6+\left(\frac{4759}{28}e^6+\left(-\frac{3115}{3}+\frac{21427}{24}e^2\right)e^4\cos 2\omega\right)\sin^2 i, \\
B_{j8} &= -\frac{235}{12}e^7+\left(\frac{235}{8}e^7+\left(-829+\frac{2265}{16}e^2\right)e^5\cos 2\omega\right)\sin^2 i, \\
B_{j9} &= -\frac{22891}{72}e^6\cos 2\omega\sin^2 i, \\
B_{j10} &= -\frac{1099}{24}e^7\cos 2\omega\sin^2 i.
\end{aligned}
\tag{5.45}
$$

These are the solutions of the $\overline{C}_{20}$ perturbations on satellite orbits. Discussions and comments will be given in the following section.

5.3 Properties of the Solutions of $\overline{C}_{20}$ Perturbations

These derived solutions of $\overline{C}_{20}$ perturbations are mathematically rigorous except for the series truncation of the transformation from f to M. It is obvious that the series of the solutions may also be truncated according to the precision requirements. The total perturbations of the orbit disturbed by $\overline{C}_{20}$ can be represented as (see (5.7), (5.19) and (5.25))

$$\Delta\sigma_j = \Delta\sigma_j(t)_0 + (\Delta\sigma_j(\omega) - \Delta\sigma_j(\omega_0))_\omega + (\Delta\sigma_j(M) - \Delta\sigma_j(M_0))_M, \quad j = 1,2,\ldots,6. \tag{5.46}$$

Adding the $\overline{C}_{20}$ perturbations (5.46) to the Keplerian orbit (i.e., the satellite orbit under the acting of the central force of the Earth), one obtains the mathematical expressions of satellite orbit under the central force field and $\overline{C}_{20}$ disturbance.

The disturbances of the $\overline{C}_{20}$ on the different Keplerian elements are inhomogeneous. There are no long–term disturbances on the (a, e, i). The order ratios of the long term disturbances on the (ω, Ω, M) are of $(1, e, e^2)$, respectively. These indicate that the disturbance on the orientation of the ellipse is stronger than that on the orbital plane. In addition to the constant motion of M, there is a small (e^2) long term change. All Keplerian elements are subjected to long periodic disturbances and they can be grouped in orders with a, (ω, M), e, and (i, Ω). The orders of the short periodic perturbations are dependent on the coefficients given in (5.26) and can be grouped with a, (ω, e, i, Ω) and M. Long and short periodic disturbances tend to change the semi-axis of the orbits very strongly, to change the orientation of the ellipse and motion of the satellite to a great extent, and to change the orbital plane and shape to a lesser extent. The order ratios of the long periodic perturbations of the zonal term of $\overline{C}_{20}$ are $(ae^2, e, 1, e^2, e^2, 1)$, respectively. The order ratios of the short periodic perturbations of zonal term $\bar{C}_{20}$ are $(ae, e, e, e, e, 1)$, respectively. These are useful for considering the truncation of the short term solutions. As it is obvious, three different truncating orders should be selected for the six series, depending on the kind of the satellite orbit (or the numbers of a, e and the coefficients of e).

It is interesting that in the solutions for the short periodic disturbance there exist long term and long periodic terms, respectively. The short periodic term is related to the variable f. The transformation from f to M given in (5.22) is not a transformation between two periodic functions. There is a small term of e which leads to the terms represented in (5.26) by the first two terms (Note that the transformation is just one of the reasons.) Using the relation (5.14) the dM (i.e., M) can be transformed to dt (i.e., t). And

$$\left(\frac{\mathrm{d}M}{\mathrm{d}t}\right)_0^{-1} \mathrm{d}M = \mathrm{d}t = \left(\frac{\mathrm{d}\omega}{\mathrm{d}t}\right)_0^{-1} \mathrm{d}\omega \tag{5.47}$$

can be used to transform easily the $d_j(\omega)M$ terms to functions of ω and the formulas are left to interested readers.

For consistence with the solutions of other order and degree geopotential disturbances given in the next chapter, the complete solution (the infinite integration of the disturbed equation of motion, to be exact) of $\overline{C}_{20}$ disturbance may be rewritten as

$$\Delta\sigma_j = b_j \left(c'_j M + d'_j(\omega)M + \sum_{k=1}^{10} A_{jk} \cos kM + \sum_{k=1}^{10} B_{jk} \sin kM \right). \tag{5.48}$$

It is obvious that this solution is the summation of the solution given in (5.12), (5.18) and (5.26). Most coefficients are the same as that of (5.26), except the c'_j and d'_j, which can be easily obtained. The relations of the integral variable transformation can be obtained as

$$\left(\frac{d\sigma_j}{dt}\right)_0 = c'_j \quad \text{and} \quad \left(\frac{d\sigma_j}{dt}\right)_\omega = d'_j. \tag{5.49}$$

5.4 Orbit Correction

When the orbit errors of GPS satellites become non–negligible for special GPS applications, a process of orbit correction is the first option. Generally, orbit correction is applied to the regional or very long baseline of GPS precise positioning. Even IGS precise GPS orbits are not homogenously precise, because they are dependent on the distribution of the IGS reference stations and the length and quality of the data used. The orbit correction is an adjustment or filtering process in which, besides the station position, the orbit errors are also modelled, determined, and corrected, based on a known orbit.

Keplerian elements also describe the orbit geometry for instantaneous time. Orbit errors can be considered geometric element errors of the orbit in general. Recalling earlier discussions on the $\overline{C}_{20}$ perturbed orbit solution, a general orbit model can be written as

$$\begin{aligned}\sigma_j(t) = {} & \sigma_{jc}(t) + \dot{\sigma}_{j0}(t-t_0) + A_{j\omega}\cos 2\omega + B_{j\omega}\sin 2\omega \\ & + \sum_{m=1}^{m(j)} [A_{jm}\cos mM + B_{jm}\sin mM],\end{aligned} \tag{5.50}$$

where $\sigma_j(t)$, $\sigma_{jc}(t)$, $\dot{\sigma}_{j0}$ are true orbit element at time t, computed element at t, element rate with respect to the initial epoch t_0, $A_{j\omega}$, $B_{j\omega}$, A_{jm}, B_{jm} are the coefficients of the long and short periodic perturbations, respectively, and $m(j)$ is the maximum integer of index m related to the jth Keplerian element. ω and M are Keplerian elements. Generally speaking, the coefficients of $A_{j\omega}$, $B_{j\omega}$, A_{jm}, B_{jm} are also functions of ω and i which can be considered in the short periodic term as constants.

For a general model, the order of the polynomial term can be raised to 2, further terms of ω (and Ω) may also be added, and $m(j)$ is selectable. The selection of the number of the order depends on the need and the situation of orbit errors.

In the GPS observation equations (see Xu, 2003/2007), the orbit state vector is represented in the range or range rate functions. It depends on the use of the GPS observables. We generally denote both the range and range rate together as ρ; their partial derivatives with respect to the orbit state vector are given in Sect. 8.3 and have the forms of

$$\frac{\partial\rho}{\partial\vec{r}} \quad \text{and} \quad \frac{\partial\rho}{\partial\dot{\vec{r}}}, \tag{5.51}$$

where the satellite state vector is $(\vec{r}, \dot{\vec{r}})$. The relations between $(\vec{r}, \dot{\vec{r}})$ and Keplerian elements σ_j are discussed in Sect. 3.4. Also, the relations between σ_j and the parameters of the orbit correction model are given in (5.50). Therefore, the orbit correction parts in the GPS observation equations are

$$\frac{\partial\rho}{\partial\vec{r}}\frac{\partial\vec{r}}{\partial\vec{\sigma}}\frac{\partial\vec{\sigma}}{\partial\vec{y}}\Delta\vec{y}^T + \frac{\partial\rho}{\partial\dot{\vec{r}}}\frac{\partial\dot{\vec{r}}}{\partial\vec{\sigma}}\frac{\partial\vec{\sigma}}{\partial\vec{y}}\Delta\vec{y}^T, \tag{5.52}$$

where $\vec{y}$ and $\Delta\vec{y}$ are the parameter vectors in model (5.50) and the parameter correction vector of the model, and $\vec{\sigma}$ is the vector of Keplerian elements. If the initial parameter vector is selected as zero, then $\vec{y} = \Delta\vec{y}$. It is obvious that

$$\vec{y} = \left(\dot{\sigma}_{j0},\ A_{j\omega},\ B_{j\omega},\ A_{jm},\ B_{jm}\right) \tag{5.53}$$

and

$$\frac{\partial\sigma_j}{\partial(\dot{\sigma}_{j0}, A_{j\omega},\ B_{j\omega},\ A_{jm},\ B_{jm})} = ((t-t_0), \cos 2\omega, \sin 2\omega, \cos mM, \sin mM)\,. \tag{5.54}$$

Here parameters A_{jm}, B_{jm} represent symbolically the unknowns of all m. For the convenience of representing the partial derivatives of the state vector with respect to the Keplerian elements, the Keplerian element vector is reordered as

$$\vec{\sigma} = (\Omega, i, \omega, a,\ e,\ M)\,. \tag{5.55}$$

This does not affect (5.54), because the right-hand side of the equation has nothing to do with index j. According to the formulas in Sect. 3.4 ((3.41)–(3.43))

$$\begin{pmatrix}\vec{r}\\ \dot{\vec{r}}\end{pmatrix} = R_3(-\Omega)R_1(-i)R_3(-\omega)\begin{pmatrix}\vec{q}\\ \dot{\vec{q}}\end{pmatrix}, \tag{5.56}$$

where

$$\vec{q} = \begin{pmatrix} a(\cos E - e)\\ a\sqrt{1-e^2}\sin E\\ 0\end{pmatrix} = \begin{pmatrix} r\cos f\\ r\sin f\\ 0\end{pmatrix} \tag{5.57}$$

and

$$\dot{\vec{q}} = \begin{pmatrix} -\sin E\\ \sqrt{1-e^2}\cos E\\ 0\end{pmatrix}\frac{na}{1-e\cos E} = \begin{pmatrix} -\sin f\\ e+\cos f\\ 0\end{pmatrix}\frac{na}{\sqrt{1-e^2}}, \tag{5.58}$$

one has

$$\frac{\partial\vec{r}}{\partial(\Omega,i,\omega)} = \frac{\partial R}{\partial(\Omega,i,\omega)}\vec{q} \quad\text{and}\quad \frac{\partial\dot{\vec{r}}}{\partial(\Omega,i,\omega)} = \frac{\partial R}{\partial(\Omega,i,\omega)}\dot{\vec{q}}, \tag{5.59}$$

where $(\vec{q}, \dot{\vec{q}})$ are position and velocity vectors of the satellite in the orbital plane coordinate system, and

$$R = R_3(-\Omega)R_1(-i)R_3(-\omega) \tag{5.60}$$

and

$$\frac{\partial R}{\partial(\Omega,i,\omega)} = \left(\frac{\partial R_3(-\Omega)}{\partial\Omega}R_1(-i)R_3(-\omega), R_3(-\Omega)\frac{\partial R_1(-i)}{\partial i}R_3(-\omega), R_3(-\Omega)R_1(-i)\frac{\partial R_3(-\omega)}{\partial\omega}\right),$$

where

$$\frac{\partial R_1(-i)}{\partial i} = \begin{pmatrix} 0 & 0 & 0 \\ 0 & -\sin i & -\cos i \\ 0 & \cos i & -\sin i \end{pmatrix},$$

$$\frac{\partial R_3(-\Omega)}{\partial\Omega} = \begin{pmatrix} -\sin\Omega & -\cos\Omega & 0 \\ \cos\Omega & -\sin\Omega & 0 \\ 0 & 0 & 0 \end{pmatrix}$$

and

$$\frac{\partial R_3(-\omega)}{\partial\omega} = \begin{pmatrix} -\sin\omega & -\cos\omega & 0 \\ \cos\omega & -\sin\omega & 0 \\ 0 & 0 & 0 \end{pmatrix}.$$

For the Keplerian elements in the orbital plane (a, e, M), one has

$$\frac{\partial\vec{r}}{\partial(a,e,M)} = R\frac{\partial\vec{q}}{\partial(a,e,M)} \quad \text{and} \quad \frac{\partial\dot{\vec{r}}}{\partial(a,e,M)} = R\frac{\partial\dot{\vec{q}}}{\partial(a,e,M)}, \tag{5.61}$$

where

$$\frac{\partial\vec{q}}{\partial(a,e,M)} = \begin{pmatrix} \cos E - e & \dfrac{-a\sin^2 E}{1-e\cos E} - a & \dfrac{-a\sin E}{1-e\cos E} \\ \sqrt{1-e^2}\sin E & a\sqrt{1-e^2}\left(\dfrac{\sin 2E}{2(1-e\cos E)} - \dfrac{e\sin E}{1-e^2}\right) & \dfrac{a\sqrt{1-e^2}\cos E}{1-e\cos E} \\ 0 & 0 & 0 \end{pmatrix}$$

and

$$\frac{\partial\dot{\vec{q}}}{\partial(a,e,M)} = \begin{pmatrix} \dfrac{n\sin E}{2(1-e\cos E)} & \dfrac{na\sin E(e-2\cos E+e\cos^2 E)}{(1-e\cos E)^3} & \dfrac{na(e-\cos E)}{(1-e\cos E)^3} \\ \dfrac{-n\sqrt{1-e^2}\cos E}{2(1-e\cos E)} & \dfrac{na\left[1+e^2-2e\cos E+\sin^2 E(e\cos E-2)\right]}{\sqrt{1-e^2}(1-e\cos E)^3} & \dfrac{-na\sqrt{1-e^2}\sin E}{(1-e\cos E)^3} \\ 0 & 0 & 0 \end{pmatrix}.$$

The partial derivative formulas given in Sect. 4.1 and the relation in (3.32) between n and a (mean angular velocity and semi–major axis of the satellite) given in Chap. 3 are used, i.e.

$$\frac{\partial E}{\partial(e,M)} = \left(\frac{a}{r}\sin E,\ \frac{a}{r}\right)$$

and

$$n^2 = \mu/a^3. \tag{5.62}$$

Chapter 6
Solutions of Geopotential Perturbations

The principle of the derivation of geopotential perturbations will be discussed first. As special examples, the solutions of the $\overline{C}_{30}$, $\overline{C}_{21}$ and $\overline{S}_{21}$, $\overline{C}_{22}$ and $\overline{S}_{22}$ are derived and given. The aim is to give solutions up to 6×6 order and degrees, which are necessary for orbit determination of satellites similar to that of GPS. The general solution of the perturbations of $\overline{C}_{\mathrm{lm}}$ and $\overline{S}_{lm}$ is derived.

6.1 Principle of the Derivations

From the solution process of the equation of satellite motion perturbed by the geopotential term $\overline{C}_{20}$ given in Chap. 4, one notices that the derivation is very complicated, even if the potential function of the perturbation is relatively simple. An alternative method is to use symbolic mathematical operation software such as Mathematica, Maple, etc. However, the principle and strategy of the derivation have still to be carefully created.

For simplification, geopotential disturbance function of l order and m degree can be written as (see (4.35))

$$R_{lm} = \frac{\mu}{r}\left(\frac{a_e}{r}\right)^l \overline{P}_{lm}(\sin\varphi)\left[\overline{C}_{lm}\cos m\lambda + \overline{S}_{lm}\sin m\lambda\right]. \tag{6.1}$$

Let

$$\begin{aligned}\overline{C}_{lm} &= D_{lm}\cos m\lambda_{lm},\\ \overline{S}_{lm} &= D_{lm}\sin m\lambda_{lm},\\ \overline{\lambda} &= \lambda - \lambda_{lm},\end{aligned} \tag{6.2}$$

where

$$D_{lm} = \sqrt{\overline{C}^2_{lm} + \overline{S}^2_{lm}},$$

$$\cos m\lambda_{lm} = \frac{\overline{C}_{lm}}{D_{lm}},$$
$$\sin m\lambda_{lm} = \frac{\overline{S}_{lm}}{D_{lm}}, \tag{6.3}$$

then (6.1) is

$$R_{lm} = \frac{b_{lm}}{r^{l+1}}\overline{P}_{lm}(\sin\varphi)\cos(m\bar{\lambda}), \tag{6.4}$$

where

$$b_{lm} = \mu a_e^l D_{lm}.$$

To transform the geographic coordinates into the Keplerian variables, the following relations are needed (see (5.2)):

$$\begin{aligned}
\sin\varphi &= \sin i \sin u, \\
\overline{\lambda} &= \alpha - \Theta - \lambda_{lm} = (\Omega - \Theta - \lambda_{lm}) + (\alpha - \Omega), \\
\cos(\alpha - \Omega) &= \frac{\cos u}{\cos\varphi}, \\
\sin(\alpha - \Omega) &= \frac{\sin u \cos i}{\cos\varphi}.
\end{aligned} \tag{6.5}$$

Because (see Wang et al., 1979)

$$\begin{aligned}
\cos(my) &= \sum_{j=0}^{m-1}(-1)^j \binom{m}{2j} (\cos y)^{m-2j}(\sin y)^{2j}, \\
\sin(my) &= \sum_{j=0}^{m-1}(-1)^j \binom{m}{2j+1} (\cos y)^{m-2j-1}(\sin y)^{2j+1},
\end{aligned} \tag{6.6}$$

where the binomial form has the well-known expression of

$$\binom{m}{k} = \frac{m!}{k!(n-k)!}. \tag{6.7}$$

Let

$$\begin{aligned}
\overline{\Omega} &= \Omega - \Theta - \overline{\lambda}_{lm}, \\
y &= \alpha - \Omega,
\end{aligned} \tag{6.8}$$

then

$$\cos m\overline{\lambda} = \cos(m\overline{\Omega} + my) = \cos m\overline{\Omega} \cos my - \sin m\overline{\Omega} \sin my$$
$$= \sum_{j=0}^{m-1} (-1)^j \begin{pmatrix} \cos m\overline{\Omega} \binom{m}{2j} (\cos u)^{m-2j} (\sin u \cos i)^{2j} - \\ \sin m\overline{\Omega} \binom{m}{2j+1} (\cos u)^{m-2j-1} (\sin u \cos i)^{2j+1} \end{pmatrix} \frac{1}{\cos^m \varphi}.$$

Note that in the definition of (4.28) there is a factor of $\cos^m \varphi$ in expression of $\overline{P}_{lm}(\sin \varphi)$; therefore, let

$$q(\Omega, u, i) = \sum_{j=0}^{m-1} (-1)^j \begin{pmatrix} \cos m\overline{\Omega} \binom{m}{2j} (\cos u)^{m-2j} (\sin u \cos i)^{2j} - \\ \sin m\overline{\Omega} \binom{m}{2j+1} (\cos u)^{m-2j-1} (\sin u \cos i)^{2j+1} \end{pmatrix}, \quad (6.9)$$

$$Q_{lm}(x) = \overline{P}_{lm}(x)/(1-x^2)^{m/2}$$
$$= N_{lm} \sum_{k=0}^{K} T_{lmk} x^{l-m-2k}, \quad (6.10)$$

where K is the integer part of $(l-m)/2$, and the factors are

$$N_{lm} = \sqrt{\frac{(l-m)!(2l+1)(2-\delta_{0m})}{(l+m)!}},$$
$$T_{lmk} = \frac{(-1)^k (2l-2k)}{2^l k!(l-k)!(l-m-2k)!}. \quad (6.11)$$

One has

$$R_{lm} = \frac{b_{lm}}{r^{l+1}} Q_{lm}(x) q(\Omega, u, i), \quad (6.12)$$

and then

$$\frac{\partial R_{lm}}{\partial a} = \frac{\partial R_{lm}}{\partial r} \frac{\partial r}{\partial a} = \frac{-(l+1)}{a} R_{lm},$$

$$\frac{\partial R_{lm}}{\partial \Omega} = \frac{b_{lm}}{r^{l+1}} Q_{lm}(x) \frac{\partial q(\Omega, u, i)}{\partial \Omega},$$

$$\frac{\partial R_{lm}}{\partial i} = \frac{b_{lm}(-l-1)}{r^{l+2}} \frac{\partial r}{\partial i} Q_{lm}(x) q(\Omega, u, i) + \frac{b_{lm}}{r^{l+1}} \frac{\partial Q_{lm}(x)}{\partial x} \frac{\partial x}{\partial i} q(\Omega, u, i)$$
$$+ \frac{b_{lm}}{r^{l+1}} Q_{lm}(x) \frac{\partial q(\Omega, u, i)}{\partial i},$$

$$\frac{\partial R_{lm}}{\partial \omega} = \frac{b_{lm}}{r^{l+1}} \frac{\partial Q_{lm}(x)}{\partial x} \frac{\partial x}{\partial u} q(\Omega, u, i) + \frac{b_{lm}}{r^{l+1}} Q_{lm}(x) \frac{\partial q(\Omega, u, i)}{\partial u},$$

$$\frac{\partial R_{lm}}{\partial e} = \frac{b_{lm}(-l-1)}{r^{l+2}}\frac{\partial r}{\partial e}Q_{lm}(x)q(\Omega,u,i) + \frac{b_{lm}}{r^{l+1}}\frac{\partial Q_{lm}(x)}{\partial x}\frac{\partial x}{\partial u}\frac{\partial u}{\partial e}q(\Omega,u,i)$$
$$+ \frac{b_{lm}}{r^{l+1}}Q_{lm}(x)\frac{\partial q(\Omega,u,i)}{\partial u}\frac{\partial u}{\partial e}$$

and

$$\frac{\partial R_{lm}}{\partial M} = \frac{b_{lm}(-l-1)}{r^{l+2}}\frac{\partial r}{\partial M}Q_{lm}(x)q(\Omega,u,i) + \frac{b_{lm}}{r^{l+1}}\frac{\partial Q_{lm}(x)}{\partial x}\frac{\partial x}{\partial u}\frac{\partial u}{\partial M}q(\Omega,u,i)$$
$$+ \frac{b_{lm}}{r^{l+1}}Q_{lm}(x)\frac{\partial q(\Omega,u,i)}{\partial u}\frac{\partial u}{\partial M}, \tag{6.13}$$

where

$$\frac{\partial q(\Omega,u,i)}{\partial \Omega} = \sum_{j=0}^{m-1}(-1)^j m\begin{pmatrix} b_1(m,j)\sin m\overline{\Omega}(\cos u)^{m-2j}(\sin u\cos i)^{2j} - \\ b_2(m,j)\cos m\overline{\Omega}(\cos u)^{m-2j-1}(\sin u\cos i)^{2j+1}\end{pmatrix},$$
$$b_1(m,j) = -\binom{m}{2j}, \tag{6.14}$$
$$b_2(m,j) = \binom{m}{2j+1},$$

$$\frac{\partial q(\Omega,u,i)}{\partial u} = \sum_{j=0}^{m-1}(-1)^j\begin{pmatrix} \cos m\overline{\Omega}\begin{pmatrix} b_3(m,j)(\cos u)^{m-2j-1}(\sin u)^{2j+1} \\ +b_4(m,j)(\cos u)^{m-2j+1}(\sin u)^{2j-1}\end{pmatrix}(\cos i)^{2j} - \\ \\ \sin m\overline{\Omega}\begin{pmatrix} b_5(m,j)(\cos u)^{m-2j-2}(\sin u)^{2j+2} \\ +b_6(m,j)(\cos u)^{m-2j}(\sin u)^{2j}\end{pmatrix}(\cos i)^{2j+1}\end{pmatrix},$$
$$b_3(m,j) = -\binom{m}{2j}(m-2j)(1-\delta_{0(m-2j)}),$$
$$b_4(m,j) = \binom{m}{2j}2j(1-\delta_{0(2j)}),$$
$$b_5(m,j) = -\binom{m}{2j+1}(m-2j-1)(1-\delta_{0(m-2j-1)}),$$
$$b_6(m,j) = \binom{m}{2j+1}(2j+1), \tag{6.15}$$

$$\frac{\partial q(\Omega,u,i)}{\partial i} = \sum_{j=0}^{m-1}(-1)^j\begin{pmatrix} -b_4(m,j)\cos m\overline{\Omega}(\cos u)^{m-2j}(\sin u)^{2j}(\cos i)^{2j-1}\sin i \\ +b_6(m,j)\sin m\overline{\Omega}(\cos u)^{m-2j-1}(\sin u)^{2j+1}(\cos i)^{2j}\sin i\end{pmatrix}, \tag{6.16}$$

$$\frac{\partial Q_{lm}(x)}{\partial x} = N_{lm}\sum_{k=0}^{K}W_{lmk}x^{l-m-2k-1}, \tag{6.17}$$
$$W_{lmk} = T_{lmk}(l-m-2k)\delta_{0(l-m-2k)},$$

$$x = \sin\varphi = \sin u \sin i, \tag{6.18}$$
$$\frac{\partial x}{\partial u} = \cos u \sin i,$$
$$\frac{\partial x}{\partial i} = \sin u \cos i,$$

$$\begin{aligned} \sin u &= \sin(f+\omega) = \sin f \cos\omega + \cos f \sin\omega, \\ \cos u &= \cos(f+\omega) = \cos f \cos\omega - \sin f \sin\omega. \end{aligned} \tag{6.19}$$

These derivations lead to simplified formulae for the perturbation function and are necessary and enough to transform the differential equations of motion into functions of Keplerian variables. They are used to derive the solutions of perturbations of geopotential function in order and degrees of 6×6 and are the basis for deriving the general solution of the perturbation of l order and m degree as described at the end of this chapter.

6.2 Solutions of $\overline{C}_{30}$ Perturbation

Similar to the solutions of $\overline{C}_{20}$ perturbation, the indefinite integration of equations of motion can be given as shown below. Define

$$(\Delta\sigma_j(M))_M = b_j\left(c_j M + d_j(\omega)M + \sum_{k=1}^{12} A_{jk}\cos kM + \sum_{k=1}^{12} B_{jk}\sin kM\right), \tag{6.20}$$

where j is the index of Keplerian elements. The first two terms are the long term and long periodic term perturbations, respectively. There are

$$b_0 = \frac{\sqrt{7}\mu a_e^3 \overline{C}_{30}}{512 a^6 e (1-e^2)^{11/2} n}, \quad b_1 = 2aeb_0, \quad b_2 = (1-e^2)b_0,$$
$$b_3 = b_2, \quad b_4 = 12eb_2, \quad b_5 = b_4, \quad b_6 = -(1-e^2)^{1/2} b_2. \tag{6.21}$$

For $j = 1$, there are

$$c_j = 0, \tag{6.22}$$
$$\begin{aligned} d_j = &\left(1200 + 1800e^2 + \frac{295}{2}e^4\right) e^5 \cos 3\omega \sin^3 i \\ &+ \cos\omega\left(\left(480 + 1680e^2 + 150e^4\right) e^5 \sin i\right. \\ &\left. - \left(600 + 2100e^2 + \frac{375}{2}e^4\right) e^5 \sin^3 i\right), \end{aligned}$$

$$
\begin{aligned}
A_{j1} &= \left(-40-1610e^2+\frac{11625}{2}e^4+\frac{9500}{3}\right)e^2\sin 3\omega\sin^3 i \\
&+\left(\left(-768+1632e^2-400e^4-13110e^6+717e^8\right)\sin i\right. \\
&+\left.\left(960-2040e^2+500e^4+\frac{32775}{2}e^6-\frac{3585}{4}e^8\right)\sin^3 i\right)\sin\omega, \\
A_{j2} &= \left(320-2160e^2-14510e^4-\frac{10180}{3}e^6+\frac{1145}{12}e^8\right)e\sin 3\omega\sin^3 i \\
&+\left(\left(-2304+9088e^2-8920e^4-19104e^6-865e^8\right)e\sin i\right. \\
&+\left.\left(2800-11360e^2+11150e^4+23880e^6+\frac{4325}{4}e^8\right)e\sin^3 i\right)\sin\omega, \\
A_{j3} &= \left(-320-3680e^2-10560e^4-\frac{179665}{6}e^6-\frac{20295}{4}e^8\right)\sin 3\omega\sin^3 i \\
&+\left(\left(-5088+23640e^2-2682e^4-7646e^6\right)e^2\sin i\right. \\
&+\left.\left(6360-29550e^2+\frac{6705}{2}e^4+\frac{19115}{2}e^6\right)e^2\sin^3 i\right)\sin\omega, \\
A_{j4} &= \left(-1600+15840e^2-\frac{27050}{3}e^4-\frac{62060}{3}e^6-\frac{3365}{3}e^8\right)e\sin 3\omega\sin^3 i \\
&+\left(\left(-9856+40936e^2+10696e^4-575e^6\right)e^3\sin i\right. \\
&+\left.\left(12320-51170e^2-13370e^4+\frac{2875}{4}e^6\right)e^3\sin^3 i\right)\sin\omega, \\
A_{j5} &= \left(-5080+\frac{123772}{3}e^2+\frac{75245}{6}e^4-\frac{101875}{24}e^6\right)e^2\sin 3\omega\sin^3 i \\
&+\left(\left(-17480+39426e^2+7954e^4\right)e^4\sin i\right. \\
&+\left.\left(21850-\frac{98565}{2}e^2-\frac{19885}{2}e^4\right)e^4\sin^3 i\right)\sin\omega,
\end{aligned}
$$

$$
\begin{aligned}
A_{j6} = & \left(-13040 + 60150e^2 + 20880e^4 + \frac{2255}{24}e^6\right) e^3 \sin 3\omega \sin^3 i \\
& + \left(\left(-25544 + 18336e^2 + \frac{2925}{2}e^4\right) e^5 \sin i \right.\\
& \left. + \left(31930 - 22920e^2 - \frac{14625}{8}e^4\right) e^5 \sin^3 i\right) \sin\omega, \\
A_{j7} = & \left(-\frac{561970}{21} + \frac{959960}{21}e^2 + \frac{794555}{84}e^4\right) e^4 \sin 3\omega \sin^3 i \\
& + \left(\left(-\frac{176574}{7} + \frac{46321}{14}e^2\right) e^6 \sin i \right.\\
& \left. + \left(\frac{441435}{14} - \frac{231605}{56}e^2\right) e^6 \sin^3 i\right) \sin\omega, \\
A_{j8} = & \left(-\frac{113605}{3} + \frac{49595}{3}e^2 + \frac{64535}{48}e^4\right) e^5 \sin 3\omega \sin^3 i \\
& + \left(\left(-14674 + \frac{215}{4}e^2\right) e^7 \sin i + \left(\frac{36685}{2} - \frac{1075}{16}e^2\right) e^7 \sin^3 i\right) \sin\omega, \\
A_{j9} = & \left(-\frac{598325}{18} + \frac{24215}{12}e^2\right) e^6 \sin 3\omega \sin^3 i \\
& + \left(-\frac{26843}{6}e^8 \sin i + \frac{134215}{24}e^8 \sin^3 i\right) \sin\omega, \\
A_{j10} = & \left(-\frac{51020}{3} - \frac{2401}{24}e^2\right) e^7 \sin 3\omega \sin^3 i \\
& + \left(-\frac{1099}{2}e^9 \sin i + \frac{5495}{8}e^9 \sin^3 i\right) \sin\omega, \\
A_{j11} = & -\frac{1221305}{264}e^8 \sin 3\omega \sin^3 i, \\
A_{j12} = & -\frac{2065}{4}e^9 \sin 3\omega \sin^3 i, \qquad (6.23)
\end{aligned}
$$

$$
\begin{aligned}
B_{j1} = & \left(-40 - \frac{5050}{3}e^2 + \frac{32875}{6}e^4 + \frac{18485}{6}e^6\right) e^2 \cos 3\omega \sin^3 i \\
& + \left(\left(-768 + 3744e^2 - 7280e^4 - 5290e^6 + 1503e^8\right) \sin i \right.\\
& \left. + \left(960 - 4680e^2 + 9100e^4 + \frac{13225}{2}e^6 - \frac{7515}{4}e^8\right) \sin^3 i\right) \cos\omega,
\end{aligned}
$$

$$B_{j2} = \left(320 - 2160e^2 - \frac{43270}{3}e^4 - \frac{9620}{3}e^6 + \frac{1915}{12}e^8\right) e\cos 3\omega \sin^3 i$$

$$+ \left(\left(-2304 + 12032e^2 - 19240e^4 - 11376e^6 - 215e^8\right) e\sin i \right.$$

$$\left. + \left(2880 - 15040e^2 + 24050e^4 + 14220e^6 + \frac{1075}{4}e^8\right) e\sin^3 i \right) \cos\omega,$$

$$B_{j3} = \left(-320 + 3680e^2 - 10560e^4 - \frac{179935}{6}e^6 - \frac{20805}{4}e^8\right) \cos 3\omega \sin^3 i$$

$$+ \left(\left(-5088 + 27720e^2 - 17658e^4 - 6296e^6\right) e^2\sin i \right.$$

$$\left. + \left(6360 - 34650e^2 + \frac{44145}{2}e^4 + 7870e^6\right) e^2\sin^3 i \right) \cos\omega,$$

$$B_{j4} = \left(-1600 + 15480e^2 - \frac{27050}{3}e^4 - \frac{62140}{3}e^6 - \frac{14125}{12}e^8\right) e\cos 3\omega \sin^3 i$$

$$+ \left(\left(-9856 + 46664e^2 - 1792e^4 - 850e^6\right) e^3\sin i \right.$$

$$\left. + \left(12320 - 58330e^2 + 2240e^4 + \frac{2125}{2}e^6\right) e^3\sin^3 i \right) \cos\omega,$$

$$B_{j5} = \left(-5080 + \frac{123772}{3}e^2 + \frac{75245}{6}e^4 - \frac{100625}{24}e^6\right) e^2\cos 3\omega \sin^3 i$$

$$+ \left((-17480 + 46814e^2 + 3604^4)\, e^4\sin i \right.$$

$$\left. + \left(21850 - \frac{117035}{2}e^2 - 4505e^4\right) e^4\sin^3 i \right) \cos\omega,$$

$$B_{j6} = \left(-13040 + 60150e^2 + 20880e^4 + \frac{2725}{24}e^6\right) e^3\cos 3\omega \sin^3 i$$

$$+ \left(\left(-25544 + 24272e^2 + \frac{1975}{2}e^4\right) e^4\sin i \right.$$

$$\left. + \left(31930 - 30340e^2 - \frac{9875}{8}e^4\right) e^5\sin^3 i \right) \cos\omega,$$

$$
\begin{aligned}
B_{j7} &= \left(-\frac{561970}{21}+\frac{959960}{21}e^2+\frac{794555}{84}e^4\right)e^4\cos 3\omega\sin^3 i\\
&+\left(\left(-\frac{176574}{7}+\frac{79259}{14}e^2\right)e^6\sin i\right.\\
&+\left.\left(\frac{441435}{14}-\frac{396295}{56}e^2\right)e^6\sin^3 i\right)\cos\omega,\\
B_{j8} &= \left(-\frac{113605}{3}+\frac{49595}{3}e^2+\frac{64535}{48}e^4\right)e^5\cos 3\omega\sin^3 i\\
&+\left(\left(-14674+\frac{1625}{4}e^2\right)e^7\sin i+\left(\frac{36685}{2}-\frac{8125}{16}e^2\right)e^7\sin^3 i\right)\cos\omega,\\
B_{j9} &= \left(-\frac{598325}{18}+\frac{24215}{12}e^2\right)e^6\cos 3\omega\sin^3 i\\
&+\left(-\frac{26843}{6}e^8\sin i+\frac{134215}{24}e^8\sin^3 i\right)\cos\omega,\\
B_{j10} &= \left(-\frac{51020}{3}-\frac{2401}{24}e^2\right)e^7\cos 3\omega\sin^3 i\\
&+\left(-\frac{1099}{2}e^9\sin i+\frac{5495}{8}e^9\sin^3 i\right)\cos\omega,\\
B_{j11} &= -\frac{1221305}{264}e^8\cos 3\omega\sin^3 i,\\
B_{j12} &= -\frac{2065}{4}e^9\cos 3\omega\sin^3 i.
\end{aligned}
\tag{6.24}
$$

For $j=2$, there are

$$
\begin{aligned}
c_j &= 0,\\
d_j &= \left(540+2280e^2+\frac{655}{2}e^4\right)e^5\cos 3\omega\sin^3 i\\
&+\cos\omega\left(\left(768-1920e^2+1728e^4+1464e^6+270e^8\right)e\sin i\right.\\
&+\left.\left(-960+2400e^2-2160e^4-1830e^6-\frac{675}{2}e^8\right)e\sin^3 i\right),\\
A_{j1} &= \left(80-370e^2+3560e^4+\frac{24335}{6}e^6\right)e^2\sin 3\omega\sin^3 i\\
&+\left(\left(-1728+5328e^2-15760e^4+231e^6\right)e^2\sin i\right.\\
&+\left.\left(2160-6660e^2+19700e^4-\frac{1155}{4}e^6\right)e^2\sin^3 i\right)\sin\omega,
\end{aligned}
\tag{6.25}
$$

$$A_{j2} = \left(-160 + 840e^2 - 11060e^4 - \frac{26800}{3}e^6 - \frac{4015}{12}e^8\right) e \sin 3\omega \sin^3 i$$

$$+ \left(\left(-1152 + 3648e^2 - 1200e^4 - 22188e^6 - 1213e^8\right) e \sin i\right.$$

$$\left. + \left(1440 - 4560e^2 + 1500e^4 + 27735e^6 + \frac{6065}{4}e^8\right) e \sin^3 i\right) \sin\omega,$$

$$A_{j3} = \left(160 - 1720e^2 - \frac{92900}{3}e^4 - \frac{38765}{4}e^6\right) e^2 \sin 3\omega \sin^3 i$$

$$+ \left(\left(-3392 + 15512e^2 + 4656e^4 - 8552e^6\right) e^2 \sin i\right.$$

$$\left. + \left(4240 - 19390e^2 - 5820e^4 + 10690e^6\right) e^2 \sin^3 i\right) \sin\omega,$$

$$A_{j4} = \left(-400 + 4560e^2 + \frac{5080}{3}e^4 - \frac{61850}{3}e^6 - \frac{5465}{3}e^8\right) e \sin 3\omega \sin^3 i$$

$$+ \left(\left(-7392 + 32376e^2 + 16692e^4 - 475e^6\right) e^3 \sin i\right.$$

$$\left. + \left(9240 - 40470e^2 - 20865e^4 + \frac{2375}{4}e^6\right) e^3 \sin^3 i\right) \sin\omega,$$

$$A_{j5} = \left(-2032 + \frac{62620}{3}e^2 + \frac{80500}{3}e^4 - \frac{28831}{24}e^6\right) e^2 \sin 3\omega \sin^3 i$$

$$+ \left(\left(-14184 + \frac{167824}{5}e^2 + \frac{52596}{5}e^4\right) e^4 \sin i\right.$$

$$\left. + \left(17730 - 41956e^2 - 13149e^4\right) e^4 \sin^3 i\right) \sin\omega,$$

$$A_{j6} = \left(-6520 + 40140e^2 + \frac{66685}{2}e^4 + \frac{26915}{24}e^6\right) e^3 \sin 3\omega \sin^3 i$$

$$+ \left(\left(-22576 + 15020e^2 + \frac{3621}{2}e^4\right) e^5 \sin i\right.$$

$$\left. + \left(28220 - 18775e^2 - \frac{18015}{8}e^4\right) e^5 \sin^3 i\right) \sin\omega,$$

$$A_{j7} = \left(-\frac{50590}{3} + \frac{664190}{21}e^2 + \frac{1146275}{84}e^4\right) e^4 \sin 3\omega \sin^3 i$$
$$+ \left(\left(-\frac{167056}{7} + \frac{27285}{14}e^2\right) e^6 \sin i\right.$$
$$\left. + \left(\frac{208820}{7} - \frac{136425}{56}e^2\right) e^6 \sin^3 i\right) \sin\omega,$$
$$A_{j8} = \left(-\frac{176765}{6} + \frac{23330}{3}e^2 + \frac{81215}{48}e^4\right) e^5 \sin 3\omega \sin^3 i$$
$$+ \left(\left(-14439 - \frac{725}{4}e^2\right) e^7 \sin i + \left(\frac{72195}{4} + \frac{3625}{16}e^2\right) e^7 \sin^3 i\right) \sin\omega,$$
$$A_{j9} = \left(-\frac{267790}{9} - \frac{17615}{12}e^2\right) e^6 \sin 3\omega \sin^3 i$$
$$+ \left(-\frac{26843}{6}e^8 \sin i + \frac{134215}{24}e^8 \sin^3 i\right) \sin\omega,$$
$$A_{j10} = \left(-\frac{98743}{6} - \frac{15589}{24}e^2\right) e^7 \sin 3\omega \sin^3 i$$
$$+ \left(-\frac{1099}{2}e^9 \sin i + \frac{5495}{8}e^9 \sin^3 i\right) \sin\omega,$$
$$A_{j11} = -\frac{1221305}{264}e^8 \sin 3\omega \sin^3 i,$$
$$A_{j12} = -\frac{2065}{4}e^9 \sin 3\omega \sin^3 i, \tag{6.26}$$

$$B_{j1} = \left(80 - \frac{1090}{3}e^2 + \frac{9620}{3}e^4 + \frac{11740}{3}e^6\right) e^2 \cos 3\omega \sin^3 i$$
$$+ \left(\left(2496 - 7440e^2 - 2832e^4 - 315e^6\right) e^2 \sin i\right.$$
$$\left. + \left(-3120 + 9300e^2 + 3540e^4 + \frac{1575}{4}e^6\right) e^2 \sin^3 i\right) \cos\omega,$$
$$B_{j2} = \left(-160 + 840e^2 - \frac{33100}{3}e^4 - \frac{26345}{3}e^6 - \frac{2105}{12}e^8\right) e \cos 3\omega \sin^3 i$$
$$+ \left(\left(-1152 + 8064e^2 - 16112e^4 - 11208e^6 - 695e^8\right) e \sin i\right.$$
$$\left. + \left(1440 - 10080e^2 + 20140e^4 + 14010e^6 + \frac{3475}{4}e^8\right) e \sin^3 i\right) \cos\omega,$$

$$B_{j3} = \left(160 - 1720e^2 - \frac{92900}{3}e^4 - \frac{39455}{4}e^6\right) e^2 \cos 3\omega \sin^3 i$$
$$+ \left(\left(-3392 + 21000e^2 - 13600e^4 - 5330e^6\right) e^2 \sin i\right.$$
$$\left.+ \left(4240 - 26250e^2 + 17000e^4 + \frac{13325}{2}e^6\right) e^2 \sin^3 i\right) \cos\omega,$$

$$B_{j4} = \left(-400 + 4560e^2 + \frac{5080}{3}e^4 - \frac{61810}{3}e^6 - \frac{23005}{12}e^8\right) e \cos 3\omega \sin^3 i$$
$$+ \left(\left(-7392 + 39568e^2 + 2400e^4 - 410e^6\right) e^3 \sin i\right.$$
$$\left.+ \left(9240 - 49460e^2 - 3000e^4 + \frac{1025}{2}e^6\right) e^3 \sin^3 i\right) \cos\omega,$$

$$B_{j5} = \left(-2032 + \frac{62620}{3}e^2 + \frac{80500}{3}e^4 - \frac{27581}{24}e^6\right) e^2 \cos 3\omega \sin^3 i$$
$$+ \left(\left(-14184 + \frac{209152}{5}e^2 + \frac{26458}{5}e^4\right) e^4 \sin i\right.$$
$$\left.+ \left(17730 - 52288e^2 - \frac{13229}{2}e^4\right) e^4 \sin^3 i\right) \cos\omega,$$

$$B_{j6} = \left(-6520 + 40140e^2 + \frac{66685}{2}e^4 + \frac{27385}{24}e^6\right) e^3 \cos 3\omega \sin^3 i$$
$$+ \left(\left(-22576 + 21144e^2 + \frac{2295}{2}e^4\right) e^5 \sin i\right.$$
$$\left.+ \left(28220 - 26430e^2 - \frac{11475}{8}e^4\right) e^5 \sin^3 i\right) \cos\omega,$$

$$B_{j7} = \left(-\frac{50590}{3} + \frac{664190}{21}e^2 + \frac{1146275}{84}e^4\right) e^4 \cos 3\omega \sin^3 i$$
$$+ \left(\left(-\frac{167056}{7} + \frac{60223}{14}e^2\right) e^6 \sin i\right.$$
$$\left.+ \left(\frac{208820}{7} - \frac{301115}{56}e^2\right) e^6 \sin^3 i\right) \cos\omega,$$

$$B_{j8} = \left(-\frac{176765}{6} + \frac{23330}{3}e^2 + \frac{81215}{48}e^4\right) e^5 \cos 3\omega \sin^3 i$$
$$+ \left(\left(-14439 + \frac{685}{4}e^2\right) e^7 \sin i + \left(\frac{72195}{4} - \frac{3425}{16}e^2\right) e^7 \sin^3 i\right) \cos\omega,$$

$$B_{j9} = \left(-\frac{267790}{9} - \frac{17615}{12}e^2\right) e^6 \cos 3\omega \sin^3 i$$

$$+ \left(-\frac{26843}{6} e^8 \sin i + \frac{134215}{24} e^8 \sin^3 i\right) \cos\omega,$$

$$B_{j10} = \left(-\frac{98743}{6} - \frac{15589}{24}e^2\right) e^7 \cos 3\omega \sin^3 i$$

$$+ \left(-\frac{1099}{2} e^9 \sin i + \frac{5495}{8} e^9 \sin^3 i\right) \cos\omega,$$

$$B_{j11} = -\frac{1221305}{264} e^8 \cos 3\omega \sin^3 i,$$

$$B_{j12} = -\frac{2065}{4} e^9 \cos 3\omega \sin^3 i. \tag{6.27}$$

For $j = 3$, there are

$$c_j = 0, \tag{6.28}$$

$$d_j = \sin 3\omega \left(\left(-660 - 180e^2\right) e^6 \sin i + \left(460 + 2100e^2 + \frac{765}{2}e^4\right) e^4 \sin^3 i \right)$$

$$+ \sin\omega \left(\left(768 - 1152e^2 + 96e^4 - 120e^6\right) e^2 \Big/ \sin i \right.$$

$$+ \left(-768 - 5568e^2 + 10080e^4 + 1008e^6 + 1080e^8\right) \sin i$$

$$\left. + \left(960 + 5280e^2 - 10080e^4 - 1470e^6 - \frac{2175}{2}e^8\right) \sin^3 i \right),$$

$$A_{j1} = \cos 3\omega \left(\left(-120 - 1360e^2 + \frac{1785}{2}e^4\right) e^3 \sin i \right.$$

$$\left. + \left(80 - \frac{770}{3}e^2 - 4720e^4 - \frac{11960}{3}e^6\right) \sin^3 i \right)$$

$$\cos\omega \left(\left(-768 + 2592e^2 - 3136e^4 - 486e^6\right) e \Big/ \sin i \right.$$

$$+ \left(4608 - 15048e^2 + 27520e^4 + \frac{3183}{2}e^6\right) e \sin i$$

$$\left. + \left(-4080 + 13140e^2 - 27540e^4 - \frac{3705}{4}e^6\right) e \sin^3 i \right),$$

$$A_{j2} = \cos 3\omega\Bigg(\left(480 - 2520e^2 - 5970e^4 - 430e^6\right) e^2 \sin i$$

$$+ \left(-160 + 920e^2 + \frac{24820}{3}e^4 + \frac{24850}{3}e^6 + \frac{1915}{4}e^8\right)\sin^3 i\Bigg)$$

$$\cos\omega\Bigg(\left(-1152 + 4288e^2 - 3432e^4 - 348e^6\right) e^2 \Big/ \sin i$$

$$+ \left(1152 + 672e^2 - 14880e^4 + 35058e^6 + 2518e^8\right)\sin i$$

$$+ \left(-1440 + 1680e^2 + 9220e^4 - 36315e^6 - \frac{9545}{4}e^8\right)\sin^3 i\Bigg),$$

$$A_{j3} = \cos 3\omega\Bigg(\left(-320 + 3200e^2 - 5640e^4 - \frac{9235}{2}e^6\right) e \sin i$$

$$+ \left(-960 + 6020e^2 + \frac{88220}{3}e^4 + \frac{121475}{12}e^6\right) e \sin^3 i\Bigg)$$

$$\cos\omega\Bigg((-1696 + 6432e^2 - 906e^4)\, e^3 \Big/ \sin i$$

$$+ \left(3392 - 9152e^2 - 28776e^4 + \frac{23899}{2}e^6\right) e \sin i$$

$$+ \left(-4240 + 15150e^2 + 21900e^4 - 12955e^6\right) e \sin^3 i\Bigg),$$

$$A_{j4} = \cos 3\omega\Bigg(\left(-1200 + 10080e^2 - 630e^4 - 700e^6\right) e^2 \sin i$$

$$+ \left(400 - 6080e^2 + \frac{14620}{3}e^4 + \frac{61360}{3}e^6 + \frac{3835}{2}e^8\right)\sin^3 i\Bigg)$$

$$\cos\omega\Bigg((-2464 + 6096e^2 + 100e^4)\, e^4 \Big/ \sin i$$

$$+ \left(7392 - 23136e^2 - 39552e^4 + 100e^6\right) e^2 \sin i$$

$$+ \left(-9240 + 34310e^2 + 36105e^4 - \frac{1375}{4}e^6\right) e^2 \sin^3 i\Bigg),$$

$$A_{j5} = \cos 3\omega\Bigg(\left(-3048 + 17336e^2 + \frac{6087}{2}e^4\right) e^3 \sin i$$

$$+ \left(2032 - \frac{70376}{3}e^2 - 22468e^4 + \frac{15191}{24}e^6\right) e \sin^3 i\Bigg)$$

$$+\cos\omega\left(\left(-3296+\frac{12826}{5}e^2\right)e^5\Big/\sin i\right.$$

$$+\left(14184-\frac{106024}{5}e^2-\frac{201387}{10}e^4\right)e^3\sin i$$

$$\left.+\left(-17730+33716e^2+19562e^4\right)e^3\sin^3 i\right),$$

$$A_{j6}=\cos 3\omega\left(\left(-6520+13490e^2+\frac{2055}{2}e^4\right)e^4\sin i\right.$$

$$\left.+\left(6520-\frac{130540}{3}e^2-\frac{65385}{2}e^4-\frac{32765}{24}e^6\right)e^2\sin^3 i\right)$$

$$+\cos\omega\left(\left(-2968+348e^2\right)e^6\Big/\sin i+\left(22576-3890e^2-\frac{6231}{2}e^4\right)e^4\sin i\right.$$

$$\left.+\left(-28220+11355e^2+\frac{25065}{8}e^4\right)e^4\sin^3 i\right),$$

$$A_{j7}=\cos 3\omega\left(\left(-\frac{69280}{7}+\frac{29310}{7}e^2\right)e^5\sin i\right.$$

$$\left.+\left(\frac{50590}{3}-\frac{237010}{7}e^2-\frac{290485}{21}e^4\right)e^3\sin^3 i\right)$$

$$+\cos\omega\left(-\frac{9518}{7}e^7/\sin i+\left(\frac{167056}{7}+3150e^2\right)e^5\sin i\right.$$

$$\left.+\left(-\frac{208820}{7}-\frac{7705}{8}e^2\right)e^5\sin^3 i\right),$$

$$A_{j8}=\cos 3\omega\left(\left(-\frac{16815}{2}+\frac{695}{2}e^2\right)e^6\sin i\right.$$

$$\left.+\left(\frac{176765}{6}-\frac{48305}{6}e^2-\frac{27215}{16}e^4\right)e^4\sin^3 i\right)$$

$$+\cos\omega\left(-235e^8/\sin i+\left(14439+\frac{2125}{2}e^2\right)e^6\sin i\right.$$

$$\left.+\left(-\frac{72195}{4}-\frac{13025}{16}e^2\right)e^6\sin^3 i\right),$$

$$
\begin{aligned}
A_{j9} &= \cos 3\omega \left(-\frac{20915}{6} e^8 \sin i + \left(\frac{267790}{9} + \frac{15970}{9} e^2 \right) e^5 \sin^3 i \right) \\
&\quad + \cos \omega \left(\frac{26843}{6} e^7 \sin i - \frac{134215}{24} e^7 \sin^3 i \right), \\
A_{j10} &= \cos 3\omega \left(-\frac{1099}{2} e^8 \sin i + \left(\frac{98743}{6} + \frac{5929}{8} e^2 \right) e^6 \sin^3 i \right) \\
&\quad + \cos \omega \left(\frac{1099}{2} e^8 \sin i - \frac{5495}{8} e^8 \sin^3 i \right), \\
A_{j11} &= \cos 3\omega \left(\frac{1221305}{264} e^7 \sin^3 i \right), \\
A_{j12} &= \cos 3\omega \left(\frac{2065}{4} e^8 \sin^3 i \right),
\end{aligned}
\tag{6.29}
$$

$$
\begin{aligned}
B_{j1} &= \sin 3\omega \Bigg(\left(120 + 1440e^2 - \frac{1665}{2} e^4 \right) e^3 \sin i \\
&\quad + \left(-80 + 250e^2 + 4420e^4 + \frac{11200}{3} e^6 \right) e \sin^3 i \Bigg) \\
&\quad \sin \omega \Bigg(\left(768 - 480e^2 - 640e^4 + 1818e^6 \right) e / \sin i \\
&\quad + \left(-8832 + 7320e^2 - 4272e^4 - \frac{26901}{2} e^6 \right) e \sin i \\
&\quad + \left(9360 - 8100e^2 + 6740e^4 + \frac{51345}{4} e^6 \right) e \sin^3 i \Bigg), \\
B_{j2} &= \sin 3\omega \Bigg(\left(-480 + 2520e^2 + 5910e^4 + 335e^6 \right) \sin i \\
&\quad + \left(160 - 920e^2 - 8300e^4 - \frac{24335}{3} e^6 - \frac{1325}{4} e^8 \right) \sin^3 i \Bigg) \\
&\quad + \sin \omega \Bigg(\left(1152 - 2816e^2 + 312e^4 + 4808e^6 \right) e^2 / \sin i \\
&\quad + \left(-1152 - 5088e^2 + 15824e^4 - 15570e^6 - 3935e^8 \right) \sin i \\
&\quad + \left(1440 + 3840e^2 - 13620e^4 + 18780e^6 + \frac{15475}{4} e^8 \right) \sin^3 i \Bigg),
\end{aligned}
$$

$$B_{j3} = \sin 3\omega \left(\left(320 - 3200e^2 + 5640e^4 + \frac{9325}{2}e^6 \right) e \sin i \right.$$
$$\left. + \left(960 - 6020e^2 - \frac{88220}{3}e^4 - \frac{121115}{12}e^6 \right) e \sin^3 i \right)$$
$$+ \sin\omega \left(\left(1696 - 5024e^2 - 966e^4\right) e^3 / \sin i \right.$$
$$+ \left(-3392 + 3664e^2 + 33304e^4 + \frac{2957}{2}e^6 \right) e \sin i$$
$$\left. + \left(4240 - 8290e^2 - 30640e^4 + 265e^6 \right) e \sin^3 i \right),$$

$$B_{j4} = \sin 3\omega \left(\left(1200 - 10080e^2 + 630e^4 + 740e^6 \right) e^2 \sin i \right.$$
$$\left. + \left(-400 + 6080e^2 - \frac{14620}{3}e^4 - \frac{61400}{3}e^6 - \frac{7795}{4}e^8 \right) \sin^3 i \right)$$
$$+ \sin\omega \left(\left(2464 - 4632e^2 - 440e^4\right) e^4 / \sin i \right.$$
$$+ \left(-7392 + 15944e^2 + 39762e^4 + 2560e^6 \right) e^2 \sin i$$
$$\left. + \left(9240 - 25320e^2 - 39570e^4 - \frac{4475}{2}e^6 \right) e^2 \sin^3 i \right),$$

$$B_{j5} = \sin 3\omega \left(\left(3048 - 17336e^2 - \frac{6087}{2}e^4 \right) e^3 \sin i \right.$$
$$\left. + \left(-2032 + \frac{70376}{3}e^2 + 22468e^4 - \frac{16441}{24}e^6 \right) e \sin^3 i \right)$$
$$+ \sin\omega \left(\left(3296 - \frac{8438}{5}e^2 \right) e^5 / \sin i \right.$$
$$+ \left(-14184 + \frac{64696}{5}e^2 + \frac{175597}{10}e^4 \right) e^3 \sin i$$
$$\left. + \left(17730 - 23384e^2 - 18258e^4\right) e^3 \sin^3 i \right),$$

$$B_{j6} = \sin 3\omega\left(\left(6520 - 13490e^2 - \frac{2055}{2}e^4\right)e^4 \sin i\right.$$
$$\left. + \left(-6520 + \frac{130540}{3}e^2 + \frac{65385}{2}e^4 + \frac{10765}{8}e^6\right)e^2 \sin^3 i\right)$$
$$+ \sin\omega\left((2968 - 160e^2)\,e^6/\sin i + \left(-22576 - 2234e^2 + \frac{4455}{2}e^4\right)e^4 \sin i\right.$$
$$\left. + \left(28220 - 3700e^2 - \frac{19475}{8}e^4\right)e^4 \sin^3 i\right),$$

$$B_{j7} = \sin 3\omega\left(\left(\frac{69280}{7} - \frac{29310}{7}e^2\right)e^5 \sin i\right.$$
$$\left. + \left(-\frac{50590}{3} + \frac{237010}{7}e^2 + \frac{290485}{21}e^4\right)e^3 \sin^3 i\right)$$
$$+ \sin\omega\left(\frac{9518}{7}e^7/\sin i + \left(-\frac{167056}{7} - \frac{38519}{7}e^2\right)e^5 \sin i\right.$$
$$\left. + \left(\frac{208820}{7} + \frac{218625}{56}e^2\right)e^5 \sin^3 i\right),$$

$$B_{j8} = \sin 3\omega\left(\left(\frac{16815}{2} - \frac{695}{2}e^2\right)e^6 \sin i\right.$$
$$\left. + \left(-\frac{176765}{6} + \frac{48305}{6}e^2 + \frac{27215}{16}e^4\right)e^4 \sin^3 i\right)$$
$$+ \sin\omega\left(235e^8/\sin i + (-14439 - 1415e^2)\,e^6 \sin i\right.$$
$$\left. + \left(\frac{72195}{4} + \frac{20075}{16}e^2\right)e^6 \sin^3 i\right),$$

$$B_{j9} = \sin 3\omega\left(\frac{20915}{6}e^7 \sin i + \left(-\frac{267790}{9} - \frac{15970}{9}e^2\right)e^5 \sin^3 i\right)$$
$$+ \sin\omega\left(-\frac{26843}{6}e^7 \sin i + \frac{134215}{24}e^7 \sin^3 i\right),$$

$$B_{j10} = \sin 3\omega\left(\frac{1099}{2}e^8 \sin i + \left(-\frac{98743}{6} - \frac{5929}{8}e^2\right)e^6 \sin^3 i\right)$$
$$+ \sin\omega\left(-\frac{1099}{2}e^8 \sin i + \frac{5495}{8}e^8 \sin^3 i\right),$$

$$B_{j11} = -\sin 3\omega\left(\frac{1221305}{264}e^7 \sin^3 i\right),$$

$$B_{j12} = -\sin 3\omega \left(\frac{2065}{4} e^8 \sin^3 i \right). \tag{6.30}$$

For $j = 4$, there are

$$c_j = 0, \tag{6.31}$$

$$d_j = \cos i \Bigg(\cos 3\omega \left(55 + 15e^2\right) e^5 \sin^2 i + \cos\omega \Bigg(\left(-64 + 96e^2 - 8e^4 + 10e^6\right) e$$

$$+ \left(80 - 120e^2 + 10e^4 - \frac{25}{2} e^6\right) e \sin^2 i \Bigg)\Bigg),$$

$$A_{j1} = \cos i \Bigg(\sin 3\omega \left(-10 - \frac{340}{3} e^2 + \frac{595}{8} e^4\right) e^2 \sin^2 i$$

$$+ \sin\omega \Bigg(\left(-64 + 216e^2 - \frac{784}{3} e^4 - \frac{81}{2} e^6\right)$$

$$+ \left(80 - 270e^2 + \frac{980}{3} e^4 + \frac{405}{8} e^6\right) \sin^2 i \Bigg)\Bigg),$$

$$A_{j2} = \cos i \Bigg(\sin 3\omega \left(40 - 210e^2 - \frac{995}{2} e^4 - \frac{215}{6} e^6\right) e \sin^2 i$$

$$+ \sin\omega \Bigg(\left(-96 + \frac{1072}{3} e^2 - 286e^4 - 29e^6\right) e$$

$$+ \left(120 - \frac{1340}{3} e^2 + \frac{715}{2} e^4 + \frac{145}{4} e^6\right) e \sin^2 i \Bigg)\Bigg),$$

$$A_{j3} = \cos i \Bigg(\sin 3\omega \left(-\frac{80}{3} + \frac{800}{3} e^2 - 470e^4 - \frac{9235}{24} e^6\right) \sin^2 i$$

$$+ \sin\omega \Bigg(\left(-\frac{424}{3} + 536e^2 - \frac{151}{2} e^4\right) e^2$$

$$+ \left(\frac{530}{3} - 670e^2 + \frac{755}{8} e^4\right) e^2 \sin^2 i \Bigg)\Bigg),$$

$$A_{j4} = \cos i \Bigg(\sin 3\omega \left(-100 + 840e^2 - \frac{105}{2} e^4 - \frac{175}{3} e^6\right) e \sin^2 i$$

$$+ \sin\omega \Bigg(\left(-\frac{616}{3} + 508e^2 + \frac{25}{3} e^4\right) e^3$$

$$+ \left(\frac{770}{3} - 635e^2 - \frac{125}{12} e^4\right) e^3 \sin^2 i \Bigg)\Bigg),$$

$$
\begin{aligned}
A_{j5} &= \cos i\left(\sin 3\omega\left(-254+\frac{4334}{3}e^2+\frac{2029}{8}e^4\right)e^2\sin^2 i\right.\\
&\quad\left.+\sin\omega\left(\left(-\frac{824}{3}+\frac{6413}{30}e^2\right)e^4+\left(\frac{1030}{3}-\frac{6413}{24}e^2\right)e^4\sin^2 i\right)\right),\\
A_{j6} &= \cos i\left(\sin 3\omega\left(-\frac{1630}{3}+\frac{6745}{6}e^2+\frac{685}{8}e^4\right)e^3\sin^2 i\right.\\
&\quad\left.+\sin\omega\left(\left(-\frac{742}{3}+29e^2\right)e^5+\left(\frac{1855}{6}-\frac{145}{4}e^2\right)e^5\sin^2 i\right)\right),\\
A_{j7} &= \cos i\left(\sin 3\omega\left(-\frac{17320}{21}+\frac{4885}{14}e^2\right)e^4\sin^2 i\right.\\
&\quad\left.+\sin\omega\left(-\frac{4759}{42}e^6+\frac{23795}{168}e^6\sin^2 i\right)\right),\\
A_{j8} &= \cos i\left(\sin 3\omega\left(-\frac{5605}{8}+\frac{695}{24}e^2\right)e^5\sin^2 i\right.\\
&\quad\left.+\sin\omega\left(-\frac{235}{12}e^7+\frac{1175}{48}e^7\sin^2 i\right)\right),\\
A_{j9} &= -\frac{20915}{72}e^6\cos i\sin 3\omega\sin^2 i,\\
A_{j10} &= -\frac{1099}{24}e^7\cos i\sin 3\omega\sin^2 i,
\end{aligned}
\tag{6.32}
$$

$$
\begin{aligned}
B_{j1} &= \cos i\left(\cos 3\omega\left(-10-120e^2+\frac{555}{8}e^4\right)e^2\sin^2 i\right.\\
&\quad+\cos\omega\left(\left(-64+40e^2+\frac{160}{3}e^4-\frac{303}{2}e^6\right)\right.\\
&\quad\left.\left.+\left(80-50e^2-\frac{200}{3}e^4+\frac{1515}{8}e^6\right)\sin^2 i\right)\right),\\
B_{j2} &= \cos i\left(\cos 3\omega\left(40-210e^2-\frac{985}{2}e^4-\frac{335}{12}e^6\right)e\sin^2 i\right.\\
&\quad+\cos\omega\left(\left(-96+\frac{704}{3}e^2-26e^4-40e^6\right)e\right.\\
&\quad\left.\left.+\left(120-\frac{880}{3}e^2+\frac{65}{2}e^4+50e^6\right)e\sin^2 i\right)\right),
\end{aligned}
$$

$$
\begin{aligned}
B_{j3} &= \cos i \Bigg(\cos 3\omega \left(-\frac{80}{3} + \frac{800}{3} e^2 - 470 e^4 - \frac{9325}{24} e^6 \right) \sin^2 i \\
&\quad + \cos\omega \Bigg(\left(-\frac{424}{3} + \frac{1256}{3} e^2 + \frac{161}{2} e^4 \right) e^2 \\
&\quad + \left(\frac{530}{3} - \frac{1570}{3} e^2 - \frac{805}{8} e^4 \right) e^2 \sin^2 i \Bigg) \Bigg), \\
B_{j4} &= \cos i \Bigg(\cos 3\omega \left(-100 + 840 e^2 - \frac{105}{2} e^4 - \frac{185}{3} e^6 \right) e \sin^2 i \\
&\quad + \cos\omega \Bigg(\left(-\frac{616}{3} + 386 e^2 + \frac{110}{3} e^4 \right) e^3 \\
&\quad + \left(\frac{770}{3} - \frac{965}{2} e^2 - \frac{275}{6} e^4 \right) e^3 \sin^2 i \Bigg) \Bigg), \\
B_{j5} &= \cos i \Bigg(\cos 3\omega \left(-254 + \frac{4334}{3} e^2 + \frac{2029}{8} e^4 \right) e^2 \sin^2 i \\
&\quad + \cos\omega \left(\left(-\frac{824}{3} + \frac{4219}{30} e^2 \right) e^4 + \left(\frac{1030}{3} - \frac{4219}{24} e^2 \right) e^4 \sin^2 i \right) \Bigg), \\
B_{j6} &= \cos i \Bigg(\cos 3\omega \left(-\frac{1630}{3} + \frac{6745}{6} e^2 + \frac{685}{8} e^4 \right) e^3 \sin^2 i \\
&\quad + \cos\omega \left(\left(-\frac{742}{3} + \frac{40}{3} e^2 \right) e^5 + \left(\frac{1855}{6} - \frac{50}{3} e^2 \right) e^5 \sin^2 i \right) \Bigg), \\
B_{j7} &= \cos i \Bigg(\cos 3\omega \left(-\frac{17320}{21} + \frac{4885}{3} e^2 \right) e^4 \sin^2 i \\
&\quad + \cos\omega \left(-\frac{4759}{42} e^6 + \frac{23795}{168} e^6 \sin^2 i \right) \Bigg), \\
B_{j8} &= \cos i \Bigg(\cos 3\omega \left(-\frac{5605}{8} + \frac{695}{24} e^2 \right) e^5 \sin^2 i \\
&\quad + \cos\omega \left(-\frac{235}{12} e^7 + \frac{1175}{48} e^7 \sin^2 i \right) \Bigg), \\
B_{j9} &= -\frac{20915}{72} e^6 \cos i \cos 3\omega \sin^2 i, \\
B_{j10} &= -\frac{1099}{24} e^7 \cos i \cos 3\omega \sin^2 i.
\end{aligned}
\tag{6.33}
$$

For $j = 5$, there are

$$
\begin{aligned}
c_j &= 0, \qquad (6.34)\\
d_j &= \cos i\left(\sin 3\omega\,(55+15e^2)\,e^5\sin i+\sin\omega\left(\left(-64+96e^2-8e^4+10e^7\right)e\Big/\sin i\right.\right.\\
&\quad\left.\left.+\left(240-360e^2+30e^4-\frac{75}{2}e^6\right)e\sin i\right)\right),\\
A_{j1} &= \cos i\left(\cos 3\omega\left(10+\frac{340}{3}e^2-\frac{595}{8}e^4\right)e^2\sin i\right.\\
&\quad+\cos\omega\left(\left(64-216e^2+\frac{784}{3}e^4+\frac{81}{2}e^6\right)\Big/\sin i\right.\\
&\quad\left.\left.+\left(-240+810e^2-980e^4-\frac{1215}{8}e^6\right)\sin i\right)\right),\\
A_{j2} &= \cos i\left(\cos 3\omega\left(-40+210e^2+\frac{995}{2}e^4+\frac{215}{6}e^6\right)e\sin i\right.\\
&\quad+\cos\omega\left(\left(96-\frac{1072}{3}e^2+286e^4+29e^6\right)e\Big/\sin i\right.\\
&\quad\left.\left.+\left(-360+1340e^2-\frac{2145}{2}e^4-\frac{435}{4}e^6\right)e\sin i\right)\right),\\
A_{j3} &= \cos i\left(\cos 3\omega\left(\frac{80}{3}-\frac{800}{3}e^2+470e^4+\frac{9235}{2}e^6\right)\sin i\right.\\
&\quad+\cos\omega\left(\left(\frac{424}{3}-536e^2+\frac{151}{2}e^4\right)e^2\Big/\sin i\right.\\
&\quad\left.\left.+\left(-530+2010e^2-\frac{2265}{8}e^4\right)e^2\sin i\right)\right),\\
A_{j4} &= \cos i\left(\cos 3\omega\left(100-840e^2+\frac{105}{2}e^4+\frac{175}{3}e^6\right)e\sin i\right.\\
&\quad+\cos\omega\left(\left(\frac{616}{3}-508e^2-\frac{25}{3}e^4\right)e^2\Big/\sin i\right.\\
&\quad\left.\left.+\left(-770+1905e^2+\frac{125}{4}e^4\right)e^3\sin i\right)\right),
\end{aligned}
$$

$$A_{j5} = \cos i\left(\cos 3\omega \left(254 - \frac{4334}{3}e^2 - \frac{2029}{8}e^4\right) e^3 \sin i \right.$$
$$\left. + \cos\omega \left(\left(\frac{824}{3} - \frac{6413}{30}e^2\right) e^4 \Big/ \sin i + \left(-1030 + \frac{6413}{8}e^2\right) e^4 \sin i \right) \right),$$
$$A_{j6} = \cos i\left(\cos 3\omega \left(\frac{1630}{3} - \frac{6745}{6}e^2 - \frac{685}{8}e^4\right) e^3 \sin i \right.$$
$$\left. + \cos\omega \left(\left(\frac{742}{3} - 29e^2\right) e^5 \Big/ \sin i + \left(-\frac{1855}{2} + \frac{435}{4}e^2\right) e^5 \sin i \right) \right),$$
$$A_{j7} = \cos i\left(\cos 3\omega \left(\frac{17320}{21} - \frac{4885}{14}e^2\right) e^4 \sin i \right.$$
$$\left. + \cos\omega \left(\frac{4759}{42}e^6 \Big/ \sin i - \frac{23795}{56}e^6 \sin i \right) \right),$$
$$A_{j8} = \cos i\left(\cos 3\omega \left(\frac{5605}{8} - \frac{695}{24}e^2\right) e^5 \sin i \right.$$
$$\left. + \cos\omega \left(\frac{235}{12}e^7 \Big/ \sin i - \frac{1175}{16}e^7 \sin i \right) \right),$$
$$A_{j9} = \frac{20915}{72}e^6 \cos i \cos 3\omega \sin i,$$
$$A_{j10} = \frac{1099}{24}e^7 \cos i \cos 3\omega \sin i, \tag{6.35}$$

$$B_{j1} = \cos i\left(\sin 3\omega \left(-10 - 120e^2 + \frac{555}{8}e^4\right) e^2 \sin i \right.$$
$$+ \sin\omega \left(\left(-64 + 406e^2 + \frac{160}{3}e^4 - \frac{303}{2}e^6\right) \Big/ \sin i \right.$$
$$\left.\left. + \left(240 - 150e^2 - 200e^4 + \frac{4545}{8}e^6\right) \sin i \right) \right),$$
$$B_{j2} = \cos i\left(\sin 3\omega \left(40 - 210e^2 - \frac{985}{2}e^4 - \frac{335}{12}e^6\right) e \sin i \right.$$
$$+ \sin\omega \left(\left(-96 + \frac{704}{3}e^2 - 26e^4 - 40e^6\right) e \Big/ \sin i \right.$$
$$\left.\left. + \left(360 - 880e^2 + \frac{195}{2}e^4 + 150e^6\right) e \sin i \right) \right),$$

$$B_{j3} = \cos i \Bigg(\sin 3\omega \left(-\frac{80}{3} + \frac{800}{3}e^2 - 470e^4 - \frac{9325}{24}e^6 \right) \sin i$$

$$+ \sin\omega \Bigg(\left(-\frac{424}{3} + \frac{1256}{3}e^2 + \frac{161}{2}e^4 \right) e^2 \Big/ \sin i$$

$$+ \left(530 - 1570e^2 - \frac{2415}{8}e^4 \right) e^2 \sin i \Bigg) \Bigg),$$

$$B_{j4} = \cos i \Bigg(\sin 3\omega \left(-100 + 840e^2 - \frac{105}{2}e^4 - \frac{185}{3}e^6 \right) e \sin i$$

$$+ \sin\omega \Bigg(\left(-\frac{616}{3} + 386e^2 + \frac{110}{3}e^4 \right) e^3 \Big/ \sin i$$

$$+ \left(770 - \frac{2895}{2}e^2 - \frac{275}{2}e^4 \right) e^3 \sin i \Bigg) \Bigg),$$

$$B_{j5} = \cos i \Bigg(\sin 3\omega \left(-254 + \frac{4334}{3}e^2 + \frac{2029}{8}e^4 \right) e^2 \sin i$$

$$+ \sin\omega \Bigg(\left(-\frac{824}{3} + \frac{4219}{30}e^2 \right) e^4 \Big/ \sin i + \left(1030 - \frac{4219}{8}e^2 \right) e^4 \sin i \Bigg) \Bigg),$$

$$B_{j6} = \cos i \Bigg(\sin 3\omega \left(-\frac{1630}{3} + \frac{6745}{6}e^2 + \frac{685}{8}e^4 \right) e^3 \sin i$$

$$+ \sin\omega \Bigg(\left(-\frac{742}{3} + \frac{40}{3}e^2 \right) e^5 \Big/ \sin i + \left(\frac{1855}{2} - 50e^2 \right) e^5 \sin i \Bigg) \Bigg),$$

$$B_{j7} = \cos i \Bigg(\sin 3\omega \left(-\frac{17320}{21} + \frac{4885}{14}e^2 \right) e^6 \sin i$$

$$+ \sin\omega \left(-\frac{4759}{42}e^6 \Big/ \sin i + \frac{23795}{56}e^7 \sin i \right) \Bigg),$$

$$B_{j8} = \cos i \Bigg(\sin 3\omega \left(-\frac{5605}{8} + \frac{695}{24}e^2 \right) e^5 \sin i$$

$$+ \sin\omega \left(-\frac{235}{12}e^7 \Big/ \sin i + \frac{1175}{16}e^7 \sin i \right) \Bigg),$$

$$B_{j9} = -\frac{20915}{72}e^6 \cos i \sin 3\omega \sin i,$$

$$B_{j10} = -\frac{1099}{24}e^7 \cos i \sin 3\omega \sin i. \tag{6.36}$$

For $j = 6$, there are

$$
\begin{aligned}
c_j &= 0, \qquad (6.37)\\
d_j &= \sin 3\omega \left(460 - 320e^2 - \frac{555}{2}e^4\right) e^4 \sin^3 i \\
&\quad + \sin\omega \Bigg(\left(-768 + 4224e^2 - 4608e^4 + 2232e^6 - 450e^8\right) \sin i \\
&\quad + \left(960 - 5280e^2 + 5760e^4 - 2790e^6 + \frac{1125}{2}e^8\right) \sin^3 i \Bigg), \\
A_{j1} &= \cos 3\omega \left(80 - \frac{2090}{3}e^2 - \frac{29120}{3}e^4 - \frac{4285}{6}e^6\right) e \sin^3 i \\
&\quad \cos\omega \Bigg(\left(-5184 + 18000e^2 - 12464e^4 - 4605e^6\right) e \sin i \\
&\quad + \left(6480 - 22500e^2 + 15580e^4 + \frac{23025}{4}e^6\right) e \sin^3 i \Bigg), \\
A_{j2} &= \cos 3\omega \left(-160 + 2680e^2 - \frac{2900}{3}e^4 - \frac{40820}{3}e^6 - \frac{13175}{12}e^8\right) \sin^3 i \\
&\quad \cos\omega \Bigg(\left(1152 - 14016e^2 + 39792e^4 - 8700e^6 - 1919e^8\right) \sin i \\
&\quad + \left(-1440 + 17520e^2 - 49740e^4 + 10875e^6 + \frac{9595}{4}e^8\right) \sin^3 i \Bigg), \\
A_{j3} &= \cos 3\omega \left(-\frac{6400}{3} + \frac{53260}{3}e^2 + \frac{26180}{3}e^4 - \frac{81695}{12}e^6\right) e \sin^3 i \\
&\quad \cos\omega \Bigg(\left(3392 - 30776e^2 + 53232e^4 + 398e^6\right) e \sin i \\
&\quad + \left(-4240 + 38470e^2 - 66540e^4 - \frac{995}{2}e^6\right) e \sin^3 i \Bigg), \\
A_{j4} &= \cos 3\omega \left(400 - 10480e^2 + \frac{125500}{3}e^4 + \frac{54430}{3}e^6 - \frac{3895}{6}e^6\right) \sin^3 i \\
&\quad \cos\omega \Bigg(\left(7392 - 54552e^2 + 38172e^4 + 1375e^6\right) e^2 \sin i \\
&\quad + \left(-9240 + 68190e^2 - 47715e^4 - \frac{6875}{4}e^6\right) e^2 \sin^3 i \Bigg),
\end{aligned}
$$

$$
\begin{aligned}
A_{j5} &= \cos 3\omega \left(2032 - \frac{103904}{3}e^2 + \frac{123292}{3}e^4 + \frac{283019}{24}e^6\right) e \sin^3 i \\
&\quad \cos\omega \left(\left(14184 - \frac{316144}{5}e^2 + \frac{62838}{5}e^4\right) e^3 \sin i \right. \\
&\quad \left. + \left(-17730 + 79036e^2 - \frac{31419}{2}e^4\right) e^3 \sin^3 i \right), \\
A_{j6} &= \cos 3\omega \left(6520 - 67420e^2 + \frac{100625}{6}e^4 + \frac{57655}{24}e^6\right) e^2 \sin^3 i \\
&\quad \cos\omega \left(\left(22576 - 41732e^2 + \frac{2643}{2}e^4\right) e^4 \sin i \right. \\
&\quad \left. + \left(-28220 + 52165e^2 - \frac{13215}{8}e^4\right) e^4 \sin^3 i \right), \\
A_{j7} &= \cos 3\omega \left(\frac{50590}{3} - \frac{1473110}{21}e^2 + \frac{31925}{21}e^4\right) e^3 \sin^3 i \\
&\quad \cos\omega \left(\left(\frac{167056}{7} - \frac{198609}{14}e^2\right) e^5 \sin i \right. \\
&\quad \left. + \left(-\frac{208820}{7} + \frac{993045}{56}e^2\right) e^5 \sin^3 i \right), \\
A_{j8} &= \cos 3\omega \left(\frac{176765}{6} - \frac{116635}{3}e^2 - \frac{20485}{48}e^4\right) e^4 \sin^3 i \\
&\quad \cos\omega \left(\left(14439 - \frac{7735}{4}e^2\right) e^6 \sin i + \left(-\frac{72195}{4} + \frac{38675}{16}e^2\right) e^6 \sin^3 i \right), \\
A_{j9} &= \cos 3\omega \left(\frac{267790}{9} - \frac{198125}{18}e^2\right) e^5 \sin^3 i \\
&\quad \cos\omega \left(\frac{26843}{6}e^7 \sin i - \frac{134215}{24}e^7 \sin^3 i\right), \\
A_{j10} &= \cos 3\omega \left(\frac{98743}{6} - \frac{30569}{24}e^2\right) e^6 \sin^3 i \\
&\quad \cos\omega \left(\frac{1099}{2}e^8 \sin i - \frac{5495}{8}e^8 \sin^3 i\right), \\
A_{j11} &= \frac{1221305}{264}e^7 \cos 3\omega \sin^3 i, \\
A_{j12} &= \frac{2065}{4}e^8 \cos 3\omega \sin^3 i,
\end{aligned}
\tag{6.38}
$$

$$B_{j1} = \sin 3\omega \left(-80 + 690e^2 + 9700e^4 + \frac{4085}{6} e^6 \right) e \sin^3 i$$

$$\sin\omega \left(\left(960 + 1200e^2 - 12432e^4 + 9729e^6 \right) e \sin i \right.$$

$$\left. + \left(-1200 - 1500e^2 + 15540e^4 - \frac{48645}{4} e^6 \right) e \sin^3 i \right),$$

$$B_{j2} = \sin 3\omega \left(160 - 2680e^2 + 940e^4 + \frac{40675}{3} e^6 + \frac{10765}{12} e^8 \right) \sin^3 i$$

$$\sin\omega \left(\left(-1152 + 9600e^2 - 20080e^4 - 11592e^6 + 2185e^8 \right) \sin i \right.$$

$$\left. + \left(1440 - 12000e^2 + 25100e^4 + 14490e^6 - \frac{10925}{4} e^8 \right) \sin^3 i \right),$$

$$B_{j3} = \sin 3\omega \left(\frac{6400}{3} - \frac{53260}{3} e^2 - \frac{26180}{3} e^4 + \frac{84035}{12} e^6 \right) e \sin^3 i$$

$$\sin\omega \left(\left(-3392 + 25288e^2 - 30752e^4 - 10838e^6 \right) e \sin i \right.$$

$$\left. + \left(4240 - 31610e^2 + 38440e^4 + \frac{27095}{2} e^6 \right) e \sin^3 i \right),$$

$$B_{j4} = \sin 3\omega \left(-400 + 10480e^2 - \frac{125500}{3} e^4 - \frac{54470}{3} e^6 + \frac{9175}{12} e^8 \right) \sin^3 i$$

$$\sin\omega \left(\left(-7392 + 47360e^2 - 19296e^4 - 3050e^6 \right) e^2 \sin i \right.$$

$$\left. + \left(9240 - 59200e^2 + 24120e^4 + \frac{7625}{2} e^6 \right) e^2 \sin^3 i \right),$$

$$B_{j5} = \sin 3\omega \left(-2032 + \frac{103904}{3} e^2 - \frac{123292}{3} e^4 - \frac{284269}{24} e^6 \right) e \sin^3 i$$

$$\sin\omega \left(\left(-14184 + \frac{274816}{5} e^2 - \frac{19786}{5} e^4 \right) e^3 \sin i \right.$$

$$\left. + \left(17730 - 68704e^2 + \frac{9893}{2} e^4 \right) e^3 \sin^3 i \right),$$

$$B_{j6} = \sin 3\omega \left(-6520 + 67420e^2 - \frac{100625}{6}e^4 - \frac{19375}{8}e^6\right) e^2 \sin^3 i$$

$$\sin\omega \left(\left(-22576 + 35608e^2 + \frac{375}{2}e^4\right) e^4 \sin i \right.$$

$$\left. + \left(28220 - 44510e^2 - \frac{1875}{8}e^4\right) e^4 \sin^3 i \right),$$

$$B_{j7} = \sin 3\omega \left(-\frac{50590}{3} + \frac{1473110}{21}e^2 - \frac{31925}{21}e^4\right) e^3 \sin^3 i$$

$$\sin\omega \left(\left(-\frac{167056}{7} + \frac{165671}{14}e^2\right) e^5 \sin i \right.$$

$$\left. + \left(\frac{208820}{7} - \frac{828355}{56}e^2\right) e^5 \sin^3 i \right),$$

$$B_{j8} = \sin 3\omega \left(-\frac{176765}{6} + \frac{116635}{3}e^2 + \frac{20485}{48}e^4\right) e^4 \sin^3 i$$

$$\sin\omega \left(\left(-14439 + \frac{6325}{4}e^2\right) e^6 \sin i + \left(\frac{72195}{4} - \frac{31625}{16}e^2\right) e^6 \sin^3 i \right),$$

$$B_{j9} = \sin 3\omega \left(-\frac{267790}{9} + \frac{198125}{18}e^2\right) e^5 \sin^3 i$$

$$\sin\omega \left(-\frac{26843}{6}e^7 \sin i + \frac{134215}{24}e^8 \sin^3 i\right),$$

$$B_{j10} = \sin 3\omega \left(-\frac{98743}{6} + \frac{30569}{24}e^2\right) e^6 \sin^3 i$$

$$\sin\omega \left(-\frac{1099}{2}e^8 \sin i + \frac{5495}{8}e^8 \sin^3 i\right),$$

$$B_{j11} = -\frac{1221305}{264}e^7 \sin 3\omega \sin^3 i,$$

$$B_{j12} = -\frac{2065}{4}e^8 \sin 3\omega \sin^3 i. \tag{6.39}$$

These are the solutions of the $\overline{C}_{30}$ perturbations on the satellite orbits. Discussions and comments will be given in the summary section.

6.3 Solutions of D_{21} Perturbations

Solutions of the D_{21} perturbations are given below in the form of

$$(\Delta\sigma_j(M))_M = b_j \left(c_j M + d_j(\omega,\Omega) M + \sum_{k=1}^{10} A_{jk} \cos kM + \sum_{k=1}^{10} B_{jk} \sin kM\right), \tag{6.40}$$

where j is the index of Keplerian elements. There are

$$b_0 = \frac{\sqrt{15}\mu a_e^2 D_{21}}{64a^5 e\left(1-e^2\right)^{9/2} n}, \quad b_1 = 2aeb_0, \quad b_2 = \left(1-e^2\right) b_0,$$
$$b_3 = -b_2, \quad b_4 = 4eb_2, \quad b_5 = b_4, \quad b_6 = \left(1-e^2\right)^{1/2} b_2. \tag{6.41}$$

For $j = 1$, there are

$$\begin{aligned}
c_j &= 0, \qquad (6.42)\\
d_j &= \left(-8-84e^2-\frac{25}{2}e^4\right) e^4 \sin i \left(\cos 2\omega \cos\overline{\Omega} - \sin 2\omega \cos i \sin\overline{\Omega}\right),\\
A_{j1} &= \left(-16+\frac{224}{3}e^2+252e^4-\frac{193}{2}\right) e\sin i \left(\sin 2\omega \cos\overline{\Omega} + \cos 2\omega \cos i \sin\overline{\Omega}\right)\\
&\quad + \left(-96+324e^2-392e^4-\frac{243}{4}e^6\right) e\sin i \cos i \sin\overline{\Omega},\\
A_{j2} &= \left(32-224e^2+\frac{1592}{3}e^4+662e^6+\frac{343}{12}e^8\right) \sin i \left(\sin 2\omega \cos\overline{\Omega} + \cos 2\omega \cos i \sin\overline{\Omega}\right)\\
&\quad + \left(-114+536e^2-429e^4-\frac{87}{2}e^6\right) e^2 \sin i \cos i \sin\overline{\Omega},\\
A_{j3} &= \left(112-750e^2+720e^4+\frac{5023}{12}e^6\right) e\sin i \left(\sin 2\omega \cos\overline{\Omega} + \cos 2\omega \cos i \sin\overline{\Omega}\right)\\
&\quad + \left(-212+804e^2-\frac{453}{4}e^4\right) e^3 \sin i \cos i \sin\overline{\Omega},\\
A_{j4} &= \left(272-\frac{4994}{3}e^2+\frac{238}{3}e^4+\frac{725}{12}e^6\right) e^2 \sin i \left(\sin 2\omega \cos\overline{\Omega} + \cos 2\omega \cos i \sin\overline{\Omega}\right)\\
&\quad + \left(-308+762e^2+\frac{25}{2}e^4\right) e^4 \sin i \cos i \sin\overline{\Omega},\\
A_{j5} &= \left(\frac{1690}{3}-\frac{32552}{15}e^2-\frac{14851}{60}e^4\right) e^3 \sin i \left(\sin 2\omega \cos\overline{\Omega} + \cos 2\omega \cos i \sin\overline{\Omega}\right)\\
&\quad + \left(-412+\frac{6413}{20}e^2\right) e^5 \sin i \cos i \sin\overline{\Omega},\\
A_{j6} &= \left(\frac{2968}{3}-1414e^2-\frac{627}{8}e^4\right) e^4 \sin i \left(\sin 2\omega \cos\overline{\Omega} + \cos 2\omega \cos i \sin\overline{\Omega}\right)\\
&\quad + \left(-371+\frac{87}{2}e^2\right) e^6 \sin i \cos i \sin\overline{\Omega},\\
A_{j7} &= \left(\frac{25468}{21}-\frac{67211}{168}e^2\right) e^5 \sin i \left(\sin 2\omega \cos\overline{\Omega} + \cos 2\omega \cos i \sin\overline{\Omega}\right)\\
&\quad - \frac{4759}{28}e^7 \sin i \cos i \sin\overline{\Omega},\\
A_{j8} &= \left(\frac{5209}{6}-\frac{1625}{48}e^2\right) e^6 \sin i \left(\sin 2\omega \cos\overline{\Omega} + \cos 2\omega \cos i \sin\overline{\Omega}\right)\\
&\quad - \frac{235}{8}e^8 \sin i \cos i \sin\overline{\Omega},
\end{aligned}$$

$$A_{j9} = \frac{22891}{72} e^7 \sin i \left(\sin 2\omega \cos \overline{\Omega} + \cos i \cos 2\omega \sin \overline{\Omega}\right),$$
$$A_{j10} = \frac{1099}{24} e^8 \sin i \left(\sin 2\omega \cos \overline{\Omega} + \cos i \cos 2\omega \sin \overline{\Omega}\right), \tag{6.43}$$

$$B_{j1} = \left(-16 + \frac{220}{3} e^2 + \frac{796}{3} e^4 - \frac{405}{4} e^6\right) e \sin i \left(\cos 2\omega \cos \overline{\Omega} - \sin 2\omega \cos i \sin \overline{\Omega}\right),$$
$$B_{j2} = \left(32 - 224e^2 + \frac{1576}{3} e^4 + 658e^6 + \frac{215}{12} e^8\right) \sin i \left(\cos 2\omega \cos \overline{\Omega} - \sin 2\omega \cos i \sin \overline{\Omega}\right),$$
$$B_{j3} = \left(112 - 750e^2 + 720e^4 + \frac{1330}{3} e^6\right) e \sin i \left(\cos 2\omega \cos \overline{\Omega} - \sin 2\omega \cos i \sin \overline{\Omega}\right),$$
$$B_{j4} = \left(272 - \frac{4994}{3} e^2 + \frac{230}{3} e^4 + \frac{425}{6} e^6\right) e^2 \sin i \left(\cos 2\omega \cos \overline{\Omega} - \sin 2\omega \cos i \sin \overline{\Omega}\right),$$
$$B_{j5} = \left(\frac{1690}{3} - \frac{32552}{15} e^2 - \frac{3869}{15} e^4\right) e^3 \sin i \left(\cos 2\omega \cos \overline{\Omega} - \sin 2\omega \cos i \sin \overline{\Omega}\right),$$
$$B_{j6} = \left(\frac{2968}{3} - 1414e^2 - \frac{1975}{24} e^4\right) e^4 \sin i \left(\cos 2\omega \cos \overline{\Omega} - \sin 2\omega \cos i \sin \overline{\Omega}\right),$$
$$B_{j7} = \left(\frac{25468}{21} - \frac{67211}{168} e^2\right) e^5 \sin i \left(\cos 2\omega \cos \overline{\Omega} - \sin 2\omega \cos i \sin \overline{\Omega}\right),$$
$$B_{j8} = \left(\frac{5029}{6} - \frac{1625}{48} e^2\right) e^6 \sin i \left(\cos 2\omega \cos \overline{\Omega} - \sin 2\omega \cos i \sin \overline{\Omega}\right),$$
$$B_{j9} = \frac{22891}{72} e^7 \sin i \left(\cos 2\omega \cos \overline{\Omega} - \sin 2\omega \cos i \sin \overline{\Omega}\right),$$
$$B_{j10} = \frac{1099}{24} e^8 \sin i \left(\cos 2\omega \cos \overline{\Omega} - \sin 2\omega \cos i \sin \overline{\Omega}\right). \tag{6.44}$$

For $j = 2$, there are

$$c_j = 0, \tag{6.45}$$
$$d_j = \left(-80 - \frac{49}{2} e^2\right) e^6 \sin i \left(\cos 2\omega \cos \overline{\Omega} - \sin 2\omega \cos i \sin \overline{\Omega}\right),$$
$$A_{j1} = \left(16 - 56e^2 + \frac{833}{3} e^4 - \frac{47}{2} e^6\right) e \sin i \left(\sin 2\omega \cos \overline{\Omega} + \cos 2\omega \cos i \sin \overline{\Omega}\right)$$
$$+ \left(-96 + 324e^2 - 392e^4 - \frac{243}{4} e^6\right) e \sin i \cos i \sin \overline{\Omega},$$
$$A_{j2} = \left(-16 + \frac{308}{3} e^2 + 856e^4 + \frac{1039}{12} e^6\right) e \sin i \left(\sin 2\omega \cos \overline{\Omega} + \cos 2\omega \cos i \sin \overline{\Omega}\right)$$
$$+ \left(-114 + 536e^2 - 429e^4 - \frac{87}{2} e^6\right) e^2 \sin i \cos i \sin \overline{\Omega},$$
$$A_{j3} = \left(\frac{112}{3} - \frac{862}{3} e^2 + 229e^4 + \frac{6259}{12} e^6\right) e \sin i \left(\sin 2\omega \cos \overline{\Omega} + \cos 2\omega \cos i \sin \overline{\Omega}\right)$$
$$+ \left(-212 + 804e^2 - \frac{453}{4} e^4\right) e^3 \sin i \cos i \sin \overline{\Omega},$$

$$
\begin{aligned}
A_{j4} &= \left(136 - \frac{2960}{3}e^2 - 446e^4 + \frac{175}{4}e^6\right) e^2 \sin i \left(\sin 2\omega \cos\overline{\Omega} + \cos 2\omega \cos i \sin\overline{\Omega}\right) \\
&\quad + \left(-308 + 762e^2 + \frac{25}{2}e^4\right) e^4 \sin i \cos i \sin\overline{\Omega}, \\
A_{j5} &= \left(338 - \frac{8109}{5}e^2 - \frac{34231}{60}e^4\right) e^3 \sin i \left(\sin 2\omega \cos\overline{\Omega} + \cos 2\omega \cos i \sin\overline{\Omega}\right) \\
&\quad + \left(-412 + \frac{6413}{20}e^2\right) e^5 \sin i \cos i \sin\overline{\Omega}, \\
A_{j6} &= \left(\frac{2132}{3} - \frac{3232}{3}e^2 - \frac{1091}{8}e^4\right) e^4 \sin i \left(\sin 2\omega \cos\overline{\Omega} + \cos 2\omega \cos i \sin\overline{\Omega}\right) \\
&\quad + \left(-371 + \frac{87}{2}e^2\right) e^6 \sin i \cos i \sin\overline{\Omega}, \\
A_{j7} &= \left(\frac{3115}{3} - \frac{37907}{168}e^2\right) e^5 \sin i \left(\sin 2\omega \cos\overline{\Omega} + \cos 2\omega \cos i \sin\overline{\Omega}\right) \\
&\quad - \frac{4759}{28}e^7 \sin i \cos i \sin\overline{\Omega}, \\
A_{j8} &= \left(829 + \frac{85}{16}e^2\right) e^6 \sin i \left(\sin 2\omega \cos\overline{\Omega} + \cos 2\omega \cos i \sin\overline{\Omega}\right) \\
&\quad - \frac{235}{8}e^8 \sin i \cos i \sin\overline{\Omega}, \\
A_{j9} &= \frac{22891}{72}e^7 \sin i \left(\sin 2\omega \cos\overline{\Omega} + \cos i \cos 2\omega \sin\overline{\Omega}\right), \\
A_{j10} &= \frac{1099}{24}e^8 \sin i \left(\sin 2\omega \cos\overline{\Omega} + \cos i \cos 2\omega \sin\overline{\Omega}\right),
\end{aligned} \tag{6.46}
$$

$$
\begin{aligned}
B_{j1} &= \left(16 - 60e^2 + \frac{899}{3}e^4 - \frac{137}{4}e^6\right) e \sin i \left(\cos 2\omega \cos\overline{\Omega} - \sin 2\omega \cos i \sin\overline{\Omega}\right), \\
B_{j2} &= \left(-16 + \frac{292}{3}e^2 + 864e^4 + \frac{767}{12}e^6\right) e^2 \sin i \left(\cos 2\omega \cos\overline{\Omega} - \sin 2\omega \cos i \sin\overline{\Omega}\right), \\
B_{j3} &= \left(\frac{112}{3} - \frac{862}{3}e^2 + 223e^4 + \frac{1657}{3}e^6\right) e \sin i \left(\cos 2\omega \cos\overline{\Omega} - \sin 2\omega \cos i \sin\overline{\Omega}\right), \\
B_{j4} &= \left(136 - \frac{2960}{3}e^2 - 454e^4 + \frac{119}{2}e^6\right) e^2 \sin i \left(\cos 2\omega \cos\overline{\Omega} - \sin 2\omega \cos i \sin\overline{\Omega}\right), \\
B_{j5} &= \left(338 - \frac{8109}{5}e^2 - \frac{8714}{15}e^4\right) e^3 \sin i \left(\cos 2\omega \cos\overline{\Omega} - \sin 2\omega \cos i \sin\overline{\Omega}\right), \\
B_{j6} &= \left(\frac{2132}{3} - \frac{3232}{3}e^2 - \frac{3367}{24}e^4\right) e^4 \sin i \left(\cos 2\omega \cos\overline{\Omega} - \sin 2\omega \cos i \sin\overline{\Omega}\right), \\
B_{j7} &= \left(\frac{3115}{3} - \frac{37907}{168}e^2\right) e^5 \sin i \left(\cos 2\omega \cos\overline{\Omega} - \sin 2\omega \cos i \sin\overline{\Omega}\right), \\
B_{j8} &= \left(829 + \frac{85}{16}e^2\right) e^6 \sin i \left(\cos 2\omega \cos\overline{\Omega} - \sin 2\omega \cos i \sin\overline{\Omega}\right),
\end{aligned}
$$

$$B_{j9} = \frac{22891}{72} e^7 \sin i \left(\cos 2\omega \cos \overline{\Omega} - \sin 2\omega \cos i \sin \overline{\Omega}\right),$$

$$B_{j10} = \frac{1099}{24} e^8 \sin i \left(\cos 2\omega \cos \overline{\Omega} - \sin 2\omega \cos i \sin \overline{\Omega}\right). \tag{6.47}$$

For $j = 3$, there are

$$c_j = 0, \tag{6.48}$$

$$\begin{aligned}
d_j = {} & \sin 2\omega \cos \overline{\Omega} \left(\left(-4 - 6e^2\right) e^5 \cos^2 i \Big/ \sin i + \left(68 + \frac{35}{2} e^2\right) e^5 \sin i \right) \\
& + \sin \overline{\Omega} \Big(\left(-32 + 48e^2 - 12e^4 + 4e^6\right) e \cos^3 i \Big/ \sin i \\
& + \left(128 - 192e^2 + 24e^4 - 19e^6\right) e \sin i \cos i \Big) \\
& + \cos 2\omega \sin \overline{\Omega} \left(\left(-4 - 6e^2\right) e^5 \cos^3 i \Big/ \sin i + \left(72 + \frac{47}{2} e^2\right) e^5 \sin i \cos i \right),
\end{aligned}$$

$$\begin{aligned}
A_{j1} = {} & \cos 2\omega \cos \overline{\Omega} \Bigg(\left(16 - \frac{148}{3} e^2 - \frac{73}{2} e^4\right) e^2 \cos^2 i \Big/ \sin i \\
& + \left(-16 + 68e^2 + \frac{293}{3} e^4 - \frac{321}{4} e^6\right) \sin i \Bigg) \\
& + \sin 2\omega \sin \overline{\Omega} \Bigg(\left(-16 + \frac{148}{3} e^2 + \frac{73}{2} e^4\right) e^2 \cos^3 i \Big/ \sin i \\
& + \left(16 - 52e^2 - 147e^4 + \frac{175}{4} e^6\right) \sin i \cos i \Bigg),
\end{aligned}$$

$$\begin{aligned}
A_{j2} = {} & \cos 2\omega \cos \overline{\Omega} \Bigg(\left(-16 + 88e^2 - 126e^4 - 29e^6\right) e \cos^2 i \Big/ \sin i \\
& + \left(-80 + \frac{1108}{3} e^2 + 587e^4 + \frac{517}{12} e^6\right) e \sin i \Bigg) \\
& + \sin 2\omega \sin \overline{\Omega} \Bigg(\left(16 - 88e^2 + 126e^4 + 29e^6\right) e \cos^3 i \Big/ \sin i \\
& + \left(64 - \frac{844}{3} e^2 - 713e^4 - \frac{865}{2} e^6\right) e \sin i \cos i \Bigg),
\end{aligned}$$

$$A_{j3} = \cos 2\omega \cos\overline{\Omega}\left(\left(-\frac{112}{3} + 194e^2 - \frac{103}{2}e^4\right) e^2 \cos^2 i \Big/ \sin i \right.$$

$$\left. + \left(\frac{112}{3} - \frac{1186}{3}e^2 + 691e^4 + \frac{1297}{3}e^6\right) \sin i\right)$$

$$+ \sin 2\omega \sin\overline{\Omega}\left(\left(\frac{112}{3} - 194e^2 + \frac{103}{2}e^4\right) e^2 \cos^3 i \Big/ \sin i \right.$$

$$\left. + \left(-\frac{112}{3} + 358e^2 - 497e^4 - \frac{2903}{6}e^6\right) \sin i \cos i\right),$$

$$A_{j4} = \cos 2\omega \cos\overline{\Omega}\left(\left(-68 + 271e^2 + \frac{25}{3}e^4\right) e^3 \cos^2 i \Big/ \sin i \right.$$

$$\left. + \left(136 - \frac{3472}{3}e^2 + 79e^4 + \frac{225}{4}e^6\right) e \sin i\right)$$

$$+ \sin 2\omega \sin\overline{\Omega}\left(- \left(-68 + 271e^2 + \frac{25}{3}e^4\right) e^3 \cos^3 i \Big/ \sin i \right.$$

$$\left. + \left(-136 + \frac{3268}{3}e^2 + 192e^4 - \frac{575}{12}e^6\right) e \sin i \cos i\right),$$

$$A_{j5} = \cos 2\omega \cos\overline{\Omega}\left(\left(-\frac{338}{3} + \frac{323}{2}e^2\right) e^4 \cos^2 i \Big/ \sin i \right.$$

$$\left. + \left(338 - \frac{9359}{5}e^2 - \frac{4532}{15}e^4\right) e^2 \sin i\right)$$

$$+ \sin 2\omega \sin\overline{\Omega}\left(\left(\frac{338}{3} - \frac{323}{2}e^2\right) e^4 \cos^3 i \Big/ \sin i \right.$$

$$\left. + \left(-338 + \frac{26387}{15}e^2 + \frac{13909}{30}e^4\right) e^2 \sin i \cos i\right),$$

$$A_{j6} = \cos 2\omega \cos\overline{\Omega}\left(\left(-\frac{418}{3} + 29e^2\right) e^5 \cos^2 i \Big/ \sin i \right.$$

$$\left. + \left(\frac{2132}{3} - \frac{4021}{3}e^2 - \frac{743}{8}e^4\right) e^3 \sin i\right)$$

$$+ \sin 2\omega \sin\overline{\Omega}\left(\left(\frac{418}{3} - 29e^2\right) e^5 \cos^3 i \Big/ \sin i \right.$$

$$\left. + \left(-\frac{2132}{3} + 1201e^2 + \frac{975}{8}e^4\right) e^3 \sin i \cos i\right),$$

$$A_{j7} = \cos 2\omega \cos \overline{\Omega} \left(-\frac{1221}{14} e^6 \cos^2 i / \sin i + \left(\frac{3115}{3} - \frac{62077}{168} e^2 \right) e^4 \sin i \right)$$
$$+ \sin 2\omega \sin \overline{\Omega} \left(\frac{1221}{14} e^6 \cos^3 i / \sin i + \left(-\frac{3115}{3} + \frac{6775}{24} e^2 \right) e^4 \sin i \cos i \right),$$
$$A_{j8} = \cos 2\omega \cos \overline{\Omega} \left(-\frac{235}{12} e^7 \cos^2 i / \sin i + \left(829 - \frac{385}{16} e^2 \right) e^5 \sin i \right)$$
$$+ \sin 2\omega \sin \overline{\Omega} \left(\frac{235}{12} e^7 \cos^3 i / \sin i + \left(-829 + \frac{215}{48} e^2 \right) e^5 \sin i \cos i \right),$$
$$A_{j9} = \cos 2\omega \cos \overline{\Omega} \left(\frac{22891}{72} e^6 \sin i \right) - \sin 2\omega \sin \overline{\Omega} \left(\frac{22891}{72} e^6 \sin i \cos i \right),$$
$$A_{j10} = \cos 2\omega \cos \overline{\Omega} \left(\frac{1099}{24} e^7 \sin i \right) - \sin 2\omega \sin \overline{\Omega} \left(\frac{1099}{24} e^7 \sin i \cos i \right), \tag{6.49}$$

$$B_{j1} = \sin 2\omega \cos \overline{\Omega} \left(\left(-16 + \frac{152}{3} e^2 + \frac{67}{2} e^4 \right) e^2 \cos^2 i \Big/ \sin i \right.$$
$$\left. + \left(16 - 72e^2 - \frac{287}{3} e^4 + \frac{123}{2} e^6 \right) \sin i \right)$$
$$+ \cos 2\omega \sin \overline{\Omega} \left(\left(-16 + \frac{152}{3} e^2 + \frac{67}{2} e^4 \right) e^2 \cos^3 i \Big/ \sin i \right.$$
$$\left. + \left(16 - 56e^2 - \frac{439}{3} e^4 + 28e^6 \right) \sin i \cos i \right)$$
$$+ \sin \overline{\Omega} \left((-96 + 180e^2 - 82e^4) e^2 \cos^3 i \Big/ \sin i \right.$$
$$\left. + \left(96 + 36e^2 - 260e^4 + \frac{1237}{4} e^6 \right) \sin i \cos i \right),$$
$$B_{j2} = \sin 2\omega \cos \overline{\Omega} \left(\left(16 - 88e^2 + 126e^4 + 23e^6 \right) e \cos^2 i \Big/ \sin i \right.$$
$$\left. + \left(80 - \frac{1124}{3} e^2 - 565e^4 - \frac{455}{12} e^6 \right) e \sin i \right)$$
$$+ \cos 2\omega \sin \overline{\Omega} \left(\left(16 - 88e^2 + 126e^4 + 23e^6 \right) e \cos^3 i \Big/ \sin i \right.$$

$$+\left(64-\frac{860}{3}e^2-691e^4-\frac{731}{12}e^6\right)e\sin i\cos i\Bigg)$$

$$+\sin\overline{\Omega}\Bigg(\left(-72+160e^2-17e^4\right)e^3\cos^3 i\Big/\sin i$$

$$+\left(144-280e^2-121e^4+77e^6\right)e\sin i\cos i\Bigg),$$

$$B_{j3}=\sin 2\omega\cos\overline{\Omega}\Bigg(\left(\frac{112}{3}-194e^2+\frac{109}{2}6e^4\right)e^2\cos^2 i\Big/\sin i$$

$$+\left(-\frac{112}{3}+\frac{1186}{3}e^2-697e^4-\frac{4981}{12}e^6\right)\sin i\Bigg)$$

$$+\cos 2\omega\sin\overline{\Omega}\Bigg(\left(\frac{112}{3}-194e^2+\frac{109}{2}6e^4\right)e^2\cos^3 i\Big/\sin i$$

$$+\left(-\frac{112}{3}+358e^2-503e^4-\frac{5635}{12}e^6\right)\sin i\cos i\Bigg)$$

$$+\sin\overline{\Omega}\Bigg(\left(-\frac{212}{3}+123e^2\right)e^4\cos^3 i\Big/\sin i$$

$$+\left(212-\frac{1672}{3}e^2-\frac{975}{4}e^4\right)e^2\sin i\cos i\Bigg),$$

$$B_{j4}=\sin 2\omega\cos\overline{\Omega}\Bigg(\left(68-271e^2-\frac{17}{3}6e^4\right)e^3\cos^2 i\Big/\sin i$$

$$+\left(-136+\frac{3472}{3}e^2-87e^4-\frac{105}{2}e^6\right)e\sin i\Bigg)$$

$$+\cos 2\omega\sin\overline{\Omega}\Bigg(\left(68-271e^2-\frac{17}{3}6e^4\right)e^3\cos^3 i\Big/\sin i$$

$$+\left(-136+\frac{3268}{3}e^2+184e^4-\frac{281}{6}e^6\right)e\sin i\cos i\Bigg)$$

$$+\sin\overline{\Omega}\left(\left(-77+31e^2\right)e^5\cos^3 i/\sin i+\left(308-502e^2-86e^4\right)e^3\sin i\cos i\right),$$

$$
\begin{aligned}
B_{j5} =& \sin 2\omega \cos\overline{\Omega}\left(\left(\frac{338}{3}-\frac{323}{2}e^2\right)e^4\cos^2 i\Big/\sin i\right.\\
&\left.+\left(-338+\frac{9359}{5}e^2+\frac{17503}{60}e^4\right)e^2\sin i\right)\\
&+\cos 2\omega \sin\overline{\Omega}\left(\left(\frac{338}{3}-\frac{323}{2}e^2\right)e^4\cos^3 i\Big/\sin i\right.\\
&\left.+\left(-338+\frac{26387}{15}e^2+\frac{27193}{60}e^4\right)e^2\sin i\cos i\right)\\
&+\sin\overline{\Omega}\left(-\frac{287}{5}e^6\cos^3 i\Big/\sin i+\left(412-\frac{3071}{20}e^2\right)e^4\sin i\cos i\right),\\
B_{j6} =& \sin 2\omega \cos\overline{\Omega}\left(\left(\frac{418}{3}-29e^2\right)e^5\cos^2 i\Big/\sin i\right.\\
&\left.+\left(-\frac{2132}{3}+\frac{4021}{3}e^2+\frac{2135}{24}e^4\right)e^3\sin i\right)\\
&+\cos 2\omega \sin\overline{\Omega}\left(\left(\frac{418}{3}-29e^2\right)e^5\cos^3 i\Big/\sin i\right.\\
&\left.+\left(-\frac{2132}{3}+1201e^2+\frac{2831}{24}e^4\right)e^3\sin i\cos i\right)\\
&+\sin\overline{\Omega}\left(-\frac{47}{3}e^7\cos^3 i\Big/\sin i+\left(371-\frac{13}{3}e^2\right)e^5\sin i\cos i\right),\\
B_{j7} =& \sin 2\omega \cos\overline{\Omega}\left(\frac{1221}{14}e^6\cos^2 i\Big/\sin i+\left(-\frac{3115}{3}+\frac{62077}{168}e^2\right)e^4\sin i\right)\\
&+\cos 2\omega \sin\overline{\Omega}\left(\frac{1221}{14}e^6\cos^3 i\Big/\sin i+\left(-\frac{3115}{3}+\frac{6775}{24}e^2\right)e^4\sin i\cos i\right)\\
&+\sin\overline{\Omega}\left(\frac{4759}{28}e^6\sin i\cos i\right),\\
B_{j8} =& \sin 2\omega \cos\overline{\Omega}\left(\frac{235}{12}e^7\cos^2 i\Big/\sin i+\left(-829+\frac{385}{16}e^2\right)e^5\sin i\right)\\
&+\cos 2\omega \sin\overline{\Omega}\left(\frac{235}{12}e^7\cos^3 i\Big/\sin i+\left(-829+\frac{215}{48}e^2\right)e^5\sin i\cos i\right)\\
&+\sin\overline{\Omega}\left(\frac{235}{8}e^7\sin i\cos i\right),\\
B_{j9} =& \sin 2\omega \cos\overline{\Omega}\left(-\frac{22891}{72}e^6\sin i\right)-\cos 2\omega \sin\overline{\Omega}\left(\frac{22891}{72}e^6\sin i\cos i\right),\\
B_{j10} =& \sin 2\omega \cos\overline{\Omega}\left(-\frac{1099}{24}e^7\sin i\right)-\cos 2\omega \sin\overline{\Omega}\left(\frac{1099}{24}e^7\sin i\cos i\right).
\end{aligned} \tag{6.50}
$$

For $j = 4$, there are

$$
\begin{aligned}
c_j &= 0, \qquad (6.51)\\
d_j &= \cos 2\omega \cos\overline{\Omega}\left(-1-\frac{3}{2}e^2\right)e^4\cos i + \cos\overline{\Omega}\cos i\left(8-12e^2+3e^4-e^6\right)\\
&\quad + \sin 2\omega \sin\overline{\Omega}\left(\left(-1-\frac{3}{2}e^2\right)e^4 + \left(2+3e^2\right)e^4\cos^2 i\right),\\
A_{j1} &= \sin 2\omega \cos\overline{\Omega}\left(-4+\frac{37}{3}e^2+\frac{73}{8}e^4\right)e\cos i\\
&\quad + \cos 2\omega \sin\overline{\Omega}\left(\left(4-\frac{37}{3}e^2-\frac{73}{8}e^4\right)e + \left(-8+\frac{74}{3}e^2+\frac{73}{4}e^4\right)\cos^2 i\right),\\
A_{j2} &= \sin 2\omega \cos\overline{\Omega}\left(4-22e^2+\frac{63}{2}e^4+\frac{29}{4}e^6\right)\cos i\\
&\quad + \cos 2\omega \sin\overline{\Omega}\left(\left(-4+22e^2-\frac{63}{2}e^4-\frac{29}{4}e^6\right)\right.\\
&\quad \left. + \left(8-44e^2+63e^4+\frac{29}{2}e^6\right)\cos^2 i\right),\\
A_{j3} &= \sin 2\omega \cos\overline{\Omega}\left(\frac{28}{3}-\frac{97}{2}e^2+\frac{103}{8}e^4\right)e\cos i\\
&\quad + \cos 2\omega \sin\overline{\Omega}\left(\left(-\frac{28}{3}+\frac{97}{2}e^2-\frac{103}{8}e^4\right)e\right.\\
&\quad \left. + \left(\frac{56}{3}-97e^2+\frac{103}{4}e^4\right)e\cos^2 i\right),\\
A_{j4} &= \sin 2\omega \cos\overline{\Omega}\left(17-\frac{271}{4}e^2-\frac{25}{12}e^4\right)e^2\cos i\\
&\quad + \cos 2\omega \sin\overline{\Omega}\left(\left(-17+\frac{271}{4}e^2+\frac{25}{12}e^4\right)e^2\right.\\
&\quad \left. + \left(34-\frac{271}{2}e^2-\frac{25}{6}e^4\right)e^2\cos^2 i\right),\\
A_{j5} &= \sin 2\omega \cos\overline{\Omega}\left(\frac{169}{6}-\frac{323}{8}e^2\right)e^3\cos i\\
&\quad + \cos 2\omega \sin\overline{\Omega}\left(\left(-\frac{169}{6}+\frac{323}{8}e^2\right)e^3 + \left(\frac{169}{3}-\frac{323}{4}e^2\right)e^3\cos^2 i\right),\\
A_{j6} &= \sin 2\omega \cos\overline{\Omega}\left(\frac{209}{6}-\frac{29}{4}e^2\right)e^4\cos i\\
&\quad + \cos 2\omega \sin\overline{\Omega}\left(\left(-\frac{209}{6}+\frac{29}{4}e^2\right)e^4 + \left(\frac{209}{3}-\frac{29}{2}e^2\right)e^4\cos^2 i\right),
\end{aligned}
$$

$$A_{j7} = \sin 2\omega \cos\overline{\Omega}\left(\frac{1221}{56}e^5\cos i\right) + \cos 2\omega \sin\overline{\Omega}\left(-\frac{1221}{56}e^5 + \frac{1221}{28}e^5\cos^2 i\right),$$

$$A_{j8} = \sin 2\omega \cos\overline{\Omega}\left(\frac{235}{48}e^6\cos i\right) + \cos 2\omega \sin\overline{\Omega}\left(-\frac{235}{48}e^6 + \frac{235}{24}e^6\cos^2 i\right), \tag{6.52}$$

$$\begin{aligned}
B_{j1} = {} & \cos 2\omega \cos\overline{\Omega}\left(-4 + \frac{38}{3}e^2 + \frac{67}{8}e^4\right) e\cos i \\
& + \cos\overline{\Omega}\left(24 - 45e^2 + \frac{41}{2}e^4\right) e\cos i \\
& + \sin 2\omega \sin\overline{\Omega}\left(\left(-4 + \frac{38}{3}e^2 + \frac{67}{8}e^4\right) e + \left(8 - \frac{76}{3}e^2 - \frac{67}{4}e^4\right) e\cos^2 i\right),
\end{aligned}$$

$$\begin{aligned}
B_{j2} = {} & \cos 2\omega \cos\overline{\Omega}\left(4 - 22e^2 + \frac{63}{2}e^4 + \frac{23}{4}e^6\right)\cos i \\
& + \cos\overline{\Omega}\left(18 - 40e^2 + \frac{17}{4}e^4\right) e^2\cos i \\
& + \sin 2\omega \sin\overline{\Omega}\left(\left(4 - 22e^2 + \frac{63}{2}e^4 + \frac{23}{4}e^6\right)\right. \\
& \left. + \left(-8 + 44e^2 - 63e^4 - \frac{23}{2}e^6\right)\cos^2 i\right),
\end{aligned}$$

$$\begin{aligned}
B_{j3} = {} & \cos 2\omega \cos\overline{\Omega}\left(\frac{28}{3} - \frac{97}{2}e^2 + \frac{109}{8}e^4\right) e\cos i \\
& + \cos\overline{\Omega}\left(\frac{53}{3} - \frac{123}{4}e^2\right) e^3\cos i \\
& + \sin 2\omega \sin\overline{\Omega}\left(\left(\frac{28}{3} - \frac{97}{2}e^2 + \frac{109}{8}e^4\right) e + \left(-\frac{56}{3} + 97e^2 - \frac{109}{4}e^4\right) e\cos^2 i\right),
\end{aligned}$$

$$\begin{aligned}
B_{j4} = {} & \cos 2\omega \cos\overline{\Omega}\left(17 - \frac{271}{4}e^2 - \frac{17}{12}e^4\right) e^2\cos i \\
& + \cos\overline{\Omega}\left(\frac{77}{4} - \frac{31}{4}e^2\right) e^4\cos i \\
& + \sin 2\omega \sin\overline{\Omega}\left(\left(17 - \frac{271}{4}e^2 - \frac{17}{12}e^4\right) e^2\right. \\
& \left. + \left(-34 + \frac{271}{2}e^2 + \frac{17}{6}e^4\right) e^2\cos^2 i\right),
\end{aligned}$$

$$\begin{aligned}
B_{j5} = {} & \cos 2\omega \cos\overline{\Omega}\left(\frac{169}{6} - \frac{323}{8}e^2\right) e^3\cos i + \cos\overline{\Omega}\left(\frac{287}{20}e^5\cos i\right) \\
& + \sin 2\omega \sin\overline{\Omega}\left(\left(\frac{169}{6} - \frac{323}{8}e^2\right) e^3 + \left(-\frac{169}{3} + \frac{323}{4}e^2\right) e^3\cos^2 i\right),
\end{aligned}$$

$$B_{j6}=\cos 2\omega\cos\overline{\Omega}\left(\frac{209}{6}-\frac{29}{4}e^2\right)e^4\cos i+\cos\overline{\Omega}\left(\frac{47}{12}e^6\cos i\right)$$
$$+\sin 2\omega\sin\overline{\Omega}\left(\left(\frac{209}{6}-\frac{29}{4}e^2\right)e^4+\left(-\frac{209}{3}+\frac{29}{2}e^2\right)e^4\cos^2 i\right),$$
$$B_{j7}=\cos 2\omega\cos\overline{\Omega}\left(\frac{1221}{56}e^5\cos i\right)+\sin 2\omega\sin\overline{\Omega}\left(\frac{1221}{56}e^5-\frac{1221}{28}e^5\cos^2 i\right),$$
$$B_{j8}=\cos 2\omega\cos\overline{\Omega}\left(\frac{235}{48}e^6\cos i\right)+\sin 2\omega\sin\overline{\Omega}\left(\frac{235}{48}e^6-\frac{235}{24}e^6\cos^2 i\right). \tag{6.53}$$

For $j=5$, there are

$$c_j=0, \tag{6.54}$$
$$d_j=\sin 2\omega\cos\overline{\Omega}\left(\left(-1-\frac{3}{2}e^2\right)e^4\cos i/\sin i\right.$$
$$+\sin\overline{\Omega}\left(-8+12e^2-3e^4+e^6\right)\left(\cos^2 i/\sin i-\sin i\right)$$
$$\left.+\cos 2\omega\sin\overline{\Omega}\left(\left(-1-\frac{3}{2}e^2\right)e^4\cos^2 i/\sin i+\left(1+\frac{3}{2}e^2\right)e^4\sin i\right)\right),$$
$$A_{j1}=\cos 2\omega\cos\overline{\Omega}\left(4-\frac{37}{3}e^2-\frac{73}{8}e^4\right)e\cos i/\sin i$$
$$+\sin 2\omega\sin\overline{\Omega}\left(-4+\frac{37}{3}e^2+\frac{73}{8}e^4\right)e\left(\cos^2 i/\sin i-\sin i\right),$$
$$A_{j2}=\cos 2\omega\cos\overline{\Omega}\left(-4+22e^2-\frac{63}{2}e^4-\frac{29}{4}e^6\right)\cos i/\sin i$$
$$+\sin 2\omega\sin\overline{\Omega}\left(4-22e^2+\frac{63}{2}e^4+\frac{29}{4}e^6\right)\left(\cos^2 i/\sin i-\sin i\right),$$
$$A_{j3}=\cos 2\omega\cos\overline{\Omega}\left(-\frac{28}{3}+\frac{97}{2}e^2-\frac{103}{8}e^4\right)e\cos i/\sin i$$
$$+\sin 2\omega\sin\overline{\Omega}\left(\frac{28}{3}-\frac{97}{2}e^2+\frac{103}{8}e^6\right)\left(\cos^2 i/\sin i-\sin i\right),$$
$$A_{j4}=\cos 2\omega\cos\overline{\Omega}\left(-17+\frac{271}{4}e^2+\frac{25}{12}e^4\right)e^2\cos i/\sin i$$
$$+\sin 2\omega\sin\overline{\Omega}\left(17-\frac{271}{4}e^2-\frac{25}{12}e^4\right)e^2\left(\cos^2 i/\sin i-\sin i\right),$$
$$A_{j5}=\cos 2\omega\cos\overline{\Omega}\left(-\frac{169}{6}+\frac{323}{8}e^2\right)e^3\cos i/\sin i$$
$$+\sin 2\omega\sin\overline{\Omega}\left(\frac{169}{6}e^2-\frac{323}{8}e^4\right)e^3\left(\cos^2 i/\sin i-\sin i\right),$$

$$
\begin{aligned}
A_{j6} &= \cos 2\omega \cos \overline{\Omega} \left(-\frac{209}{6} + \frac{29}{4} e^2 \right) e^3 \cos i / \sin i \\
&\quad + \sin 2\omega \sin \overline{\Omega} \left(\frac{209}{6} - \frac{29}{4} e^2 \right) e^4 \left(\cos^2 i / \sin i - \sin i \right), \\
A_{j7} &= \cos 2\omega \cos \overline{\Omega} \left(-\frac{1221}{56} e^5 \cos i / \sin i \right) \\
&\quad + \sin 2\omega \sin \overline{\Omega} \left(\frac{1221}{56} \right) \left(\cos^2 i / \sin i - \sin i \right) e^5, \\
A_{j8} &= \cos 2\omega \cos \overline{\Omega} \left(-\frac{235}{48} e^6 \cos i / \sin i \right) \\
&\quad + \sin 2\omega \sin \overline{\Omega} \left(\frac{235}{48} \right) \left(\cos^2 i / \sin i - \sin i \right) e^5, \qquad (6.55) \\
B_{j1} &= \sin 2\omega \cos \overline{\Omega} \left(-4 + \frac{38}{3} e^2 + \frac{67}{8} e^4 \right) e \cos i / \sin i \\
&\quad + \cos 2\omega \sin \overline{\Omega} \left(-4 + \frac{38}{3} e^2 + \frac{67}{8} e^4 \right) e \left(\cos^2 i / \sin i - \sin i \right) \\
&\quad + \sin \overline{\Omega} \left(-24 + 45 e^2 - \frac{41}{2} e^4 \right) e \left(\cos^2 i / \sin i - \sin i \right), \\
B_{j2} &= \sin 2\omega \cos \overline{\Omega} \left(4 - 22 e^2 + \frac{68}{2} e^4 + \frac{23}{4} e^6 \right) \cos i / \sin i \\
&\quad + \cos 2\omega \sin \overline{\Omega} \left(4 - 22 e^2 + \frac{68}{2} e^4 + \frac{23}{4} e^6 \right) e \left(\cos^2 i / \sin i - \sin i \right), \\
&\quad + \sin \overline{\Omega} \left(-18 + 40 e^2 - \frac{17}{4} e^4 \right) e^2 \left(\cos^2 i / \sin i - \sin i \right), \\
B_{j3} &= \sin 2\omega \cos \overline{\Omega} \left(\frac{28}{3} - \frac{97}{2} e^2 + \frac{109}{8} e^4 \right) e \cos i / \sin i \\
&\quad + \sin i \sin \overline{\Omega} \left(-\frac{53}{3} + \frac{123}{4} e^2 \right) e^3 \left(\cos^2 i / \sin i - \sin i \right) \\
&\quad + \cos 2\omega \sin \overline{\Omega} \left(\frac{28}{3} - \frac{97}{2} e^2 + \frac{109}{8} e^4 \right) e \left(\cos^2 i / \sin i - \sin i \right), \\
B_{j4} &= \sin 2\omega \cos \overline{\Omega} \left(17 - \frac{271}{4} e^2 - \frac{17}{12} e^4 \right) e^2 \cos i / \sin i \\
&\quad + \cos 2\omega \sin \overline{\Omega} \left(17 - \frac{271}{4} e^2 - \frac{17}{12} e^4 \right) e^2 \left(\cos^2 i / \sin i - \sin i \right) \\
&\quad + \sin \overline{\Omega} \left(-\frac{77}{4} + \frac{31}{4} e^2 \right) e^4 \left(\cos^2 i / \sin i - \sin i \right),
\end{aligned}
$$

$$
\begin{aligned}
B_{j5} &= \sin 2\omega \cos\overline{\Omega}\left(\frac{169}{6}-\frac{323}{8}e^2\right)e^3\cos i/\sin i \\
&\quad +\cos 2\omega \sin\overline{\Omega}\left(\frac{169}{6}-\frac{323}{8}e^2\right)e^3\left(\cos^2 i/\sin i-\sin i\right) \\
&\quad +\sin\overline{\Omega}\left(-\frac{287}{20}e^5\right)\left(\cos^2 i/\sin i-\sin i\right), \\
B_{j6} &= \sin 2\omega \cos\overline{\Omega}\left(\frac{209}{6}-\frac{29}{4}e^2\right)e^4\cos i/\sin i \\
&\quad +\cos 2\omega \sin\overline{\Omega}\left(\frac{209}{6}-\frac{29}{4}e^2\right)e^4\left(\cos^2 i/\sin i-\sin i\right) \\
&\quad +\sin\overline{\Omega}\left(-\frac{47}{12}e^6\right)\left(\cos^2 i/\sin i-\sin i\right), \\
B_{j7} &= \sin 2\omega \cos\overline{\Omega}\left(\frac{1221}{56}e^5\right)\cos i/\sin i \\
&\quad +\cos 2\omega \sin\overline{\Omega}\left(\frac{1221}{56}e^5\right)\left(\cos^2 i/\sin i-\sin i\right), \\
B_{j8} &= \sin 2\omega \cos\overline{\Omega}\left(\frac{235}{48}e^6\right)\cos i/\sin i \\
&\quad +\cos 2\omega \sin\overline{\Omega}\left(\frac{235}{48}e^6\right)\left(\cos^2 i/\sin i-\sin i\right).
\end{aligned} \tag{6.56}
$$

For $j = 6$, there are

$$
\begin{aligned}
c_j &= 0, \\
d_j &= \sin 2\omega \cos\overline{\Omega}\left(44-\frac{37}{2}e^2\right)e^5\sin i \\
&\quad +\sin\overline{\Omega}\left(-96+144e^2-60e^4+9e^6\right)e\cos i\sin i \\
&\quad +\cos 2\omega \sin\overline{\Omega}\left(44-\frac{37}{2}e^2\right)e^5\cos i\sin i, \\
A_{j1} &= \left(-16+164e^2-\frac{595}{3}e^4-\frac{1197}{4}e^6\right)\sin i\left(\cos 2\omega\cos\overline{\Omega}-\sin 2\omega\sin\overline{\Omega}\cos i\right), \\
A_{j2} &= \left(-176+\frac{2696}{3}e^2-169e^4-\frac{1571}{12}e^6\right)\sin i\left(\cos 2\omega\cos\overline{\Omega}-\sin 2\omega\sin\overline{\Omega}\cos i\right), \\
A_{j3} &= \left(\frac{112}{3}-\frac{1858}{3}e^2+1855e^4+\frac{370}{3}e^6\right)\sin i\left(\cos 2\omega\cos\overline{\Omega}-\sin 2\omega\sin\overline{\Omega}\cos i\right), \\
A_{j4} &= \left(136-\frac{4696}{3}e^2+1705e^4+\frac{425}{4}e^6\right)e\sin i\left(\cos 2\omega\cos\overline{\Omega}-\sin 2\omega\sin\overline{\Omega}\cos i\right), \\
A_{j5} &= \left(338-\frac{12739}{5}e^2+\frac{10003}{15}e^4\right)e^2\sin i\left(\cos 2\omega\cos\overline{\Omega}-\sin 2\omega\sin\overline{\Omega}\cos i\right),
\end{aligned} \tag{6.57}
$$

$$A_{j6} = \left(\frac{2132}{3} - \frac{6529}{3}e^2 + \frac{649}{8}e^4\right) e^3 \sin i \left(\cos 2\omega \cos \overline{\Omega} - \sin 2\omega \sin \overline{\Omega} \cos i\right),$$

$$A_{j7} = \left(\frac{3115}{3} - \frac{21427}{24}e^2\right) e^4 \sin i \left(\cos 2\omega \cos \overline{\Omega} - \sin 2\omega \sin \overline{\Omega} \cos i\right),$$

$$A_{j8} = \left(829 - \frac{2265}{16}e^2\right) e^5 \sin i \left(\cos 2\omega \cos \overline{\Omega} - \sin 2\omega \sin \overline{\Omega} \cos i\right),$$

$$A_{j9} = \left(\frac{22891}{72}e^6\right) \sin i \left(\cos 2\omega \cos \overline{\Omega} - \sin 2\omega \sin \overline{\Omega} \cos i\right),$$

$$A_{j10} = \left(\frac{1099}{24}e^7\right) \sin i \left(\cos 2\omega \cos \overline{\Omega} - \sin 2\omega \sin \overline{\Omega} \cos i\right), \tag{6.58}$$

$$B_{j1} = \left(16 - 168e^2 + \frac{625}{3}e^4 + \frac{525}{2}e^6\right) \left(\sin 2\omega \cos \overline{\Omega} + \cos 2\omega \sin \overline{\Omega} \cos i\right) \sin i$$
$$+ \sin \overline{\Omega} \left(96 - 636e^2 + 1000e^4 - \frac{1059}{4}e^6\right) \cos i \sin i,$$

$$B_{j2} = \left(176 - \frac{2078}{3}e^2 + 191e^4 + \frac{1201}{12}e^6\right) e \left(\sin 2\omega \cos \overline{\Omega} + \cos 2\omega \sin \overline{\Omega} \cos i\right) \sin i$$
$$+ \sin \overline{\Omega} \left(144 - 784e^2 + 999e^4 - 42e^6\right) e \cos i \sin i,$$

$$B_{j3} = \left(-\frac{112}{3} + \frac{1858}{3}e^2 - 1861e^4 - \frac{1057}{12}e^6\right) \left(\sin 2\omega \cos \overline{\Omega} + \cos 2\omega \sin \overline{\Omega} \cos i\right) \sin i$$
$$+ \sin \overline{\Omega} \left(212 - 1052e^2 + \frac{2469}{4}e^4\right) \cos i \sin i,$$

$$B_{j4} = \left(-136 + \frac{4696}{3}e^2 - 1713e^4 - \frac{173}{2}e^6\right) e \left(\sin 2\omega \cos \overline{\Omega} + \cos 2\omega \sin \overline{\Omega} \cos i\right) \sin i$$
$$+ \sin \overline{\Omega} \left(308 - 1041e^2 + 131e^4\right) e^3 \cos i \sin i,$$

$$B_{j5} = \left(-338 + \frac{12739}{5}e^2 - \frac{40637}{60}e^4\right) e^2 \left(\sin 2\omega \cos \overline{\Omega} + \cos 2\omega \sin \overline{\Omega} \cos i\right) \sin i$$
$$+ \sin \overline{\Omega} \left(412 - \frac{11107}{20}e^2\right) e^4 \cos i \sin i,$$

$$B_{j6} = \left(-\frac{2132}{3} + \frac{6529}{3}e^2 - \frac{2041}{24}e^4\right) e^3 \left(\sin 2\omega \cos \overline{\Omega} + \cos 2\omega \sin \overline{\Omega} \cos i\right) \sin i$$
$$+ \sin \overline{\Omega} \left(371 - 114e^2\right) e^5 \cos i \sin i,$$

$$B_{j7} = \left(-\frac{3115}{3} + \frac{21427}{24}e^2\right) e^4 \left(\sin 2\omega \cos \overline{\Omega} + \cos 2\omega \sin \overline{\Omega} \cos i\right) \sin i$$
$$+ \sin \overline{\Omega} \left(\frac{4759}{8}e^6\right) \cos i \sin i,$$

$$B_{j8} = \left(-829 + \frac{2265}{16}e^2\right) e^5 \left(\sin 2\omega \cos \overline{\Omega} + \cos 2\omega \sin \overline{\Omega} \cos i\right) \sin i$$
$$+ \sin \overline{\Omega} \left(\frac{235}{8}e^7\right) \cos i \sin i,$$

$$B_{j9} = \left(-\frac{22891}{72}e^6\right)\left(\sin 2\omega \cos\overline{\Omega} + \cos 2\omega \sin\overline{\Omega} \cos i\right)\sin i,$$

$$B_{j10} = \left(-\frac{1099}{24}e^7\right)\left(\sin 2\omega \cos\overline{\Omega} + \cos 2\omega \sin\overline{\Omega} \cos i\right)\sin i. \tag{6.59}$$

These are the solutions of the D_{21} perturbations on the satellite orbits. Discussions and comments will be given in the summary section.

6.4 Solutions of D_{22} Perturbations

Solutions of the D_{22} perturbations are given below in the form of

$$(\Delta\sigma_j(M))_M = b_j\left(c_j M + d_j(\omega,\Omega)M + \sum_{k=1}^{10} A_{jk}\cos kM + \sum_{k=1}^{10} B_{jk}\sin kM\right), \tag{6.60}$$

where j is the index of Keplerian elements. Then there are

$$b_0 = \frac{\sqrt{15}\mu a_e^2 D_{22}}{128a^5 e\left(1-e^2\right)^{9/2} n}, \quad b_1 = -2aeb_0, \quad b_2 = -\left(1-e^2\right)b_0,$$

$$b_3 = b_2, \quad b_4 = 8eb_2, \quad b_5 = -b_4, \quad b_6 = -\left(1-e^2\right)^{1/2} b_2. \tag{6.61}$$

For $j = 1$, there are

$$d_j = \cos 2\omega \sin 2\overline{\Omega}\left(-16 - 168e^2 - 25e^4\right)e^4\cos i$$
$$+ \sin 2\omega \cos 2\overline{\Omega}\left(-8 - 84e^2 - \frac{25}{2}e^4\right)e^4\left(1+\cos^2 i\right), \tag{6.62}$$

$$A_{j1} = \cos 2\omega \cos 2\overline{\Omega}\left(16 - \frac{224}{3}e^2 - 252e^4 + \frac{193}{2}e^6\right)e\left(1+\cos^2 i\right)$$
$$+ \sin 2\omega \sin 2\overline{\Omega}\left(-32 + \frac{448}{3}e^2 + 504e^4 - 193e^6\right)e\cos i$$
$$+ \cos 2\overline{\Omega}\left(-96 + 324e^2 - 392e^4 - \frac{243}{4}e^6\right)e\sin^2 i,$$

$$A_{j2} = \cos 2\omega \cos 2\overline{\Omega}\left(-32 + 224e^2 - \frac{1592}{3}e^4 - 662e^6 - \frac{343}{12}e^8\right)\left(1+\cos^2 i\right)$$
$$+ \sin 2\omega \sin 2\overline{\Omega}\left(64 - 448e^2 + \frac{3184}{3}e^4 + 1324e^6 + \frac{343}{6}e^8\right)\cos i$$
$$+ \cos 2\overline{\Omega}\left(-144 + 536e^2 - 429e^4 - \frac{87}{2}e^6\right)e^2\sin^2 i,$$

$$
\begin{aligned}
A_{j3} =& \cos 2\omega \cos 2\overline{\Omega}\left(-112+750e^2-720e^4-\frac{5023}{12}e^6\right)e\left(1+\cos^2 i\right)\\
&+\sin 2\omega \sin 2\overline{\Omega}\left(224-1500e^2+1440e^4-\frac{5023}{12}e^6\right)e\cos i\\
&+\cos 2\overline{\Omega}\left(-212+804e^2-\frac{453}{4}e^4\right)e^3\sin^2 i,\\
A_{j4} =& \cos 2\omega \cos 2\overline{\Omega}\left(-272+\frac{4994}{3}e^2-\frac{238}{3}e^4-\frac{725}{12}e^6\right)e^2\left(1+\cos^2 i\right)\\
&+\sin 2\omega \sin 2\overline{\Omega}\left(544-\frac{9988}{3}e^2+\frac{476}{3}e^4+\frac{725}{6}e^6\right)e^2\cos i\\
&+\cos 2\overline{\Omega}\left(-308+762e^2+\frac{25}{2}e^4\right)e^4\sin^2 i,\\
A_{j5} =& \cos 2\omega \cos 2\overline{\Omega}\left(-\frac{1690}{3}+\frac{32552}{15}e^2-\frac{14851}{60}e^4\right)e^3\left(1+\cos^2 i\right)\\
&+\sin 2\omega \sin 2\overline{\Omega}\left(\frac{3380}{3}-\frac{65104}{15}e^2-\frac{14851}{30}e^4\right)e^3\cos i\\
&+\cos 2\overline{\Omega}\left(-412+\frac{6413}{20}e^2\right)e^5\sin^2 i,\\
A_{j6} =& \cos 2\omega \cos 2\overline{\Omega}\left(-\frac{2968}{3}+1414e^2+\frac{627}{8}e^4\right)e^4\left(1+\cos^2 i\right)\\
&+\sin 2\omega \sin 2\overline{\Omega}\left(\frac{5936}{3}-2828e^2-\frac{627}{4}e^4\right)e^4\cos i\\
&+\cos 2\overline{\Omega}\left(-371+\frac{87}{2}e^2\right)e^6\sin^2 i,\\
A_{j7} =& \cos 2\omega \cos 2\overline{\Omega}\left(-\frac{25468}{21}+\frac{67211}{168}e^2\right)e^5\left(1+\cos^2 i\right)\\
&+\sin 2\omega \sin 2\overline{\Omega}\left(\frac{50936}{21}-\frac{67211}{84}e^2\right)e^5\cos i+\cos 2\overline{\Omega}\left(\frac{4759}{28}e^7\right)\sin^2 i,\\
A_{j8} =& \cos 2\omega \cos 2\overline{\Omega}\left(-\frac{5209}{6}+\frac{1625}{48}e^2\right)e^6\left(1+\cos^2 i\right)\\
&+\sin 2\omega \sin 2\overline{\Omega}\left(\frac{5209}{3}-\frac{1625}{24}e^2\right)e^6\cos i+\cos 2\overline{\Omega}\left(-\frac{235}{8}e^8\right)\sin^2 i,\\
A_{j9} =& \cos 2\omega \cos 2\overline{\Omega}\left(-\frac{22891}{72}e^7\right)\left(1+\cos^2 i\right)+\sin 2\omega \sin 2\overline{\Omega}\left(\frac{22891}{36}e^7\right)\cos i,\\
A_{j10} =& \cos 2\omega \cos 2\overline{\Omega}\left(-\frac{1099}{24}e^8\right)\left(1+\cos^2 i\right)+\sin 2\omega \sin 2\overline{\Omega}\left(\frac{10999}{12}e^8\right)\cos i.
\end{aligned}
\tag{6.63}
$$

$$
\begin{aligned}
B_{j1} &= \cos 2\omega \sin 2\overline{\Omega}\left(-32+\frac{440}{3}e^2+\frac{1592}{3}e^4-\frac{405}{2}e^6\right)e\cos i \\
&\quad +\sin 2\omega \cos 2\overline{\Omega}\left(-16+\frac{220}{3}e^2+\frac{796}{3}e^4-\frac{405}{4}e^6\right)e\left(1+\cos^2 i\right), \\
B_{j2} &= \cos 2\omega \sin 2\overline{\Omega}\left(64-448e^2+\frac{3152}{3}e^4+1316e^6+\frac{215}{6}e^8\right)\cos i \\
&\quad +\sin 2\omega \cos 2\overline{\Omega}\left(32-224e^2+\frac{1576}{3}e^4+658e^6+\frac{215}{12}e^8\right)\left(1+\cos^2 i\right), \\
B_{j3} &= \cos 2\omega \sin 2\overline{\Omega}\left(224-1500e^2+1440e^4+\frac{2660}{3}e^6\right)e\cos i \\
&\quad +\sin 2\omega \cos 2\overline{\Omega}\left(112-750e^2+720e^4+\frac{1330}{3}e^6\right)e\left(1+\cos^2 i\right), \\
B_{j4} &= \cos 2\omega \sin 2\overline{\Omega}\left(544-\frac{9988}{3}e^2+\frac{460}{3}e^4+\frac{425}{3}e^6\right)e^2\cos i \\
&\quad +\sin 2\omega \cos 2\overline{\Omega}\left(272-\frac{4994}{3}e^2+\frac{230}{3}e^4+\frac{425}{6}e^6\right)e^2\left(1+\cos^2 i\right), \\
B_{j5} &= \cos 2\omega \sin 2\overline{\Omega}\left(\frac{3380}{3}-\frac{65104}{15}e^2-\frac{7738}{15}e^4\right)e^3\cos i \\
&\quad +\sin 2\omega \cos 2\overline{\Omega}\left(\frac{1690}{3}-\frac{32552}{15}e^2-\frac{3869}{15}e^4\right)e^3\left(1+\cos^2 i\right), \\
B_{j6} &= \cos 2\omega \sin 2\overline{\Omega}\left(\frac{5936}{3}-2828e^2-\frac{1975}{12}e^4\right)e^4\cos i \\
&\quad +\sin 2\omega \cos 2\overline{\Omega}\left(\frac{2968}{3}-1414e^2-\frac{1975}{24}e^4\right)e^4\left(1+\cos^2 i\right), \\
B_{j7} &= \cos 2\omega \sin 2\overline{\Omega}\left(\frac{50936}{21}-\frac{67211}{84}e^2\right)e^5\cos i \\
&\quad +\sin 2\omega \cos 2\overline{\Omega}\left(\frac{25468}{21}-\frac{67211}{168}e^2\right)e^5\left(1+\cos^2 i\right), \\
B_{j8} &= \cos 2\omega \sin 2\overline{\Omega}\left(\frac{5209}{3}-\frac{1625}{24}e^2\right)e^6\cos i \\
&\quad +\sin 2\omega \cos 2\overline{\Omega}\left(\frac{5209}{6}-\frac{1625}{48}e^2\right)e^6\left(1+\cos^2 i\right), \\
B_{j9} &= \cos 2\omega \sin 2\overline{\Omega}\left(\frac{22891}{36}e^7\right)\cos i+\sin 2\omega \cos 2\overline{\Omega}\left(\frac{22891}{72}e^7\right)\left(1+\cos^2 i\right), \\
B_{j10} &= \cos 2\omega \sin 2\overline{\Omega}\left(\frac{1099}{12}e^8\right)\cos i+\sin 2\omega \cos 2\overline{\Omega}\left(\frac{1099}{24}e^8\right)\left(1+\cos^2 i\right).
\end{aligned}
\tag{6.64}
$$

For $j = 2$, there are

$$\begin{aligned}
d_j &= \cos 2\omega \sin 2\overline{\Omega} \left(-160 - 49e^2\right) e^6 \cos i \\
&\quad + \sin 2\omega \cos 2\overline{\Omega} \left(-80 - \frac{49}{2}e^2\right) e^6 \left(1 + \cos^2 i\right), \qquad (6.65) \\
A_{j1} &= \cos 2\omega \cos 2\overline{\Omega} \left(-16 + 56e^2 - \frac{833}{3}e^4 + \frac{47}{2}e^6\right) e \left(1 + \cos^2 i\right) \\
&\quad + \sin 2\omega \sin 2\overline{\Omega} \left(32 - 122e^2 + \frac{1666}{3}e^4 - 47e^6\right) e \cos i \\
&\quad + \cos 2\overline{\Omega} \left(-96 + 324e^2 - 392e^4 - \frac{243}{4}e^6\right) e \sin^2 i, \\
A_{j2} &= \cos 2\omega \cos 2\overline{\Omega} \left(16 - \frac{308}{3}e^2 - 856e^4 - \frac{1039}{12}e^6\right) e^2 \left(1 + \cos^2 i\right) \\
&\quad + \sin 2\omega \sin 2\overline{\Omega} \left(-32 + \frac{616}{3}e^2 + 1712e^4 + \frac{1039}{6}e^6\right) e^2 \cos i \\
&\quad + \cos 2\overline{\Omega} \left(-144 + 536e^2 - 429e^4 - \frac{87}{2}e^6\right) e^2 \sin^2 i, \\
A_{j3} &= \cos 2\omega \cos 2\overline{\Omega} \left(-\frac{112}{3} + \frac{862}{3}e^2 - 229e^4 - \frac{6259}{12}e^6\right) e \left(1 + \cos^2 i\right) \\
&\quad + \sin 2\omega \sin 2\overline{\Omega} \left(\frac{224}{3} - \frac{1724}{3}e^2 + 458e^4 + \frac{6259}{62}e^6\right) e \cos i \\
&\quad + \cos 2\overline{\Omega} \left(-212 + 804e^2 - \frac{453}{4}e^4\right) e^3 \sin^2 i, \\
A_{j4} &= \cos 2\omega \cos 2\overline{\Omega} \left(-136 + \frac{2960}{3}e^2 + 446e^4 - \frac{175}{4}e^6\right) e^2 \left(1 + \cos^2 i\right) \\
&\quad + \sin 2\omega \sin 2\overline{\Omega} \left(272 - \frac{5920}{3}e^2 - 892e^4 + \frac{175}{2}e^6\right) e^2 \cos i \\
&\quad + \cos 2\overline{\Omega} \left(-308 + 762e^2 + \frac{25}{2}e^4\right) e^4 \sin^2 i, \\
A_{j5} &= \cos 2\omega \cos 2\overline{\Omega} \left(-338 + \frac{8109}{5}e^2 + \frac{34231}{60}e^4\right) e^3 \left(1 + \cos^2 i\right) \\
&\quad + \sin 2\omega \sin 2\overline{\Omega} \left(676 - \frac{16218}{5}e^2 - \frac{34231}{30}e^4\right) e^3 \cos i \\
&\quad + \cos 2\overline{\Omega} \left(-412 + \frac{6413}{20}e^2\right) e^5 \sin^2 i,
\end{aligned}$$

$$
\begin{aligned}
A_{j6} &= \cos 2\omega \cos 2\overline{\Omega} \left(-\frac{2132}{3} + \frac{3232}{3} e^2 + \frac{1091}{8} e^4 \right) e^4 \left(1 + \cos^2 i \right) \\
&\quad + \sin 2\omega \sin 2\overline{\Omega} \left(\frac{4264}{3} - \frac{6464}{3} e^2 - \frac{1091}{4} e^4 \right) e^4 \cos i \\
&\quad + \cos 2\overline{\Omega} \left(-371 + \frac{87}{2} e^2 \right) e^6 \sin^2 i, \\
A_{j7} &= \cos 2\omega \cos 2\overline{\Omega} \left(-\frac{3115}{3} + \frac{37907}{168} e^2 \right) e^5 \left(1 + \cos^2 i \right) \\
&\quad + \sin 2\omega \sin 2\overline{\Omega} \left(\frac{6230}{3} - \frac{37907}{84} e^2 \right) e^5 \cos i + \cos 2\overline{\Omega} \left(-\frac{4759}{28} e^7 \right) \sin^2 i, \\
A_{j8} &= \cos 2\omega \cos 2\overline{\Omega} \left(-829 - \frac{85}{16} e^2 \right) e^6 \left(1 + \cos^2 i \right) \\
&\quad + \sin 2\omega \sin 2\overline{\Omega} \left(1658 + \frac{85}{8} e^2 \right) e^6 \cos i + \cos 2\overline{\Omega} \left(-\frac{235}{8} e^8 \right) \sin^2 i, \\
A_{j9} &= \cos 2\omega \cos 2\overline{\Omega} \left(-\frac{22891}{72} e^7 \right) \left(1 + \cos^2 i \right) + \sin 2\omega \sin 2\overline{\Omega} \left(\frac{22891}{36} e^7 \right) \cos i, \\
A_{j10} &= \cos 2\omega \cos 2\overline{\Omega} \left(-\frac{1099}{24} e^8 \right) \left(1 + \cos^2 i \right) + \sin 2\omega \sin 2\overline{\Omega} \left(\frac{1099}{12} e^8 \right) \cos i,
\end{aligned}
\tag{6.66}
$$

$$
\begin{aligned}
B_{j1} &= \cos 2\omega \sin 2\overline{\Omega} \left(32 - 120 e^2 + \frac{1798}{3} e^4 - \frac{137}{2} e^6 \right) e \cos i \\
&\quad + \sin 2\omega \cos 2\overline{\Omega} \left(16 - 60 e^2 + \frac{899}{3} e^4 - \frac{137}{4} e^6 \right) e \left(1 + \cos^2 i \right), \\
B_{j2} &= \cos 2\omega \sin 2\overline{\Omega} \left(-32 + \frac{584}{3} e^2 + 1728 e^4 + \frac{676}{6} e^6 \right) e^2 \cos i \\
&\quad + \sin 2\omega \cos 2\overline{\Omega} \left(-16 + \frac{292}{3} e^2 + 864 e^4 + \frac{676}{12} e^6 \right) e^2 \left(1 + \cos^2 i \right), \\
B_{j3} &= \cos 2\omega \sin 2\overline{\Omega} \left(\frac{224}{3} - \frac{1724}{3} e^2 + 446 e^4 + \frac{3314}{3} e^6 \right) e \cos i \\
&\quad + \sin 2\omega \cos 2\overline{\Omega} \left(\frac{112}{3} - \frac{862}{3} e^2 + 223 e^4 + \frac{1657}{3} e^6 \right) e \left(1 + \cos^2 i \right), \\
B_{j4} &= \cos 2\omega \sin 2\overline{\Omega} \left(272 - \frac{5920}{3} e^2 - 908 e^4 + 119 e^6 \right) e^2 \cos i \\
&\quad + \sin 2\omega \cos 2\overline{\Omega} \left(136 - \frac{2960}{3} e^2 - 454 e^4 + \frac{119}{2} e^6 \right) e^2 \left(1 + \cos^2 i \right),
\end{aligned}
$$

$$
\begin{aligned}
B_{j5} &= \cos 2\omega \sin 2\overline{\Omega}\left(676-\frac{16218}{5}e^2-\frac{17428}{15}e^4\right)e^3\cos i \\
&\quad +\sin 2\omega \cos 2\overline{\Omega}\left(338-\frac{8109}{5}e^2-\frac{8714}{15}e^4\right)e^3\left(1+\cos^2 i\right), \\
B_{j6} &= \cos 2\omega \sin 2\overline{\Omega}\left(\frac{4264}{3}-\frac{6464}{3}e^2-\frac{3367}{12}e^4\right)e^4\cos i \\
&\quad +\sin 2\omega \cos 2\overline{\Omega}\left(\frac{2132}{3}-\frac{3232}{3}e^2-\frac{3367}{24}e^4\right)e^4\left(1+\cos^2 i\right), \\
B_{j7} &= \cos 2\omega \sin 2\overline{\Omega}\left(\frac{6230}{3}-\frac{37907}{84}e^2\right)e^5\cos i \\
&\quad +\sin 2\omega \cos 2\overline{\Omega}\left(\frac{3115}{3}-\frac{37907}{168}e^2\right)e^5\left(1+\cos^2 i\right), \\
B_{j8} &= \cos 2\omega \sin 2\overline{\Omega}\left(1658+\frac{85}{8}e^2\right)e^6\cos i \\
&\quad +\sin 2\omega \cos 2\overline{\Omega}\left(829+\frac{85}{16}e^2\right)e^6\left(1+\cos^2 i\right), \\
B_{j9} &= \cos 2\omega \sin 2\overline{\Omega}\left(\frac{22891}{36}e^7\right)\cos i+\sin 2\omega \cos 2\overline{\Omega}\left(\frac{22891}{72}e^7\right)\left(1+\cos^2 i\right), \\
B_{j10} &= \cos 2\omega \sin 2\overline{\Omega}\left(\frac{1099}{12}e^8\right)\cos i+\sin 2\omega \cos 2\overline{\Omega}\left(\frac{1099}{24}e^8\right)\left(1+\cos^2 i\right).
\end{aligned}
\tag{6.67}
$$

For $j = 3$, there are

$$
\begin{aligned}
d_j &= \sin 2\omega \sin 2\overline{\Omega}\left(-144-47e^2\right)e^5\cos i \\
&\quad +\cos 2\omega \cos 2\overline{\Omega}\left(\left(68+\frac{35}{2}e^2\right)e^5+\cos^2 i\left(76+\frac{59}{2}e^2\right)e^5\right) \\
&\quad \cos 2\overline{\Omega}\left(\left(-96+144e^2-12e^4+15e^6\right)e\right. \\
&\quad \left.+\left(160-240e^2+36e^4-23e^6\right)e\cos^2 i\right),
\end{aligned}
\tag{6.68}
$$

$$
\begin{aligned}
A_{j1} &= \cos 2\omega \sin 2\overline{\Omega}\left(32-104e^2-294e^4+\frac{175}{2}e^6\right)\cos i \\
&\quad +\sin 2\omega \cos 2\overline{\Omega}\left(\left(16-68e^2-\frac{293}{3}e^4+\frac{321}{4}e^6\right)\right. \\
&\quad \left.+\left(16-36e^2-\frac{589}{3}e^4+\frac{29}{4}e^6\right)\cos^2 i\right),
\end{aligned}
$$

$$A_{j2} = \cos 2\omega \sin 2\overline{\Omega}\left(128 - \frac{1688}{3}e^2 - 1426e^4 - \frac{865}{6}e^6\right) e\cos i$$
$$+ \sin 2\omega \cos 2\overline{\Omega}\left(\left(80 - \frac{1108}{3}e^2 - 587e^4 - \frac{517}{12}e^6\right) e\right.$$
$$\left. + \left(48 - \frac{580}{3}e^2 - 839e^4 - \frac{1213}{12}e^6\right) e\cos^2 i\right),$$

$$A_{j3} = \cos 2\omega \sin 2\overline{\Omega}\left(-\frac{224}{3} + 716e^2 - 994e^4 - \frac{2903}{3}e^6\right)\cos i$$
$$+ \sin 2\omega \cos 2\overline{\Omega}\left(\left(-\frac{112}{3} + \frac{1186}{3}e^2 - 691e^4 - \frac{1297}{3}e^6\right)\right.$$
$$\left. + \left(-\frac{112}{3} + \frac{962}{3}e^2 - 303e^4 - \frac{1606}{3}e^6\right)\cos^2 i\right),$$

$$A_{j4} = \cos 2\omega \sin 2\overline{\Omega}\left(-272 + \frac{6536}{3}e^2 + 384e^4 - \frac{575}{6}e^6\right) e\cos i$$
$$+ \sin 2\omega \cos 2\overline{\Omega}\left(\left(-136 + \frac{3472}{3}e^2 - 79e^4 - \frac{225}{4}e^6\right) e\right.$$
$$\left. + \left(-136 + \frac{3064}{3}e^2 + 463e^4 - \frac{475}{12}e^6\right) e\cos^2 i\right),$$

$$A_{j5} = \cos 2\omega \sin 2\overline{\Omega}\left(-676 + \frac{52774}{15}e^2 + \frac{13909}{15}e^4\right) e^2\cos i$$
$$+ \sin 2\omega \cos 2\overline{\Omega}\left(\left(-338 + \frac{9359}{5}e^2 + \frac{4532}{15}e^4\right) e^2\right.$$
$$\left. + \left(-338 + \frac{24697}{15}e^2 + \frac{9377}{15}e^4\right) e^2\cos^2 i\right),$$

$$A_{j6} = \cos 2\omega \sin 2\overline{\Omega}\left(-\frac{4264}{3} + 2402e^2 + \frac{975}{4}e^4\right) e^3\cos i$$
$$+ \sin 2\omega \cos 2\overline{\Omega}\left(\left(-\frac{2132}{3} + \frac{4021}{3}e^2 + \frac{743}{8}e^4\right) e^2\right.$$
$$\left. + \left(-\frac{2132}{3} + \frac{3185}{3}e^2 + \frac{1207}{8}e^4\right) e^2\cos^2 i\right),$$

$$A_{j7} = \cos 2\omega \sin 2\overline{\Omega}\left(-\frac{6230}{3}+\frac{6775}{12}e^2\right)e^4\cos i$$
$$+\sin 2\omega \cos 2\overline{\Omega}\left(\left(-\frac{3115}{3}+\frac{62077}{168}e^2\right)e^4\right.$$
$$\left.+\left(-\frac{3115}{3}+\frac{32773}{168}e^2\right)e^4\cos^2 i\right),$$
$$A_{j8} = \cos 2\omega \sin 2\overline{\Omega}\left(-1658+\frac{215}{24}e^2\right)e^5\cos i$$
$$+\sin 2\omega \cos 2\overline{\Omega}\left(\left(-829+\frac{385}{16}e^2\right)e^5+\left(-829-\frac{725}{48}e^2\right)e^5\cos^2 i\right),$$
$$A_{j9} = \cos 2\omega \sin 2\overline{\Omega}\left(-\frac{22891}{36}e^6\right)\cos i+\sin 2\omega \cos 2\overline{\Omega}\left(-\frac{22891}{72}e^6\right)\left(1+\cos^2 i\right),$$
$$A_{j10} = \cos 2\omega \sin 2\overline{\Omega}\left(-\frac{1099}{12}e^7\right)\cos i+\sin 2\omega \cos 2\overline{\Omega}\left(-\frac{1099}{24}e^7\right)\left(1+\cos^2 i\right),\tag{6.69}$$

$$B_{j1} = \sin 2\omega \sin 2\overline{\Omega}\left(-32+122e^2+\frac{878}{3}e^4-56e^6\right)\cos i$$
$$+\cos 2\omega \cos 2\overline{\Omega}\left(\left(16-72e^2-\frac{287}{3}e^4+\frac{123}{2}e^6\right)\right.$$
$$\left.+\cos^2 i\left(16-40e^2-197e^4-\frac{11}{2}e^6\right)\right)$$
$$+\cos 2\overline{\Omega}\left(\left(-96+60e^2+80e^4-\frac{909}{4}e^6\right)e\right.$$
$$\left.+\left(96+132e^2-440e^4+\frac{1565}{4}e^6\right)\cos^2 i\right),$$

$$B_{j2} = \sin 2\omega \sin 2\overline{\Omega}\left(-128+\frac{1720}{3}e^2+1382e^4+\frac{731}{6}e^6\right)\cos i$$
$$+\cos 2\omega \cos 2\overline{\Omega}\left(\left(80-\frac{1124}{3}e^2-565e^4-\frac{455}{2}e^6\right)e\right.$$
$$\left.+\cos^2 i\left(48-\frac{596}{3}e^2-817e^4-\frac{1007}{12}e^6\right)\right)$$

$$
\begin{aligned}
&+\cos 2\overline{\Omega}\Bigg(\left(-114+352e^2-39e^4-60e^6\right)e \\
&+\left(144-208e^2-281e^4+94e^6\right)e\cos^2 i\Bigg),
\end{aligned}
$$

$$
\begin{aligned}
B_{j3} =& \sin 2\omega \sin 2\overline{\Omega}\left(\frac{224}{3}-716e^2+1006e^4+\frac{5635}{6}e^6\right)\cos i \\
&+\cos 2\omega\cos 2\overline{\Omega}\Bigg(\left(-\frac{112}{3}+\frac{1186}{3}e^2-697e^4-\frac{4981}{12}e^6\right) \\
&+\cos^2 i\left(-\frac{112}{3}+\frac{962}{3}e^2-309e^4-\frac{6289}{12}e^6\right)\Bigg) \\
&+\cos 2\overline{\Omega}\Bigg(\left(-212+628e^2+\frac{483}{4}e^4\right)e^2 \\
&+\left(212-\frac{1460}{3}e^2-\frac{1467}{4}e^4\right)e^2\cos^2 i\Bigg),
\end{aligned}
$$

$$
\begin{aligned}
B_{j4} =& \sin 2\omega \sin 2\overline{\Omega}\left(272-\frac{6536}{3}e^2-368e^4+\frac{281}{3}e^6\right)e\cos i \\
&+\cos 2\omega\cos 2\overline{\Omega}\Bigg(\left(-136+\frac{3472}{3}e^2-87e^4-\frac{105}{2}e^6\right)e \\
&+\cos^2 i\left(-136+\frac{3064}{3}e^2+455e^4-\frac{247}{6}e^6\right)e\Bigg) \\
&+\cos 2\overline{\Omega}\left(\left(-308+579e^2+55e^4\right)e^3+\left(308-425e^2-117e^4\right)e^3\cos^2 i\right),
\end{aligned}
$$

$$
\begin{aligned}
B_{j5} =& \sin 2\omega \sin 2\overline{\Omega}\left(676-\frac{52774}{15}e^2-\frac{27193}{30}e^4\right)e^2\cos i \\
&+\cos 2\omega\cos 2\overline{\Omega}\Bigg(\left(-338+\frac{9359}{5}e^2+\frac{17503}{60}e^4\right)e^2 \\
&+\cos^2 i\left(-338+\frac{24697}{15}e^2+\frac{36883}{60}e^4\right)e^2\Bigg) \\
&+\cos 2\overline{\Omega}\left(\left(-412+\frac{4219}{20}e^2\right)e^4+\left(412-\frac{1923}{20}e^2\right)e^4\cos^2 i\right),
\end{aligned}
$$

$$
\begin{aligned}
B_{j6} =& \sin 2\omega \sin 2\overline{\Omega}\left(\frac{4264}{3}-2402e^2-\frac{2831}{12}e^4\right)e^3\cos i \\
&+\cos 2\omega\cos 2\overline{\Omega}\Bigg(\left(-\frac{2132}{3}+\frac{4021}{3}e^2+\frac{2135}{24}e^4\right)e^3
\end{aligned}
$$

$$
\begin{aligned}
&\quad +\cos^2 i\left(-\frac{2132}{3}+\frac{3185}{3}e^2+\frac{3527}{24}e^4\right)e^3\Bigg)\\
&\quad +\cos 2\overline{\Omega}\left(\left(-371+20e^2\right)e^5+\left(371+\frac{34}{3}e^2\right)e^5\cos^2 i\right),\\
B_{j7} &= \sin 2\omega\sin 2\overline{\Omega}\left(\frac{6230}{3}-\frac{6775}{12}e^2\right)e^4\cos i\\
&\quad +\cos 2\omega\cos 2\overline{\Omega}\left(\left(-\frac{3115}{3}+\frac{62077}{168}e^2\right)e^4\right.\\
&\quad \left.+\cos^2 i\left(-\frac{3155}{3}+\frac{32773}{168}e^2\right)e^4\right)\\
&\quad +\cos 2\overline{\Omega}\left(-\frac{4759}{28}e^6\right)\sin^2 i,\\
B_{j8} &= \sin 2\omega\sin 2\overline{\Omega}\left(1658-\frac{215}{24}e^2\right)e^5\cos i\\
&\quad +\cos 2\omega\cos 2\overline{\Omega}\left(\left(-829+\frac{385}{16}e^2\right)e^5+\cos^2 i\left(-829-\frac{725}{48}e^2\right)e^5\right)\\
&\quad +\cos 2\overline{\Omega}\left(-\frac{235}{8}e^7\right)\sin^2 i,\\
B_{j9} &= \sin 2\omega\sin 2\overline{\Omega}\left(\frac{22891}{36}e^6\right)\cos i+\cos 2\omega\cos 2\overline{\Omega}\left(-\frac{22891}{72}e^6\right)\left(1+\cos^2 i\right),\\
B_{j10} &= \sin 2\omega\sin 2\overline{\Omega}\left(\frac{1099}{12}e^7\right)\cos i+\cos 2\omega\cos 2\overline{\Omega}\left(-\frac{1099}{24}e^7\right)\left(1+\cos^2 i\right).
\end{aligned}
\tag{6.70}
$$

For $j=4$, there are

$$
\begin{aligned}
d_j &= \sin 2\omega\cos 2\overline{\Omega}\left(1+\frac{3}{2}e^2\right)e^4\cos i\sin i+\cos 2\omega\sin 2\overline{\Omega}\left(1+\frac{3}{2}e^2\right)e^4\sin i\\
&\quad +\sin 2\overline{\Omega}\left(-8+12e^2-3e^4+e^6\right)\sin i,
\end{aligned}
\tag{6.71}
$$

$$
\begin{aligned}
A_{j1} &= \left(-4+\frac{37}{3}e^2+\frac{73}{8}e^4\right)e\sin i\left(\cos 2\omega\cos 2\overline{\Omega}\cos i-\sin 2\omega\sin 2\overline{\Omega}\right),\\
A_{j2} &= \left(4-22e^2+\frac{63}{2}e^4+\frac{29}{4}e^6\right)\sin i\left(\cos 2\omega\cos 2\overline{\Omega}\cos i-\sin 2\omega\sin 2\overline{\Omega}\right),\\
A_{j3} &= \left(\frac{28}{3}-\frac{97}{2}e^2+\frac{103}{8}e^4\right)e\sin i\left(\cos 2\omega\cos 2\overline{\Omega}\cos i-\sin 2\omega\sin 2\overline{\Omega}\right),\\
A_{j4} &= \left(-17+\frac{271}{4}e^2+\frac{25}{12}e^4\right)e^2\sin i\left(\cos 2\omega\cos 2\overline{\Omega}\cos i-\sin 2\omega\sin 2\overline{\Omega}\right),
\end{aligned}
$$

$$A_{j5} = \left(\frac{169}{6} - \frac{323}{8}e^2\right) e^3 \sin i \left(\cos 2\omega \cos 2\overline{\Omega} \cos i - \sin 2\omega \sin 2\overline{\Omega}\right),$$
$$A_{j6} = \left(\frac{209}{6} - \frac{29}{4}e^2\right) e^4 \sin i \left(\cos 2\omega \cos 2\overline{\Omega} \cos i - \sin 2\omega \sin 2\overline{\Omega}\right),$$
$$A_{j7} = \left(\frac{1221}{56}e^5\right) \sin i \left(\cos 2\omega \cos 2\overline{\Omega} \cos i - \sin 2\omega \sin 2\overline{\Omega}\right),$$
$$A_{j8} = \left(\frac{235}{48}e^6\right) \sin i \left(\cos 2\omega \cos 2\overline{\Omega} \cos i - \sin 2\omega \sin 2\overline{\Omega}\right), \tag{6.72}$$

$$\begin{aligned}
B_{j1} =& \left(4 - \frac{38}{3}e^2 - \frac{67}{8}e^4\right) e \sin i \left(\sin 2\omega \cos 2\overline{\Omega} \cos i - \cos 2\omega \sin 2\overline{\Omega}\right) \\
&+ \left(-24 + 45e^2 - \frac{41}{2}e^4\right) e \sin i \sin 2\overline{\Omega}, \\
B_{j2} =& \left(-4 + 22e^2 - \frac{63}{2}e^4 - \frac{23}{4}e^6\right) \sin i \left(\sin 2\omega \cos 2\overline{\Omega} \cos i - \cos 2\omega \sin 2\overline{\Omega}\right) \\
&+ \left(-18 + 40e^2 - \frac{17}{4}e^4\right) e^2 \sin i \sin 2\overline{\Omega}, \\
B_{j3} =& \left(-\frac{28}{3} + \frac{97}{2}e^2 - \frac{109}{8}e^4\right) e \sin i \left(\sin 2\omega \cos 2\overline{\Omega} \cos i - \cos 2\omega \sin 2\overline{\Omega}\right) \\
&+ \left(-\frac{53}{3} + \frac{123}{4}e^2\right) e^3 \sin i \sin 2\overline{\Omega}, \\
B_{j4} =& \left(-17 + \frac{271}{4}e^2 + \frac{17}{12}e^4\right) e^2 \sin i \left(\sin 2\omega \cos 2\overline{\Omega} \cos i - \cos 2\omega \sin 2\overline{\Omega}\right) \\
&+ \left(-\frac{77}{4} + \frac{31}{4}e^2\right) e^4 \sin i \sin 2\overline{\Omega}, \\
B_{j5} =& \left(-\frac{169}{6} + \frac{323}{8}e^2\right) e^3 \sin i \left(\sin 2\omega \cos 2\overline{\Omega} \cos i - \cos 2\omega \sin 2\overline{\Omega}\right) \\
&+ \left(-\frac{287}{20}e^5\right) \sin i \sin 2\overline{\Omega}, \\
B_{j6} =& \left(-\frac{209}{6} + \frac{29}{4}e^2\right) e^4 \sin i \left(\sin 2\omega \cos 2\overline{\Omega} \cos i - \cos 2\omega \sin 2\overline{\Omega}\right) \\
&+ \left(-\frac{47}{12}e^6\right) \sin i \sin 2\overline{\Omega}, \\
B_{j7} =& \left(-\frac{1221}{56}e^5\right) \sin i \left(\sin 2\omega \cos 2\overline{\Omega} \cos i - \cos 2\omega \sin 2\overline{\Omega}\right), \\
B_{j8} =& \left(-\frac{235}{48}e^6\right) \sin i \left(\sin 2\omega \cos 2\overline{\Omega} \cos i - \cos 2\omega \sin 2\overline{\Omega}\right).
\end{aligned} \tag{6.73}$$

For $j=5$, there are

$$d_j = \sin 2\omega \sin 2\overline{\Omega}\left(-1-\frac{3}{2}e^2\right)e^4 + \cos 2\omega \cos 2\overline{\Omega}\left(1+\frac{3}{2}e^2\right)e^4 \cos i$$
$$+\cos 2\overline{\Omega}\left(8-12e^2+3e^4-e^6\right)\cos i, \tag{6.74}$$

$$A_{j1} = \left(4-\frac{37}{3}e^2-\frac{73}{8}e^4\right)e\left(\cos 2\omega \sin 2\overline{\Omega}+\sin 2\omega \cos 2\overline{\Omega}\cos i\right),$$

$$A_{j2} = \left(-4+22e^2-\frac{63}{2}e^4-\frac{29}{4}e^6\right)\left(\cos 2\omega \sin 2\overline{\Omega}+\sin 2\omega \cos 2\overline{\Omega}\cos i\right),$$

$$A_{j3} = \left(-\frac{28}{3}+\frac{97}{2}e^2-\frac{103}{8}e^4\right)e\left(\cos 2\omega \sin 2\overline{\Omega}+\sin 2\omega \cos 2\overline{\Omega}\cos i\right),$$

$$A_{j4} = \left(-17+\frac{271}{4}e^2+\frac{25}{12}e^4\right)e^2\left(\cos 2\omega \sin 2\overline{\Omega}+\sin 2\omega \cos 2\overline{\Omega}\cos i\right),$$

$$A_{j5} = \left(-\frac{169}{6}+\frac{323}{8}e^2\right)e^3\left(\cos 2\omega \sin 2\overline{\Omega}+\sin 2\omega \cos 2\overline{\Omega}\cos i\right),$$

$$A_{j6} = \left(-\frac{209}{6}+\frac{29}{4}e^2\right)e^4\left(\cos 2\omega \sin 2\overline{\Omega}+\sin 2\omega \cos 2\overline{\Omega}\cos i\right),$$

$$A_{j7} = \left(-\frac{1221}{56}e^5\right)\left(\cos 2\omega \sin 2\overline{\Omega}+\sin 2\omega \cos 2\overline{\Omega}\cos i\right),$$

$$A_{j8} = \left(-\frac{235}{48}e^6\right)\left(\cos 2\omega \sin 2\overline{\Omega}+\sin 2\omega \cos 2\overline{\Omega}\cos i\right), \tag{6.75}$$

$$B_{j1} = \left(-4+\frac{38}{3}e^2+\frac{67}{8}e^4\right)e\left(\sin 2\omega \sin 2\overline{\Omega}-\cos 2\omega \cos 2\overline{\Omega}\cos i\right)$$
$$+\cos 2\overline{\Omega}\left(24-45e^2+\frac{41}{2}e^4\right)e\cos i,$$

$$B_{j2} = \left(4-22e^2+\frac{63}{2}e^4+\frac{23}{4}e^6\right)\left(\sin 2\omega \sin 2\overline{\Omega}-\cos 2\omega \cos 2\overline{\Omega}\cos i\right)$$
$$+\cos 2\overline{\Omega}\left(18-40e^2+\frac{17}{4}e^4\right)e^2\cos i,$$

$$B_{j3} = \left(\frac{28}{3}-\frac{97}{2}e^2+\frac{109}{8}e^4\right)e\left(\sin 2\omega \sin 2\overline{\Omega}-\cos 2\omega \cos 2\overline{\Omega}\cos i\right)$$
$$+\cos 2\overline{\Omega}\left(\frac{53}{3}-\frac{123}{4}e^2\right)e^3\cos i,$$

$$B_{j4} = \left(17-\frac{271}{4}e^2-\frac{17}{12}e^4\right)e^2\left(\sin 2\omega \sin 2\overline{\Omega}-\cos 2\omega \cos 2\overline{\Omega}\cos i\right)$$
$$+\cos 2\overline{\Omega}\left(\frac{77}{4}-\frac{31}{4}e^2\right)e^4\cos i,$$

$$\begin{aligned}
B_{j5} &= \left(\frac{169}{6} - \frac{323}{8}e^2\right) e^3 \left(\sin 2\omega \sin 2\overline{\Omega} - \cos 2\omega \cos 2\overline{\Omega} \cos i\right) \\
&\quad + \cos 2\overline{\Omega} \left(\frac{287}{20} e^5\right) \cos i, \\
B_{j6} &= \left(\frac{209}{6} - \frac{29}{4}e^2\right) e^4 \left(\sin 2\omega \sin 2\overline{\Omega} - \cos 2\omega \cos 2\overline{\Omega} \cos i\right) \\
&\quad + \cos 2\overline{\Omega} \left(\frac{47}{12} e^6\right) \cos i, \\
B_{j7} &= \left(\frac{1221}{56} e^5\right) \left(\sin 2\omega \sin 2\overline{\Omega} - \cos 2\omega \cos 2\overline{\Omega} \cos i\right), \\
B_{j8} &= \left(\frac{235}{48} e^6\right) \left(\sin 2\omega \sin 2\overline{\Omega} - \cos 2\omega \cos 2\overline{\Omega} \cos i\right).
\end{aligned} \tag{6.76}$$

For $j = 6$, there are

$$\begin{aligned}
d_j &= \sin 2\omega \sin 2\overline{\Omega} \left(-88 + 37e^2\right) e^5 \cos i \\
&\quad + \cos 2\omega \cos 2\overline{\Omega} \left(44 - \frac{37}{2} e^2\right) e^5 \left(1 + \cos^2 i\right) \\
&\quad + \cos 2\overline{\Omega} \left(96 - 144e^2 + 60e^4 - 9e^6\right) e \sin^2 i,
\end{aligned} \tag{6.77}$$

$$\begin{aligned}
A_{j1} &= \left(16 - 164e^2 + \frac{595}{3} e^4 + \frac{1197}{4} e^6\right) \\
&\quad \left(2\cos 2\omega \sin 2\overline{\Omega} \cos i + \sin 2\omega \cos 2\overline{\Omega} \left(1 + \cos^2 i\right)\right), \\
A_{j2} &= \left(176 - \frac{2692}{3} e^2 + 169e^4 + \frac{1571}{12} e^6\right) e \\
&\quad \left(2\cos 2\omega \sin 2\overline{\Omega} \cos i + \sin 2\omega \cos 2\overline{\Omega} \left(1 + \cos^2 i\right)\right), \\
A_{j3} &= \left(-\frac{112}{3} + \frac{1858}{3} e^2 - 1855e^4 - \frac{370}{3} e^6\right) \\
&\quad \left(2\cos 2\omega \sin 2\overline{\Omega} \cos i + \sin 2\omega \cos 2\overline{\Omega} \left(1 + \cos^2 i\right)\right), \\
A_{j4} &= \left(-136 + \frac{4696}{3} e^2 - 1705e^4 - \frac{425}{4} e^6\right) e \\
&\quad \left(2\cos 2\omega \sin 2\overline{\Omega} \cos i + \sin 2\omega \cos 2\overline{\Omega} \left(1 + \cos^2 i\right)\right), \\
A_{j5} &= \left(-338 + \frac{12739}{5} e^2 - \frac{10003}{15} e^4\right) e^2 \\
&\quad \left(2\cos 2\omega \sin 2\overline{\Omega} \cos i + \sin 2\omega \cos 2\overline{\Omega} \left(1 + \cos^2 i\right)\right), \\
A_{j6} &= \left(-\frac{2132}{3} + \frac{6529}{3} e^2 - \frac{649}{8} e^4\right) e^3 \\
&\quad \left(2\cos 2\omega \sin 2\overline{\Omega} \cos i + \sin 2\omega \cos 2\overline{\Omega} \left(1 + \cos^2 i\right)\right),
\end{aligned}$$

$$
\begin{aligned}
A_{j7} &= \left(-\frac{3115}{3}+\frac{21427}{24}e^2\right)e^4 \\
&\quad \left(2\cos 2\omega \sin 2\overline{\Omega}\cos i+\sin 2\omega \cos 2\overline{\Omega}\left(1+\cos^2 i\right)\right), \\
A_{j8} &= \left(-829+\frac{2265}{16}e^2\right)e^5 \\
&\quad \left(2\cos 2\omega \sin 2\overline{\Omega}\cos i+\sin 2\omega \cos 2\overline{\Omega}\left(1+\cos^2 i\right)\right), \\
A_{j9} &= \left(-\frac{22891}{72}e^6\right)\left(2\cos 2\omega \sin 2\overline{\Omega}\cos i+\sin 2\omega \cos 2\overline{\Omega}\left(1+\cos^2 i\right)\right), \\
A_{j10} &= \left(-\frac{1099}{24}e^7\right)\left(2\cos 2\omega \sin 2\overline{\Omega}\cos i+\sin 2\omega \cos 2\overline{\Omega}\left(1+\cos^2 i\right)\right), \qquad (6.78) \\
B_{j1} &= \left(16-168e^2+\frac{625}{3}e^4+\frac{525}{2}e^6\right)\left(\cos 2\omega \cos 2\overline{\Omega}\left(1+\cos^2 i\right)\right. \\
&\quad \left.-2\sin 2\omega \sin 2\overline{\Omega}\cos i\right)+\left(-96+636e^2-1000e^4+\frac{1059}{4}e^6\right)\sin^2 i\cos 2\overline{\Omega}, \\
B_{j2} &= \left(176-\frac{2708}{3}e^2+191e^4+\frac{1201}{12}e^6\right)e\left(\cos 2\omega \cos 2\overline{\Omega}\left(1+\cos^2 i\right)\right. \\
&\quad \left.-2\sin 2\omega \sin 2\overline{\Omega}\cos i\right)+\left(-144+784e^2-999e^4+42e^6\right)e\sin^2 i\cos 2\overline{\Omega}, \\
B_{j3} &= \left(-\frac{112}{3}+\frac{1858}{3}e^2-1861e^4-\frac{1057}{12}e^6\right)\left(\cos 2\omega \cos 2\overline{\Omega}\left(1+\cos^2 i\right)\right. \\
&\quad \left.-2\sin 2\omega \sin 2\overline{\Omega}\cos i\right)+\left(-212+1052e^2-\frac{2469}{4}e^4\right)e^2\sin^2 i\cos 2\overline{\Omega}, \\
B_{j4} &= \left(-136+\frac{4696}{3}e^2-1713e^4-\frac{173}{2}e^6\right)e\left(\cos 2\omega \cos 2\overline{\Omega}\left(1+\cos^2 i\right)\right. \\
&\quad \left.-2\sin 2\omega \sin 2\overline{\Omega}\cos i\right)+\left(-308+1041e^2-131e^4\right)e^3\sin^2 i\cos 2\overline{\Omega}, \\
B_{j5} &= \left(-338+\frac{12739}{5}e^2-\frac{40637}{60}e^4\right)e^2\left(\cos 2\omega \cos 2\overline{\Omega}\left(1+\cos^2 i\right)\right. \\
&\quad \left.-2\sin 2\omega \sin 2\overline{\Omega}\cos i\right)+\left(-412+\frac{11107}{20}e^2\right)e^4\sin^2 i\cos 2\overline{\Omega}, \\
B_{j6} &= \left(-\frac{2132}{3}+\frac{6529}{3}e^2-\frac{2041}{24}e^4\right)e^3\left(\cos 2\omega \cos 2\overline{\Omega}\left(1+\cos^2 i\right)\right. \\
&\quad \left.-2\sin 2\omega \sin 2\overline{\Omega}\cos i\right)+\left(-317+114e^2\right)e^5\sin^2 i\cos 2\overline{\Omega},
\end{aligned}
$$

$$B_{j7} = \left(-\frac{3115}{3} + \frac{21427}{24}e^2\right)e^4\left(\cos 2\omega \cos 2\overline{\Omega}\left(1+\cos^2 i\right) - 2\sin 2\omega \sin 2\overline{\Omega}\cos i\right)$$
$$+\left(-\frac{4759}{28}e^6\right)\sin^2 i \cos 2\overline{\Omega},$$
$$B_{j8} = \left(-829 + \frac{2265}{16}e^2\right)e^5\left(\cos 2\omega \cos 2\overline{\Omega}\left(1+\cos^2 i\right) - 2\sin 2\omega \sin 2\overline{\Omega}\cos i\right)$$
$$+\left(-\frac{235}{8}e^7\right)\sin^2 i \cos 2\overline{\Omega},$$
$$B_{j9} = \left(-\frac{22891}{72}e^6\right)\left(\cos 2\omega \cos 2\overline{\Omega}\left(1+\cos^2 i\right) - 2\sin 2\omega \sin 2\overline{\Omega}\cos i\right),$$
$$B_{j10} = \left(-\frac{1099}{24}e^7\right)\left(\cos 2\omega \cos 2\overline{\Omega}\left(1+\cos^2 i\right) - 2\sin 2\omega \sin 2\overline{\Omega}\cos i\right). \tag{6.79}$$

6.5 Properties of the Solutions of Geopotential Perturbations

The properties of the solutions of geopotential disturbances of lower order and degrees may be summarised from the solutions given in Sects. 5.2, 6.2, 6.3 and 6.4.

Long Term Perturbations

Only the disturbances of the $\overline{C}_{20}$ have long term components and only the Keplerian elements of (ω, Ω, M) are long term perturbed. Long terms are functions of (a, e, i), which are considered constants here. The order ratios of the perturbations are of $(1, e, e^2)$, respectively. Because of slow changes of (ω, Ω), long periodic terms are periodic functions of (ω, Ω). Short periodic terms are functions of M.

Long Periodic Perturbations

All geopotential disturbances have long periodic components on all Keplerian elements $(a, e, \omega, i, \Omega, M)$. The order ratios of the long periodic perturbations of the zonal term of $\overline{C}_{20}$ and $\overline{C}_{30}$ are $(ae^2, e, 1, e^2, e^2, 1)$ and $(ae^6, e, 1, e^2, e^2, 1)$, respectively. The order ratios of the long periodic perturbations of the tesseral term of D_{21} and D_{22} are $(ae^5, e^6, e, e, e, 1)$ and $(ae^4, e^5, 1, 1, 1, 1)$, respectively. This is important information for truncation in computing practice.

Short Periodic Perturbations

The order ratios of the short periodic perturbations of the zonal term of $\overline{C}_{20}$ and $\overline{C}_{30}$ are $(ae, e, e, e, e, 1)$ and $(ae, e, 1, e, e, 1)$, respectively. The order ratios of the short periodic perturbations of the tesseral term of D_{21} and D_{22} are (ae, e, 1, e, e, 1) and (ae, e, 1, e, e, 1), respectively.

Truncation of the Short Periodic Perturbations

Generally speaking, the amplitudes of the short periodic terms are of orders of $e^{|k-2|}$ or $e^{|k-3|}$, where k is the index of the series in (5.26), (6.20), (6.40) and (6.60). These facts are not allowed to be used directly for truncation of series. As seen in the formulas of the solution, the coefficients of potent of e may some times be very big. Truncation has to be done carefully by taking account of individual satellite characters.

Transformation of the Long Periodic Perturbations

As mentioned already, the terms with a factor of M in the solution have to be transformed back to functions of t, ω, Ω, or (ω and Ω). The transformation may be carried out by using relations of

$$\left(\frac{\mathrm{d}M}{\mathrm{d}t}\right)_0^{-1}\mathrm{d}M = \mathrm{d}t = \left(\frac{\mathrm{d}\omega}{dt}\right)_0^{-1}\mathrm{d}\omega = \left(\frac{d\Omega}{dt}\right)_0^{-1}d\Omega. \tag{6.80}$$

All functions of $\sin\omega$, $\cos\omega$, $\sin\Omega$, $\cos\Omega$ should be reduced to sine and cosine functions of $n\omega + m\Omega$.

Singularity of the Solutions

The short periodic perturbations of the zonal term of $\overline{C}_{20}$ are singular if $e = 0$. The long periodic perturbations of the zonal term of $\overline{C}_{30}$ are singular if $e = 0$. The short periodic perturbations of the zonal term of $\overline{C}_{30}$ are singular if $e = 0$ and/or $\sin i = 0$. The long and short periodic perturbations of the tesseral terms of D_{21} and D_{22} are singular if $e = 0$. The singularity problem will be discussed further in Chap. 10.

6.6 Solutions of Geopotential Perturbations of 6 × 6 Order and Degrees

Similarly, the solutions of the geopotential perturbations up to order and degrees of 6×6 can be derived. However, because of the length of the formulas, the solutions will not be given here. Interested readers may visit the homepage of the author for more information about the development.

6.7 Solutions of Geopotential Perturbations of l × m Order and Degrees

For geopotential perturbation function of $l \times m$ order and degree one has

$$R_{lm} = \frac{b_{lm}}{r^{l+1}} Q_{lm}(x)\, q(\Omega, u, i), \tag{6.81}$$

where

$$\frac{1}{r^N} = \frac{1}{a^N (1-e^2)^N} (1 + e\cos f)^N = \frac{1}{a^N (1-e^2)^N} \sum_{n=0}^{N} \binom{N}{n} e^n \cos^n f. \tag{6.82}$$

Then (6.13) can be written as

$$\begin{aligned}
\frac{\partial R_{lm}}{\partial a} &= \frac{-(l+1)}{a} \frac{b_{lm}}{r^{l+1}} Q_{lm}(x)\, q(\Omega, u, i), \\
\frac{\partial R_{lm}}{\partial \Omega} &= \frac{b_{lm}}{r^{l+1}} Q_{lm}(x) \frac{\partial q(\Omega, u, i)}{\partial \Omega}, \\
\frac{\partial R_{lm}}{\partial i} &= \frac{b_{lm}}{r^{l+1}} \frac{\partial Q_{lm}(x)}{\partial x} q(\Omega, u, i) \sin u \cos i + \frac{b_{lm}}{r^{l+1}} Q_{lm}(x) \frac{\partial q(\Omega, u, i)}{\partial i}, \\
\frac{\partial R_{lm}}{\partial \omega} &= \frac{b_{lm}}{r^{l+1}} \frac{\partial Q_{lm}(x)}{\partial x} q(\Omega, u, i) \cos u \sin i + \frac{b_{lm}}{r^{l+1}} Q_{lm}(x) \frac{\partial q(\Omega, u, i)}{\partial u}, \\
\frac{\partial R_{lm}}{\partial e} &= \frac{b_{lm}(-l-1)}{r^{l+2}} (-a\cos f)\, Q_{lm}(x)\, q(\Omega, u, i) \\
&\quad + \frac{b_{lm}}{r^{l+1}} \frac{\partial Q_{lm}(x)}{\partial x} \left(\cos u \sin i \frac{2 + e\cos f}{1-e^2} \sin f \right) q(\Omega, u, i) \\
&\quad + \frac{b_{lm}}{r^{l+1}} Q_{lm}(x) \frac{\partial q(\Omega, u, i)}{\partial u} \left(\frac{2 + e\cos f}{1-e^2} \sin f \right)
\end{aligned}$$

and

$$\begin{aligned}
\frac{\partial R_{lm}}{\partial M} &= \frac{b_{lm}(-l-1)}{r^{l+2}} \left(\frac{ae}{\sqrt{1-e^2}} \sin f \right) Q_{lm}(x)\, q(\Omega, u, i) \\
&\quad + \frac{b_{lm} a^2}{r^{l+3}} \frac{\partial Q_{lm}(x)}{\partial x} \left(\cos u \sin i \sqrt{1-e^2} \right) q(\Omega, u, i) \\
&\quad + \frac{b_{lm} a^2}{r^{l+3}} Q_{lm}(x) \frac{\partial q(\Omega, u, i)}{\partial u} \left(\cos u \sin i \sqrt{1-e^2} \right).
\end{aligned} \tag{6.83}$$

By substituting equations (6.83) into (4.11) the disturbed equations of motion can be obtained and they can be generally written as (for $j = 1, \ldots, 6$)

$$\frac{\mathrm{d}\sigma_j}{\mathrm{d}t} = F\left(\frac{1}{r^N}, Q(x), \frac{\partial Q(x)}{\partial x}, q(\Omega, u, i), \frac{\partial q(\Omega, u, i)}{\partial \Omega}, \frac{\partial q(\Omega, u, i)}{\partial u}, \frac{\partial q(\Omega, u, i)}{\partial i} \right), \tag{6.84}$$

where N is an integer. Notice

$$\begin{aligned}
\sin^N u &= (\sin f \cos\omega + \cos f \sin\omega)^N \\
&= \sum_{k1=0}^{N} \binom{N}{k1} \cos^{k1}\omega \sin^{k1} f \cos^{N-k1} f \sin^{N-k1}\omega, \\
\cos^N u &= (\cos f \cos\omega - \sin f \sin\omega)^N \\
&= \sum_{k1=0}^{N} (-1)^{N-k1} \binom{N}{k1} \cos^{k1} f \cos^{k1}\omega \sin^{N-k1} f \sin^{N-k1}\omega,
\end{aligned} \tag{6.85}$$

then (6.78) is the potential function of $\sin f$ and $\cos f$. Using transformation (5.22), disturbed equations of motion can be transformed into functions of M. Integrating the differential equations with respect to M, and transforming the M–related long periodic terms back into functions of (ω, Ω), the solutions of the disturbed equations of motion can be derived.

In principle, the solutions of the general geopotential perturbation of any $l \times m$ order and degree can be obtained.

Chapter 7
Solutions of Extraterrestrial Disturbances

Solutions of the extraterrestrial disturbances of the attracting forces of the sun, the moon, and planets, the drag force of the atmosphere, and solar radiation pressure are given in this chapter. For convenience, the ephemeris of the sun and the moon, as well as planets, are described.

7.1 Key Notes to the Problems

As mentioned in Chap. 1, the Keplerian motions of the satellite under the influence of the centre force of the Earth are perfect, exact and of mathematical beauty. As soon as it is found by derivation that the satellite is moving in an orbital plane, the equations of motion are re-represented in the plane and the Keplerian motion is then derived. Note that even in the centre force field, it would be nearly impossible to derive the solution without the step of coordinate transformation. This indicates the importance of the selection of the coordinate system. The transformation of the coordinate system is allowed because the frame remained an inertial one after a series of constant rotations.

The use of an alternative coordinate system is the first key to the solution of the equation of motion influenced by extraterrestrial disturbances. Xu (2004) introduced the so-called disturbance coordinate system by proposing an adjustment model of solar radiation (see Sect. 4.2.4). However, the coordinate system is not an orthogonal Cartesian one and its axis changes direction with time and therefore the coordinate system is not an inertial one. An approximation of the expression of the solar radiation model is the second key to the solution. The approximation allows the position of the satellite with respect to the Earth to be neglected in case of solar radiation under special conditions. For a properly selected time interval, the disturbance coordinate system may be considered a frame that has constant rotational relations with respect to the inertial one. In such a case, the coordinate system can be considered approximately "inertial". Then Newton's second law can be used and the orbital disturbance of the solar radiation can be solved. The approximation can be made as precise as required.

The orbits of the satellite can be considered a central motion (Keplerian motion) plus a series of disturbances. According to the order estimation discussed in

G. Xu, *Orbits*,

Sect. 4.2.7, extraterrestrial perturbations are of second order. These are important for the approximation measure taken during the derivation.

For convenience during later discussions, the definition of the so-called disturbance coordinate system is given again (see Sect. 4.2.4). The origin is the geocentre, and the three axes are defined by $\vec{r}$ (radial vector of the satellite), $\vec{n}$ (the sun-satellite identity vector) and $\vec{p}$ (the atmospheric drag identity vector). These three axes are always in the main disturbance directions of the indirect solar radiation (reflected from the Earth's surface), direct solar radiation and atmospheric drag, respectively.

7.2 Solutions of Disturbance of Solar Radiation Pressure

Solar radiation pressure is a force caused by sunlight acting on the satellite's surface. The radiation force can be represented as (see (4.70))

$$\vec{f}_{\text{solar}} = m\gamma P_s C_r \frac{S}{m} \frac{r_{\text{sun}}^2}{|\vec{r}-\vec{r}_{\text{sun}}|^2} \vec{n}_{\text{sun}}, \tag{7.1}$$

where

$$\vec{e}_z = -\frac{\vec{r}}{|\vec{r}|}, \quad \vec{e}_y = \frac{\vec{e}_z \times \vec{n}_{\text{sun}}}{|\vec{e}_z \times \vec{n}_{\text{sun}}|}, \quad \vec{e}_x = \vec{e}_y \times \vec{e}_z \quad \text{and} \quad \vec{n}_{\text{sun}} = \frac{\vec{r}-\vec{r}_{\text{sun}}}{|\vec{r}-\vec{r}_{\text{sun}}|}. \tag{7.2}$$

where the meanings of all symbols are the same as that of (4.63).

Three Approximations

The solar radiation force vector is pointed from the sun to the satellite. If the shadow factor is known exactly, and the luminosity of the sun and the surface reflectivity of the satellite are considered constants, then the length of the solar force vector can be considered a constant, because (see (4.73))

$$\frac{r_{\text{sun}}^2}{(r_{\text{sun}}+r)^2} \le \frac{r_{\text{sun}}^2}{|\vec{r}-\vec{r}_{\text{sun}}|^2} \le \frac{r_{\text{sun}}^2}{(r_{\text{sun}}-r)^2}. \tag{7.3}$$

For GPS and GEO satellites there are

$$\frac{r_{\text{sun}}^2}{(r_{\text{sun}}\pm r)^2} = \left(\frac{r_{\text{sun}}}{r_{\text{sun}}\pm r}\right)^2 \approx \left(1 \mp \frac{r}{r_{\text{sun}}} \pm \cdots\right)^2 \approx 1 \mp \frac{2r}{r_{\text{sun}}} \approx 1 \mp 3\times 10^{-5}$$

and

$$\frac{r_{\text{sun}}^2}{(r_{\text{sun}}\pm r)^2} = \left(\frac{r_{\text{sun}}}{r_{\text{sun}}\pm r}\right)^2 \approx \left(1 \mp \frac{r}{r_{\text{sun}}} \pm \cdots\right)^2 \approx 1 \mp \frac{2r}{r_{\text{sun}}} \approx 1 \mp 5\times 10^{-5}, \tag{7.4}$$

respectively. That is, the solar radiation force vector can be considered approximately a vector, with constant length and changing direction. The approximation has a precision of better than 3rd order and is precise enough for our purposes. For convenience, this approximation is called the first approximation later on.

The identity solar vector of the satellite $\vec{n}_{\text{sun}}$ can be approximated by

$$\vec{n}_{se} = \frac{\vec{r}_{se}}{|\vec{r}_{se}|}, \tag{7.5}$$

where index *se* denotes that the vector is pointing from the sun to the centre of the Earth. For GPS and GEO satellites the maximal angles between the above two identity vectors are 1.5×10^{-5} and 2.5×10^{-5} (rad), respectively. Therefore, such an approximation (called the second approximation) is allowed and is precise enough.

The third approximation is made for suitable time duration of $\Delta t = t'_k - t'_{k-1}$ by

$$\vec{n}_{se}(t) \approx \vec{n}_{se}(t_k), \quad t_k = (t'_k + t'_{k-1})/2, \quad t \in [t'_{k-1}, t'_k]. \tag{7.6}$$

The discrete vector in this equation may be called an average vector of the time duration Δt. For $\Delta t = 5\,\text{min}$, the third approximation has a precision of 3×10^{-5} (rad).

Note that the order of the solar radiation disturbance on a GPS satellite is about 50 m. For GPS satellite, all the three approximations will lead to a precision of millimetre level. For the other satellite, the precision of the approximations should be individually estimated.

Discretization and Solution

Denote the satellite period as T, and shadow access and exit points as t_{a} and t_{e}, respectively. The local noon is selected as the starting point of counting (see Fig. 7.1). A so-called sign function can be defined as

$$\delta(t) = \begin{cases} 1, & 0 \le t < T/2, \\ -1, & T/2 \le t \le T. \end{cases} \tag{7.7}$$

The sign function shows that the solar radiation accelerates the satellite during the first half period and decelerates it during the second half period with respect to the nominal motion of the satellite. Then the duration of one period of $0 \sim T$ can be equally divided by Δt, i.e., by $t'_0, t'_1, \ldots, t'_k, \ldots, T$. The acceleration of the solar radiation of (7.1) is then discretized as

$$\vec{a}_{\text{solar}}(t) = \gamma P_s C_r \frac{S}{m} \vec{n}_{\text{se}}(t_k). \tag{7.8}$$

The disturbed velocity caused by the solar radiation is then

$$\vec{v}_{\text{solar}}(t) = \sum_{i=1}^{k} \gamma P_s C_r \frac{S}{m} \vec{n}_{\text{se}}(t_i) \delta(t_i) \Delta t. \tag{7.9}$$

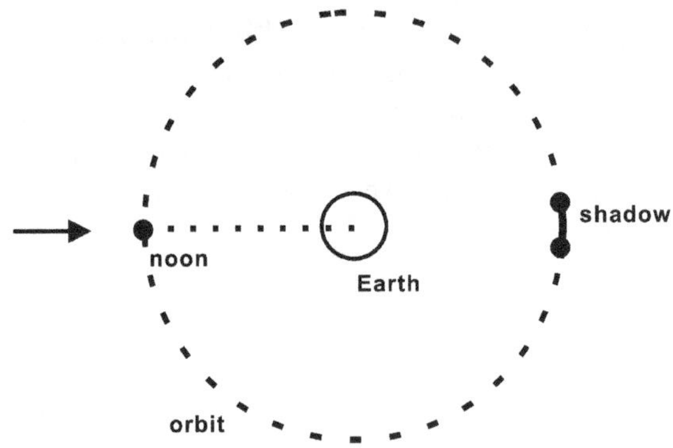

Fig. 7.1 Solar radiation pressure

It is obvious that, the disturbed velocity of the satellite is not zero during the passing of the shadow. The disturbed position caused by the solar radiation is then

$$\vec{\rho}_{\text{solar}}(t) = \sum_{j=1}^{k} \vec{v}_{\text{solar}}(t_j)\Delta t. \tag{7.10}$$

Equation (7.10) is the solution of the solar radiation disturbance on the orbit of the satellite.

Properties of the Solution

The integration (or summation) of the acceleration of the solar radiation within a period T is nearly zero. However, the position disturbed by the solar radiation during a period T is not zero. In other words, the disturbance of the solar radiation has the non–conservative behaviour. The disturbance may not be a periodic function of the orbit. The parameters of the force model, if they are not well known, can be determined using the expressions of the solution.

7.2.1 Solutions via Gaussian Perturbed Equations

Gaussian Perturbed Equations

Equation (7.8) is the approximated solar radiation force (acceleration) vector with constant length, which can be written as

$$\vec{f}_{\text{solar}}(t) = m\gamma P_s C_r \frac{S}{m} \vec{n}_{\text{se}}(t) \tag{7.11}$$

or

$$\vec{f}_{\text{solar}}(t) = \begin{pmatrix} f_x \\ f_y \\ f_z \end{pmatrix} = \xi \begin{pmatrix} n_x \\ n_y \\ n_z \end{pmatrix}, \tag{7.12}$$

where solar-Earth identity vector (7.5) in ECSF frame can be computed by the theory given in Sect. 7.8; ξ represents the constant length of the solar radiation force vector.

The force vector in the ECSF frame can be transformed to the orbital coordinate system (see (4.16)) using

$$\begin{pmatrix} f_r \\ f_\alpha \\ f_h \end{pmatrix} = R_3(f)R_3(\omega)R_1(i)R_3(\Omega) \begin{pmatrix} f_x \\ f_y \\ f_z \end{pmatrix}, \tag{7.13}$$

where

$$R_3(\omega)R_1(i)R_3(\Omega)$$
$$= \begin{pmatrix} \cos\omega\cos\Omega - \sin\omega\cos i\sin\Omega & \cos\omega\sin\Omega + \sin\omega\cos i\cos\Omega & \sin\omega\sin i \\ -\sin\omega\cos\Omega - \cos\omega\cos i\sin\Omega & -\sin\omega\sin\Omega + \cos\omega\cos i\cos\Omega & \cos\omega\sin i \\ \sin i\sin\Omega & -\sin i\cos\Omega & \cos i \end{pmatrix}.$$

Denote these elements of the matrix with R_{ij}, and

$$\begin{pmatrix} n_1 \\ n_2 \\ n_3 \end{pmatrix} = \xi \begin{pmatrix} R_{11}n_x + R_{12}n_y + R_{13}n_z \\ R_{21}n_x + R_{22}n_y + R_{23}n_z \\ R_{31}n_x + R_{32}n_y + R_{33}n_z \end{pmatrix}, \tag{7.14}$$

then one has

$$\begin{pmatrix} f_r \\ f_\alpha \\ f_h \end{pmatrix} = R_3(f) \begin{pmatrix} n_1 \\ n_2 \\ n_3 \end{pmatrix} = \begin{pmatrix} n_1\cos f + n_2\sin f \\ -n_1\sin f + n_2\cos f \\ n_3 \end{pmatrix}. \tag{7.15}$$

Using (4.23) one has

$$\frac{(1-e^2)}{1+e\cos f} = (1 - e\cos E) \tag{7.16}$$

and

$$\cos E = \frac{(e + \cos f)}{1 + e\cos f}. \tag{7.17}$$

Putting all these formulas into (4.26), the Gaussian disturbed equations are

$$\frac{\mathrm{d}a}{\mathrm{d}t} = \frac{2}{n\sqrt{1-e^2}} \begin{bmatrix} e\cos f(n_1\cos f + n_2\sin f) \\ +(1+e\cos f)(-n_1\sin f + n_2\cos f) \end{bmatrix},$$

$$\frac{de}{dt} = \frac{\sqrt{1-e^2}}{na}\left[\begin{array}{l}\sin f(n_1\cos f + n_2\sin f) \\ +\left(\dfrac{e+\cos f}{1+e\cos f} + \cos f\right)(-n_1\sin f + n_2\cos f)\end{array}\right],$$

$$\frac{di}{dt} = \frac{\sqrt{1-e^2}\cos u}{na(1+e\cos f)}n_3,$$

$$\frac{d\Omega}{dt} = \frac{\sqrt{1-e^2}\sin u}{na\sin i(1+e\cos f)}n_3, \tag{7.18}$$

$$\frac{d\omega}{dt} = \frac{\sqrt{1-e^2}}{nae}\left[\begin{array}{l}-\cos f(n_1\cos f + n_2\sin f) \\ +\dfrac{2+e\cos f}{1+e\cos f}\sin f(-n_1\sin f + n_2\cos f)\end{array}\right] - \cos i\frac{d\Omega}{dt},$$

$$\frac{dM}{dt} = -\frac{1-e^2}{nae}\left[\begin{array}{l}-\left(\cos f - \dfrac{2e}{1+e\cos f}\right)(n_1\cos f + n_2\sin f) \\ +\dfrac{2+e\cos f}{1+e\cos f}\sin f(-n_1\sin f + n_2\cos f)\end{array}\right].$$

Characters of Gaussian Perturbed Equations

1. There exist long and short periodic perturbations

Note that

$$\begin{aligned}\sin^2 f &= \frac{1-\cos 2f}{2}, \\ \cos^2 f &= \frac{1+\cos 2f}{2},\end{aligned} \tag{7.19}$$

$$\frac{1}{1+e\cos f} \approx 1 - e\cos f + \cdots \tag{7.20}$$

and

$$\begin{aligned}\cos u &= \cos\omega\cos f - \sin\omega\sin f, \\ \sin u &= \cos\omega\sin f + \cos f\sin\omega.\end{aligned} \tag{7.21}$$

Obviously, all the six Gaussian perturbed equations include the long periodic term perturbations, which are formed by terms without f (in other words, constant terms are created by terms of $\sin^2 f$ and $\cos^2 f$). And the remaining terms are short periodic terms. Remember that by integration variable transformation from t to f or M for solving the short periodic $\overline{C}_{20}$ perturbations, long periodic terms will also be created (see Sect. 5.2). Therefore, no effort will be made to separate the long and short periodic terms.

2. Concerning time variable (n_x, n_y and n_z)

In (7.18) the variables n_1, n_2 and n_3 are functions of (ω, Ω, i) and (n_x, n_y, n_z). (ω, Ω, i) are long periodic variables and they are considered constants in short periodic integrations. However, the identity vector (n_x, n_y, n_z) of solar-Earth is also time variable. In the discussion in Sect. 7.2 the (n_x, n_y, n_z) can be considered constants within 5 min. The maximum change of the identity vector around its average is ca. 0.5 degrees per day, that is, the maximum of change rate is about 0.0086 rad/day. In other words, the identity vector (n_x, n_y, n_z) can be represented by an average plus a drift term, and the drift term compared with the average term is about one order smaller and in some cases is allowed to be neglected. As soon as the vector (n_x, n_y, n_z) is considered constant, (7.18) can be solved by integration as shown in Chaps. 5 and 6.

In cases where change of the identity vector is not allowed to be neglected, the integration interval has to be made shorter so that the assumption will be valid and then the integrated solution should be summated to obtain the complete solutions.

Solutions of Gaussian Perturbed Equations

For simplifying the disturbed equations, denote

$$n_4 = \frac{n_3 e}{\sin i}, \quad n_5 = n_4 \cos i, \quad n_6 = \frac{n_5}{n_4}, \quad g_1 = \frac{2}{n\sqrt{1-e^2}}, \quad g_2 = \frac{\sqrt{1-e^2}}{na}, \tag{7.22}$$
$$g_3 = g_2, \quad g_4 = \frac{\sqrt{1-e^2}}{nae}, \quad g_5 = g_4, \quad g_6 = -\frac{1-e^2}{nae}.$$

Omitting the factors g_j ($j = 1,\ldots,6$) in the disturbing equations (of course, after the equations are solved, the factors shall be multiplied back), one has

$$\begin{aligned}
\frac{da}{dt} &= \begin{bmatrix} e\cos f(n_1\cos f + n_2\sin f) \\ +(1+e\cos f)(-n_1\sin f + n_2\cos f) \end{bmatrix}, \\
\frac{de}{dt} &= \begin{bmatrix} \sin f(n_1\cos f + n_2\sin f) \\ +\left(\dfrac{e+\cos f}{1+e\cos f} + \cos f\right)(-n_1\sin f + n_2\cos f) \end{bmatrix}, \\
\frac{di}{dt} &= \frac{\cos u}{1+e\cos f} n_3, \\
\frac{d\Omega}{dt} &= \frac{\sin u}{1+e\cos f} n_4, \\
\frac{d\omega}{dt} &= \begin{bmatrix} -\cos f(n_1\cos f + n_2\sin f) \\ +\dfrac{2+e\cos f}{1+e\cos f}\sin f(-n_1\sin f + n_2\cos f) \end{bmatrix} - \frac{d\Omega}{dt} n_6, \\
\frac{dM}{dt} &= \begin{bmatrix} -\left(\cos f - \dfrac{2e}{1+e\cos f}\right)(n_1\cos f + n_2\sin f) \\ +\dfrac{2+e\cos f}{1+e\cos f}\sin f(-n_1\sin f + n_2\cos f) \end{bmatrix}.
\end{aligned} \tag{7.23}$$

Let

$$\omega_1 = \omega + n_6\Omega, \quad M_1 = M - \omega_1, \quad \frac{1}{1+e\cos f} \approx 1 - e\cos f, \tag{7.24}$$

and equation (7.23) can be further simplified as

$$\begin{aligned}
\frac{da}{dt} &= \begin{bmatrix} e\cos f(n_1\cos f + n_2\sin f) \\ +(1+e\cos f)(-n_1\sin f + n_2\cos f) \end{bmatrix}, \\
\frac{de}{dt} &= \begin{bmatrix} \sin f(n_1\cos f + n_2\sin f) \\ +(e+(2-e^2)\cos f - e\cos^2 f)(-n_1\sin f + n_2\cos f) \end{bmatrix}, \\
\frac{di}{dt} &= \cos u(1-e\cos f)n_3, \\
\frac{d\Omega}{dt} &= \sin u(1-e\cos f)n_4, \\
\frac{d\omega_1}{dt} &= \begin{bmatrix} -\cos f(n_1\cos f + n_2\sin f) \\ +(2+e\cos f)(1-e\cos f)\sin f(-n_1\sin f + n_2\cos f) \end{bmatrix}, \\
\frac{dM_1}{dt} &= [-2e(1-e\cos f)(n_1\cos f + n_2\sin f)].
\end{aligned} \tag{7.25}$$

Simplified Gaussian perturbed equations (7.25) may be solved using symbolic computational software. The infinite integrations of the differential equations can be represented by

$$(\Delta\sigma_j(M))_M = b_j\left(d_j(\omega,\Omega)M + \sum_{k=1}^{16} A_{jk}\cos kM + \sum_{k=1}^{16} B_{jk}\sin kM\right), \tag{7.26}$$

where j is the index of Keplerian elements. b_j includes the omitted factors g_j and the factor caused by the variable transformation from t to M (see (5.24)) as well as the factors h_j given below:

$$\begin{aligned}
&h_1 = (1152\times 210)^{-1}, \quad h_2 = (55296\times 2310)^{-1}, \quad h_3 = h_1, \\
&h_4 = h_1, \quad h_5 = (2654208\times 60060)^{-1}, \quad h_6 = (576\times 210)^{-1},
\end{aligned} \tag{7.27}$$

h_j–factors are introduced to simplify the derivations of (7.26). The first term on the right-hand side of (7.26) is symbolic and represents the long periodic perturbation of

$$\int d_j(\omega,\Omega)\mathrm{d}M. \tag{7.28}$$

$\mathrm{d}M$ can be transformed to $d(n\omega + m\Omega)$ depending on the form of d_j according to (7.28).

For j= 1, there are

$$\begin{aligned}
d_j &= 120960e(n_1-n_2) + (90720e^3 - 16380e^5 + 430080e^7)(n_1+n_2), \\
A_{j1} &= 241920(n_1+e^2n_2) - 453600e^2n_1 + (201600e^4 - 5040e^6)(n_1-n_2),
\end{aligned} \tag{7.29}$$

$$
\begin{aligned}
A_{j2} &= (181440e - 383040e^3)n_1 + (-60480e + 241920e^3)n_2 \\
&\quad + (183645e^5 - 13440e^7)(n_1 - n_2), \\
A_{j3} &= 171360e^2 n_1 - 80640e^2 n_2 - (272160e^4 - 86520e^6)(n_1 - n_2), \\
A_{j4} &= 178920e^3 n_1 - 98280e^3 n_2 - (224280e^5 - 47040e^7)(n_1 - n_2), \\
A_{j5} &= (118944e^4 - 132552e^6)(n_1 - n_2), \\
A_{j6} &= (79275e^5 - 67200e^7)(n_1 - n_2), \\
A_{j7} &= 51840e^6(n_1 - n_2), \\
A_{j8} &= 26880e^7(n_1 - n_2),
\end{aligned} \tag{7.30}
$$

$$
\begin{aligned}
B_{j1} &= 241920(n_2 - e^2 n_1) - 514080e^2 n_2 + (221760e^4 + 362880e^6)(n_1 + n_2), \\
B_{j2} &= (60480e - 241920e^3)n_1 + (181440e - 403200e^3)n_2 \\
&\quad + (407295e^5 - 215040e^7)(n_1 + n_2), \\
B_{j3} &= 80640e^2 n_1 + 171360e^2 n_2 - 272160e^4(n_1 + n_2), \\
B_{j4} &= 178920e^3 n_2 + 98280e^3 n_1 - (318465e^5 - 53760e^7)(n_1 + n_2), \\
B_{j5} &= (118944e^4 - 145152e^6)(n_1 + n_2), \\
B_{j6} &= (79275e^5 - 71680e^7)(n_1 + n_2), \\
B_{j7} &= 51840e^6(n_1 + n_2), \\
B_{j8} &= 26880e^7(n_1 + n_2).
\end{aligned} \tag{7.31}
$$

For $j = 2$, there are

$$
\begin{aligned}
d_j &= (191600640 - 47900160e^2 - 68523840e^4 + 757254960e^6 \\
&\quad - 584836560e^8 + 74511360e^{10})n_2,
\end{aligned} \tag{7.32}
$$

$$
\begin{aligned}
A_{j1} &= (-31933440e + 66528000e^3 - 220374000e^5 \\
&\quad + 1193886540e^7 - 975633120e^9)n_1, \\
A_{j2} &= (31933440 - 63866880e^2 + 37089360e^4 + 153901440e^6 \\
&\quad + 162189720e^8 - 274760640e^{10})n_1, \\
A_{j3} &= (31933440e - 54552960e^3 + 93749040e^5 \\
&\quad - 75121200e^7 + 162660960e^9)n_1, \\
A_{j4} &= (27941760e^2 + 17297280e^4 - 51586920e^6 \\
&\quad - 3090780e^8 + 143700480e^{10})n_1, \\
A_{j5} &= (22087296e^3 + 41746320e^5 - 79767072e^7 + 46236960e^9)n_1, \\
A_{j6} &= (-20679120e^4 + 99792000e^6 - 52751160e^8 - 24664640e^{10})n_1, \\
A_{j7} &= (-41841360e^5 + 122551110e^7 - 79500960e^9)n_1,
\end{aligned}
$$

$$
\begin{aligned}
A_{j8} &= (-52855110e^6 + 72179415e^8 - 21732480e^{10})n_1, \\
A_{j9} &= (-55908930e^7 + 53813760e^9)n_1, \\
A_{j10} &= (-33197472e^8 + 19869696e^{10})n_1, \\
A_{j11} &= -17418240e^9 n_1, \\
A_{j12} &= -6307840e^{10} n_1, \qquad (7.33) \\
B_{j1} &= (-95800320e + 125072640e^3 + 525072240e^5 \\
&\quad - 262383660e^7 - 166209120e^9)n_2, \\
B_{j2} &= (31933440 - 95800320e^2 + 402660720e^4 - 534829680e^6 \\
&\quad + 295162560e^8 - 104670720e^{10})n_2, \\
B_{j3} &= (31933440e - 78503040e^3 - 146860560e^5 \\
&\quad + 88787160e^7 + 53037600e^9)n_2, \\
B_{j4} &= (27941760e^2 - 153180720e^4 + 15744960e^6 \\
&\quad + 19015920e^8 + 47900160e^{10})n_2, \\
B_{j5} &= (22087296e^3 + 41679792e^5 - 98855064e^7 + 35681184e^9)n_2, \\
B_{j6} &= (-20679120e^4 + 112044240e^6 - 80905440e^8 + 197120e^{10})n_2, \\
B_{j7} &= (-41841360e^5 + 141925410e^7 - 70598880e^9)n_2, \\
B_{j8} &= (-52855110e^6 + 91981890e^8 - 23802240e^{10})n_2, \\
B_{j9} &= (-55908930e^7 + 57164800e^9)n_2, \\
B_{j10} &= (-33197472e^8 + 20815872e^{10})n_2, \\
B_{j11} &= -17418240e^9 n_2, \\
B_{j12} &= -6307840e^{10} n_2. \qquad (7.34)
\end{aligned}
$$

For $j = 3$, there are

$$
d_j = (-362880e - 90720e^3 + 16380e^5 - 430080e^7)n_3 \cos\omega, \qquad (7.35)
$$

$$
\begin{aligned}
A_{j1} &= (241920 + 30240e^2 - 201600e^4 + 5040e^6)n_3 \sin\omega, \\
A_{j2} &= (60480e + 100800e^3 - 183645e^5 + 13440e^7)n_3 \sin\omega, \\
A_{j3} &= (10080e^2 + 272160e^4 - 86520e^6)n_3 \sin\omega, \\
A_{j4} &= (-17640e^3 n_1 + 224280e^5 - 47040e^7)n_3 \sin\omega, \\
A_{j5} &= (-118944e^4 + 132552e^6)n_3 \sin\omega, \\
A_{j6} &= (-79275e^5 + 67200e^7)n_3 \sin\omega, \\
A_{j7} &= -51840e^6 n_3 \sin\omega, \\
A_{j8} &= -26880e^7 n_3 \sin\omega, \qquad (7.36) \\
B_{j1} &= (241920 - 30240e^2 - 221760e^4 - 362880e^6)n_3 \cos\omega,
\end{aligned}
$$

$$
\begin{aligned}
B_{j2} &= (60480e + 80640e^3 - 407295e^5 + 215040e^7)n_3 \cos\omega, \\
B_{j3} &= (10080e^2 + 272160e^4)n_3 \cos\omega, \\
B_{j4} &= (-17640e^3 + 318465e^5 - 53760e^7)n_3 \cos\omega, \\
B_{j5} &= (-118944e^4 + 145152e^6)n_3 \cos\omega, \\
B_{j6} &= (-79275e^5 + 71680e^7)n_3 \cos\omega, \\
B_{j7} &= -51840e^6 n_3 \cos\omega, \\
B_{j8} &= -26880e^7 n_3 \cos\omega.
\end{aligned} \tag{7.37}
$$

For $j = 4$, there are

$$
d_j = (-362880e - 90720e^3 + 16380e^5 - 430080e^7)n_4 \sin\omega, \tag{7.38}
$$

$$
\begin{aligned}
A_{j1} &= -(241920 + 30240e^2 - 201600e^4 + 5040e^6)n_4 \cos\omega, \\
A_{j2} &= -(60480e + 100800e^3 - 183645e^5 + 13440e^7)n_4 \cos\omega, \\
A_{j3} &= -(100800e^2 + 272160e^4 - 86520e^6)n_4 \cos\omega, \\
A_{j4} &= -(-17640e^3 + 224280e^5 - 47040e^7)n_4 \cos\omega, \\
A_{j5} &= -(-118944e^4 + 132552e^6)n_4 \cos\omega, \\
A_{j6} &= -(-79275e^5 + 67200e^7)n_4 \cos\omega, \\
A_{j7} &= 51840e^6 n_4 \cos\omega, \\
A_{j8} &= 26880e^7 n_4 \cos\omega,
\end{aligned} \tag{7.39}
$$

$$
\begin{aligned}
B_{j1} &= (241920 - 30240e^2 - 221760e^4 - 362880e^6)n_4 \sin\omega, \\
B_{j2} &= (60480e + 80640e^3 - 407295e^5 + 215040e^7)n_4 \sin\omega, \\
B_{j3} &= (10080e^2 + 272160e^4)n_4 \sin\omega, \\
B_{j4} &= (-17640e^3 n_2 + 318465e^5 - 53760e^7)n_4 \sin\omega, \\
B_{j5} &= (-118944e^4 + 145152e^6)n_4 \sin\omega, \\
B_{j6} &= (-79275e^5 + 71680e^7)n_4 \sin\omega, \\
B_{j7} &= -51840e^6 n_4 \sin\omega, \\
B_{j8} &= -26880e^7 n_4 \sin\omega.
\end{aligned} \tag{7.40}
$$

For $j = 5$, there are

$$
\begin{aligned}
d_j = (&-239117598720 + 39852933120e^2 + 78875596800e^4 - 687047961600e^6 \\
&- 280700259840e^8 + 264212708760e^{10} - 139810130560e^{12} \\
&+ 333583810560e^{14})n_1,
\end{aligned} \tag{7.41}
$$

$$
\begin{aligned}
A_{j1} = (&199264665600e - 142806343680e^3 - 148410662400e^5 + 1111921050240e^7 \\
&- 2852805795840e^9 + 2015768795040e^{11} - 261258117120e^{13})n_2,
\end{aligned}
$$

$$
\begin{aligned}
A_{j2} = {} & (-39852933120+139485265920e^2-4774049280e^4-48639951360e^6\\
& +196151155200e^8-1068285418200e^{10}+777063006720e^{12}\\
& -74785751040e^{14})n_2,\\
A_{j3} = {} & (-39852933120e+74724249600e^3+293431057920e^5-414339045120e^7\\
& +497781123840e^9-818554857120e^{11}+298896998400e^{13})n_2,\\
A_{j4} = {} & (-29889699840e^2-80536135680e^4+493525992960e^6-455895760320e^8\\
& +447162390675e^{10}-560126286720e^{12}+116114718720e^{14})n_2,\\
A_{j5} = {} & (-11623772160e^3-310313203200e^5+737853532416e^7\\
& -502012038528e^9+33185797448e^{11}-262881755136e^{13})n_2,\\
A_{j6} = {} & (60678858240e^4-499891392000e^6+844928884800e^8\\
& -47343700700e^{10}+242311829760e^{12}-118082764800e^{14})n_2,\\
A_{j7} = {} & (117216253440e^5-634880600640e^7+820172770560e^9\\
& -340771345200e^{11}+118188195840e^{13})n_2,\\
A_{j8} = {} & (160847406720e^6-641571250320e^8\\
& +670415635890e^{10}-204615290880e^{12}+69865635840e^{14})n_2,\\
A_{j9} = {} & (188641252800e^7-570422012160e^9\\
& +462878816400e^{11}-72571699200e^{13})n_2,\\
A_{j10} = {} & (171528477120e^8-433911602124e^{10}\\
& +261564855552e^{12}-36212047872e^{14})n_2,\\
A_{j11} = {} & (139990032000e^9-284139732240e^{11}+96546078720e^{13})n_2,\\
A_{j12} = {} & (101723306685e^{10}-155150835840e^{12}+29520691200e^{14})n_2,\\
A_{j13} = {} & (62960325120e^{11}-62958551040e^{13})n_2,\\
A_{j14} = {} & (32709980160e^{12}-17431265280e^{14})n_2,\\
A_{j15} = {} & 14169931776e^{13}n_2,\\
A_{j16} = {} & 3936092160e^{14}n_2,
\end{aligned}
\tag{7.42}
$$

$$
\begin{aligned}
B_{j1} = {} & (-199558799360e-3321077760e^3-437344427520e^5-22745923200e^7\\
& -153357684480e^9+129582573120e^{11}+316609413120e^{13})n_1,\\
B_{j2} = {} & (39852933120-79705866240e^2-508747599360e^4+588937789440e^6\\
& +79757758080e^8-195407372160e^{10}+317154277440e^{12}\\
& -228293345280e^{14})n_1,\\
B_{j3} = {} & (39852933120e-18265927680e^3-163355512320e^5+320034274560e^7\\
& -38665186560e^9+41147346240e^{11}-147234447360e^{13})n_1,
\end{aligned}
$$

$$
\begin{aligned}
B_{j4} &= (29889699840e^2 + 325673187840e^4 - 551575664640e^6 + 398736898560e^8 \\
&\quad - 18850656825e^{10} - 166889923200e^{12} + 34686812160e^{14})n_1, \\
B_{j5} &= (11623772160e^3 + 349584947712e^5 - 699910219008e^7 \\
&\quad + 442322584704e^9 - 155644673184e^{11} + 30517014528e^{13})n_1, \\
B_{j6} &= (-60678858240e^4 + 500375715840e^6 - 756216901440e^8 \\
&\quad + 340157737920e^{10} - 24104240160e^{12} - 328007680e^{14})n_1, \\
B_{j7} &= (-117216253440e^5 + 608566495680e^7 - 735828762240e^9 \\
&\quad + 300759544800e^{11} - 27148492800e^{13})n_1, \\
B_{j8} &= (-160847406720e^6 + 597352916160e^8 \\
&\quad - 579512888955e^{10} + 171441270000e^{12} - 9717227520e^{14})n_1, \\
B_{j9} &= (-188641252800e^7 + 527603556480e^9 \\
&\quad - 404266742880e^{11} + 72694702080e^{13})n_1, \\
B_{j10} &= (-171528477120e^8 + 401784102720e^{10} \\
&\quad - 229893247584e^{12} + 22238920704e^{14})n_1, \\
B_{j11} &= (-139990032000e^9 + 261524259360e^{11} - 96898314240e^{13})n_1, \\
B_{j12} &= (-101723306685e^{10} + 141910728960e^{12} - 27634647040e^{14})n_1, \\
B_{j13} &= (-62960325120e^{11} + 60517416960e^{13})n_1, \\
B_{j14} &= (-32709980160e^{12} + 16868966400e^{14})n_1, \\
B_{j15} &= -14169931776e^{13}n_1, \\
B_{j16} &= -3936092160e^{14}n_1.
\end{aligned} \tag{7.43}
$$

For $j = 6$, there are

$$
d_j = (362880e^2 + 90720e^4 - 16380e^6 + 430080e^8)n_1, \tag{7.44}
$$

$$
\begin{aligned}
A_{j1} &= (241920e + 30240e^3 - 201600e^5 + 5040e^7)n_2, \\
A_{j2} &= (60480e^2 + 100800e^4 - 183645e^6 + 13440e^8)n_2, \\
A_{j3} &= (10080e^3 + 272160e^5 - 86520e^7)n_2, \\
A_{j4} &= (-17640e^4 + 224280e^6 - 47040e^8)n_2, \\
A_{j5} &= (-118944e^5 + 132552e^7)n_2, \\
A_{j6} &= (-79275e^6 + 67200e^8)n_2, \\
A_{j7} &= -51840e^7 n_2, \\
A_{j8} &= -26880e^8 n_2, \\
B_{j1} &= (-241920e + 30240e^3 + 221760e^5 + 362880e^7)n_1, \\
B_{j2} &= (-60480e^2 - 80640e^4 + 407295e^6 - 215040e^8)n_1,
\end{aligned} \tag{7.45}
$$

$$
\begin{aligned}
B_{j3} &= (-10080e^3 - 272160e^5)n_1, \\
B_{j4} &= (17640e^4 - 318465e^6 + 53760e^8)n_1, \\
B_{j5} &= (118944e^5 - 145152e^7)n_1, \\
B_{j6} &= (79275e^6 - 71680e^8)n_1, \\
B_{j7} &= 51840e^7 n_1, \\
B_{j8} &= 26880e^8 n_1.
\end{aligned}
\tag{7.46}
$$

Properties of the Solution

Disturbances of the solar radiation consist of both the long periodic and short periodic terms. The orientation of the orbital ellipse is subjected to higher frequency disturbance than that of the other Keplerian elements.

7.3 Solutions of Disturbance of Atmospheric Drag

Atmospheric drag, caused by the air, is the disturbance force acting on the satellite's surface. Air drag force can be represented as (see (4.75))

$$
\vec{f}_{\text{drag}} = -m\frac{1}{2}\left(\frac{C_{\text{d}}S}{m}\right)\sigma\left|\dot{\vec{r}} - \dot{\vec{r}}_{\text{air}}\right|^2 \vec{n}_a, \quad \vec{n}_a = \frac{\dot{\vec{r}} - \dot{\vec{r}}_{\text{air}}}{\left|\dot{\vec{r}} - \dot{\vec{r}}_{\text{air}}\right|}, \tag{7.47}
$$

where the meanings of the symbols are the same as given in (4.75); and identity vector $\vec{n}_a$ is the direction of the air drag force. For CHAMP satellite, with an orbit height of 400 km, the air drag force identity vector $\vec{n}_a$ changes its direction about 1.2×10^{-3} (rad) per second. The amount of $\left|\dot{\vec{r}} - \dot{\vec{r}}_{\text{air}}\right|^2$ changes slower than the direction. In such a case the acceleration of the air drag can be discretized by

$$
\vec{a}_{\text{drag}} = -\frac{1}{2}\left(\frac{C_{\text{d}}S}{m}\right)\sigma\left|\dot{\vec{r}}(t_k) - \dot{\vec{r}}_{\text{air}}(t_k)\right|^2 \vec{n}_a(t_k).
$$

The disturbed velocity caused by the atmospheric drag is then

$$
\vec{v}_{\text{airdrag}}(t) = \sum_{i=1}^{k} -\frac{C_d S}{2m}\sigma\left|\dot{\vec{r}}(t_i) - \dot{\vec{r}}_{\text{air}}(t_i)\right|^2 \vec{n}_{\text{a}}(t_i)\Delta t. \tag{7.48}
$$

The disturbed position caused by the solar radiation is then

$$
\vec{\rho}_{\text{air}}(t) = \sum_{j=1}^{k} \vec{v}_{\text{airdrag}}(t_j)\Delta t. \tag{7.49}
$$

Equation (7.49) is the solution of the solar radiation disturbance on the orbit of the satellite.

For all satellites, with an orbit height higher than 1,000 km, the atmospheric drag is nearly zero; therefore this effect does not need to be taken into account.

7.3.1 Solutions via Gaussian Perturbed Equations

Air Drag Force Vector for Gaussian Perturbed Equations

Air drag force is given in (7.47) (using ξ to represent the coefficient part of the air drag force vector)

$$\vec{f}_{\text{drag}} = \xi \left|\dot{\vec{r}} - \dot{\vec{r}}_{\text{air}}\right| (\dot{\vec{r}} - \dot{\vec{r}}_{\text{air}}). \tag{7.50}$$

Using (4.16) the air drag force vector can be rotated from the ECSF to the orbital coordinate frame by

$$\begin{pmatrix} f_r \\ f_\alpha \\ f_h \end{pmatrix} = R_3(f)R_3(\omega)R_1(i)R_3(\Omega) \begin{pmatrix} f_x \\ f_y \\ f_z \end{pmatrix}. \tag{7.51}$$

Satellite position and velocity vectors in orbital frame are given in (3.41) and (3.42)

$$\vec{q} = \begin{pmatrix} a(\cos E - e) \\ a\sqrt{1-e^2}\sin E \\ 0 \end{pmatrix} = \begin{pmatrix} r\cos f \\ r\sin f \\ 0 \end{pmatrix}, \tag{7.52}$$

$$\dot{\vec{q}} = \begin{pmatrix} -\sin E \\ \sqrt{1-e^2}\cos E \\ 0 \end{pmatrix} \frac{na}{1 - e\cos E} = \begin{pmatrix} -\sin f \\ e + \cos f \\ 0 \end{pmatrix} \frac{na}{\sqrt{1-e^2}}. \tag{7.53}$$

They can be rotated from the orbital frame to the ECSF frame using (3.43)

$$\begin{pmatrix} \vec{r} \\ \dot{\vec{r}} \end{pmatrix} = R_3(-\Omega)R_1(-i)R_3(-\omega) \begin{pmatrix} \vec{q} \\ \dot{\vec{q}} \end{pmatrix}. \tag{7.54}$$

Air velocity in the ECSF frame is given in (4.76)

$$\dot{\vec{r}}_{\text{air}} = k\vec{\omega}_e \times \vec{r} = k\omega_e \begin{pmatrix} -y \\ x \\ 0 \end{pmatrix} = k\omega_e \begin{pmatrix} 0 & -1 & 0 \\ 1 & 0 & 0 \\ 0 & 0 & 0 \end{pmatrix} \begin{pmatrix} x \\ y \\ z \end{pmatrix} = k\omega_e R_4 \vec{r}, \tag{7.55}$$

where ω_e is the angle velocity of the Earth's rotation. Thus in the ECSF frame there is

$$\dot{\vec{r}} - \dot{\vec{r}}_{air} = R_3(-\Omega)R_1(-i)R_3(-\omega)\dot{\vec{q}} - k\omega_e R_4 R_3(-\Omega)R_1(-i)R_3(-\omega)\vec{q}. \tag{7.56}$$

Denote following matrix as R

$$R_3(\omega)R_1(i)R_3(\Omega)R_4R_3(-\Omega)R_1(-i)R_3(-\omega)=R_3(\omega)R_1(i)R_4R_1(-i)R_3(-\omega)$$
$$=R_3(\omega)\begin{pmatrix}0 & -\cos i & \sin i\\ \cos i & 0 & 0\\ -\sin i & 0 & 0\end{pmatrix}R_3(-\omega)=\begin{pmatrix}0 & -\cos i & \sin i\cos\omega\\ \cos i & 0 & -\sin i\sin\omega\\ -\sin i\cos\omega & \sin i\sin\omega & 0\end{pmatrix}, \tag{7.57}$$

and note that the length of a vector is invariable under rotational transformations, one has

$$\begin{pmatrix}f_r\\ f_\alpha\\ f_h\end{pmatrix}=\xi\left|\dot{\vec{r}}-\dot{\vec{r}}_{air}\right|R_3(f)\left(\dot{\vec{q}}-k\omega_eR\vec{q}\right)=\xi\left|\dot{\vec{q}}-k\omega_eR\vec{q}\right|R_3(f)\left(\dot{\vec{q}}-k\omega_eR\vec{q}\right). \tag{7.58}$$

The force vector (7.58) is represented completely in Keplerian elements.

Gaussian Perturbed Equations and the Solutions

The air drag force vector (7.58) has to be further simplified. Denote the elements of the matrix R with R_{ij}, then one has approximately

$$\begin{aligned}\dot{\vec{q}}-k\omega_eR\vec{q}&=\frac{na}{\sqrt{1-e^2}}\begin{pmatrix}-\sin f\\ e+\cos f\\ 0\end{pmatrix}-k\omega R(1-e^2)\begin{pmatrix}\cos f\\ \sin f\\ 0\end{pmatrix}(1-e\cos f)\\ &=\begin{pmatrix}b_{11}\sin f+b_{13}\sin f\cos f\\ b_{22}\cos f+b_{24}\cos^2 f+b_{25}\\ b_{31}\sin f+b_{32}\cos f+b_{33}\sin f\cos f+b_{34}\cos^2 f\end{pmatrix},\end{aligned} \tag{7.59}$$

where coefficients b_{ij} can be obtained by comparison.

For convenience, the simplified Gaussian disturbed equations of motion can be written as shown below (see (7.25), (7.24) and (7.22))

$$\begin{aligned}\frac{\mathrm{d}a}{\mathrm{d}t}&=\left[e\cos f\cdot f_r+(1+e\cos f)\cdot f_\alpha\right],\\ \frac{\mathrm{d}e}{\mathrm{d}t}&=\left[\sin f\cdot f_r+(e\sin^2 f+(2-e^2)\cos f)\cdot f_\alpha\right],\\ \frac{\mathrm{d}i}{\mathrm{d}t}&=\cos u(1-e\cos f)\cdot f_h,\\ \frac{\mathrm{d}\Omega}{\mathrm{d}t}&=\sin u(1-e\cos f)\frac{e}{\sin i}\cdot f_h,\\ \frac{\mathrm{d}\omega_1}{\mathrm{d}t}&=\left[-\cos f\cdot f_r+(2+e\cos f)(1-e\cos f)\sin f\cdot f_\alpha\right],\\ \frac{\mathrm{d}M_1}{\mathrm{d}t}&=\left[-2e(1-e\cos f)\cdot f_r\right].\end{aligned} \tag{7.60}$$

Putting the air drag force vector and other mathematical relations into the simplified Gaussian disturbed equations (7.60), theoretically, the equations could be solved. It is obvious that numerical solutions may be computed; however, there are still problems that have to be solved by deriving the analytic solutions.

7.4 Solutions of Disturbance of the Sun

The solutions of the disturbance of the sun (see Fig. 7.2) may be similarly derived by the discretization demonstrated in Sect. 7.2. However, analytic solutions are preferred in theoretical and practical aspects.

Potential Function of the Sun

The disturbance forces of multiple point-masses acting on the satellite are (see (4.50))

$$\vec{f}_{\text{mul}} = -m\sum_j Gm(j)\left[\frac{\vec{r}-\vec{r}_{m(j)}}{\left|\vec{r}-\vec{r}_{m(j)}\right|^3} + \frac{\vec{r}_{m(j)}}{r^3_{m(j)}}\right], \qquad (7.61)$$

where $Gm(j)$ are the gravitational constants of the sun and the moon as well as the planets. The disturbance acceleration of the sun is then

$$\vec{f}_s = -m\mu_s\left(\frac{1}{|\vec{r}-\vec{r}_s|^2}\vec{n}_{ss} + \frac{1}{|\vec{r}_s|^2}\vec{n}_s\right), \quad \vec{n}_{ss} = \frac{\vec{r}-\vec{r}_s}{|\vec{r}-\vec{r}_s|}, \quad \vec{n}_s = \frac{\vec{r}_s}{|\vec{r}_s|}. \qquad (7.62)$$

The identity vectors $\vec{n}_{ss}$ and $\vec{n}_s$ represent the vectors from the satellite to the sun and the geocentric vector of the sun, respectively. The maximum difference between the two identity vectors is about 1.7×10^{-4} (rad) for the GPS satellite (except the sign difference). For CHAMP satellite the difference is about 4.5×10^{-5} (rad). That is, for most satellites the difference between the two identity vectors can be neglected. Then (7.62) can be approximated by

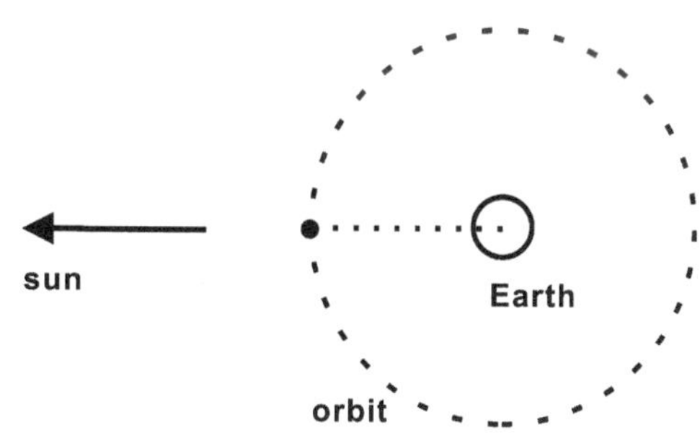

Fig. 7.2 Attracting disturbance force of the sun

$$\vec{f}_s = -m\mu_s \left(\frac{-1}{|\vec{r}-\vec{r}_s|^2} + \frac{1}{|\vec{r}_s|^2} \right) \vec{n}_s. \tag{7.63}$$

Because (see (7.3))

$$\frac{1}{(r_s+r)^2} \le \frac{1}{|\vec{r}-\vec{r}_s|^2} \le \frac{1}{(r_s-r)^2}, \tag{7.64}$$

there are (for GPS or GEO satellite)

$$\frac{1}{(r_s \pm r)^2} = \frac{1}{r_s^2}\left(1 \mp \frac{2r}{r_s} \pm \frac{3r^2}{r_s^2} \mp \cdots \right) \approx \frac{1}{r_s^2}\left(1 \mp \frac{2r}{r_s} \pm 2.4 \times 10^{-7} \cdots \right). \tag{7.65}$$

Then (7.63) turns out to be

$$\vec{f}_s = m\mu_s \frac{2r}{r_s^3} \vec{n}_s. \tag{7.66}$$

Neglecting the change of the geocentric distance of the satellite r, the potential function of the disturbing force of the sun is

$$V_s = -m\mu_s \frac{r}{r_s^2}. \tag{7.67}$$

Disturbed Equation of Motion and the Solutions

Note that the potential function is the only function of the three Keplerian elements $(a,\ M,\ e)$. The derivatives of the potential function with respect to Keplerian elements are then

$$\frac{\partial V_s}{\partial a} = \frac{\partial V_s}{\partial r}\frac{\partial r}{\partial a} = \frac{1}{a}V_s\,, \qquad \frac{\partial V_s}{\partial \Omega} = \frac{\partial V_s}{\partial i} = \frac{\partial V_s}{\partial \omega} = 0,$$

$$\frac{\partial V_s}{\partial e} = \frac{V_s}{r}\frac{\partial r}{\partial e} = \frac{-a\cos f}{r}V_s$$

and

$$\frac{\partial V_s}{\partial M} = \frac{V_s}{r}\frac{\partial r}{\partial M} = \frac{ae\sin f}{r\sqrt{1-e^2}}V_s. \tag{7.68}$$

Substituting the above derivatives and V_s into the equation of motion (4.11), one has

$$\frac{\mathrm{d}a}{\mathrm{d}t} = \frac{-2m\mu_s}{nr_s^2}\frac{e\sin f}{\sqrt{1-e^2}},$$

$$\frac{\mathrm{d}e}{\mathrm{d}t} = \frac{-m\mu_s\sqrt{1-e^2}}{na}\frac{\sin f}{r_s^2},$$

$$\frac{\mathrm{d}\omega}{\mathrm{d}t} = \frac{m\mu_s\sqrt{1-e^2}}{nae}\frac{\cos f}{r_s^2},$$

$$
\begin{aligned}
&\frac{\mathrm{d}i}{\mathrm{d}t}=0, \quad \frac{\mathrm{d}\Omega}{\mathrm{d}t}=0, \\
&\frac{\mathrm{d}M}{\mathrm{d}t}=\frac{2}{na}\frac{m\mu_s}{r_s^2}\frac{(1-e^2)}{1+e\cos f}-\frac{1-e^2}{nae}\frac{m\mu_s\cos f}{r_s^2} \\
&\quad=\frac{m\mu_s(1-e^2)}{naer_s^2}\left(2e\frac{1}{1+e\cos f}-\cos f\right) \\
&\quad=\frac{m\mu_s(1-e^2)}{naer_s^2}\left(2e-(1+2e^2)\cos f\right).
\end{aligned} \tag{7.69}
$$

According to (5.22) there are

$$
\begin{aligned}
\sin f &= \left(1-\frac{7}{8}e^2\right)\sin M+e\left(1-\frac{7}{6}e^2\right)\sin 2M+\frac{9}{8}e^2\sin 3M+\frac{4}{3}e^3\sin 4M, \\
\cos f+e &= \left(1-\frac{9}{8}e^2\right)\cos M+e\left(1-\frac{4}{3}e^2\right)\cos 2M+\frac{9}{8}e^2\cos 3M+\frac{4}{3}e^3\cos 4M.
\end{aligned} \tag{7.70}
$$

Denote

$$
\begin{aligned}
\delta S &= \int \sin f\mathrm{d}t=\int \sin f\left(\frac{\mathrm{d}M}{\mathrm{d}t}\right)_0^{-1}\mathrm{d}M=\left(\frac{\mathrm{d}M}{\mathrm{d}t}\right)_0^{-1} \\
&\times\left(-\left(1-\frac{7}{8}e^2\right)\cos M-\frac{e}{2}\left(1-\frac{7}{6}e^2\right)\cos 2M-\frac{3}{8}e^2\cos 3M-\frac{1}{3}e^3\cos 4M\right), \\
\delta C &= \int (\cos f+e)\mathrm{d}t=\int (\cos f+e)\left(\frac{\mathrm{d}M}{\mathrm{d}t}\right)_0^{-1}\mathrm{d}M=\left(\frac{\mathrm{d}M}{\mathrm{d}t}\right)_0^{-1} \\
&\times\left(\left(1-\frac{9}{8}e^2\right)\sin M+\frac{e}{2}\left(1-\frac{4}{3}e^2\right)\sin 2M+\frac{3}{8}e^2\sin 3M+\frac{1}{3}e^3\sin 4M\right).
\end{aligned} \tag{7.71}
$$

Then the solutions are

$$
\begin{aligned}
\Delta a &= \frac{-2m\mu_s}{nr_s^2}\frac{e}{\sqrt{1-e^2}}\delta S, \\
\Delta e &= \frac{-m\mu_s\sqrt{1-e^2}}{nar_s^2}\delta S, \\
\Delta\omega &= \frac{m\mu_s\sqrt{1-e^2}}{naer_s^2}(-et+\delta C), \\
\Delta i &= \text{const.}, \quad \Delta\Omega=\text{const.}, \\
\Delta M &= \frac{m\mu_s(1-e^2)}{naer_s^2}\left(e(3-2e^2)t-(1+2e^2)\delta C\right).
\end{aligned} \tag{7.72}
$$

The orbital parameters (i, Ω) are not affected by the sun perturbation. Ω and M are partly linearly perturbed by the sun. The remaining terms are all short periodic ones. The solutions are called potential function approximated solutions.

7.4.1 Solutions via Gaussian Perturbed Equations

Equation (7.66) is the disturbing force of the sun and can be simplified by

$$\vec{f}_s = m\mu_s \frac{2r}{r_s^3}\vec{n}_s = \xi\, r \begin{pmatrix} n_x \\ n_y \\ n_z \end{pmatrix}. \tag{7.73}$$

Using (7.51), one has

$$\begin{pmatrix} f_r \\ f_\alpha \\ f_h \end{pmatrix} = R_3(f) r \xi R_3(\omega) R_1(i) R_3(\Omega) \begin{pmatrix} n_x \\ n_y \\ n_z \end{pmatrix} = R_3(f) \frac{1}{1+e\cos f} \begin{pmatrix} n_1 \\ n_2 \\ n_3 \end{pmatrix} \tag{7.74}$$

or approximately

$$\begin{pmatrix} f_r \\ f_\alpha \\ f_h \end{pmatrix} = (1 - e\cos f) \begin{pmatrix} n_1\cos f + n_2 \sin f \\ -n_1 \sin f + n_2 \cos f \\ n_3 \end{pmatrix}, \tag{7.75}$$

where coefficients n_1, n_2, n_3 can be obtained by comparing with (7.74) and it is notable that they are different from those in (7.15). Comparing (7.75) with (7.15) one notices that the solutions of the simplified Gaussian perturbed equations can be derived in principle without great problem; however, they are rather complicated because of the factor $(1 - e\cos f)$. Then the simplified Gaussian disturbed equations (7.60) can be used to derive the solution.

7.5 Solutions of Disturbance of the Moon

The disturbance acceleration of the moon is (see (7.61))

$$\vec{f}_m = -m\,\mu_m \left(\frac{1}{|\vec{r}-\vec{r}_m|^2}\vec{n}_{sm} + \frac{1}{|\vec{r}_m|^2}\vec{n}_m \right), \quad \vec{n}_{sm} = \frac{\vec{r}-\vec{r}_m}{|\vec{r}-\vec{r}_m|}, \quad \vec{n}_m = \frac{\vec{r}_m}{|\vec{r}_m|}. \tag{7.76}$$

The identity vectors $\vec{n}_{sm}$ and $\vec{n}_m$ represent the vectors from the satellite to the moon and the geocentric vector of the moon, respectively. The maximum difference between the two identity vectors is about 7×10^{-2} (rad) for the GPS satellite (except the sign difference). For CHAMP satellite the difference is about 1.7×10^{-2} (rad).

It is obvious that for precise purpose the difference between the two identity vectors should be taken into account and the solutions can be derived via discretezation as in Sect. 7.4. Suppose the difference of the two identity vectors can be neglected. Then (7.76) can be approximated by

$$\vec{f}_m = -m\,\mu_m \left(\frac{-1}{|\vec{r}-\vec{r}_m|^2} + \frac{1}{|\vec{r}_m|^2} \right) \vec{n}_m. \tag{7.77}$$

Because (see (7.3))

$$\frac{1}{(r_m+r)^2} \le \frac{1}{|\vec{r}-\vec{r}_{\rm m}|^2} \le \frac{1}{(r_m-r)^2}, \tag{7.78}$$

there are (for GPS satellite)

$$\frac{1}{(r_m \pm r)^2} = \frac{1}{r_m^2}\left(1 \mp \frac{2r}{r_m} \pm \frac{3r^2}{r_m^2} \mp \ldots\right) \approx \frac{1}{r_m^2}\left(1 \mp \frac{2r}{r_m} \pm 1.4\times 10^{-2}\ldots\right). \tag{7.79}$$

Then (7.77) turns out to be

$$\vec{f}_m = m\mu_m \frac{2r}{r_m^3}\vec{n}_m. \tag{7.80}$$

Neglecting the change of the geocentric distance of the satellite r, the potential function of the disturbing force of the moon is

$$V_m = -m\mu_m \frac{r}{r_m^2}. \tag{7.81}$$

The only difference between (7.67) and (7.81) is the index; instead of "s" for the sun, "m" is used for the moon. Therefore the solutions of the disturbance of the moon are similar to that of (7.72); one just needs to change the index "s" to "m".

Discretization and Solution

Denote the satellite period as T. The local noon of the moon is selected as the starting point of counting (see Fig. 7.3). A so-called sign function can be defined as

$$\delta(t) = \begin{cases} -1, & 0 \le t < T/2, \\ 1, & T/2 \le t \le T. \end{cases} \tag{7.82}$$

The sign function shows that the attracting force of the moon decelerates the satellite during the first half period and accelerates during the second half period with respect to the nominal motion of the satellite. Then the duration of one period of $0 \sim T$ can be equally divided by Δt, i.e., by $t'_0, t'_1, \ldots, t'_k, \ldots, T$. The acceleration of the disturbance of the moon (7.76) is then discretized as

$$\vec{a}_m(t) = -\mu_m \left(\frac{1}{|\vec{r}(t_k)-\vec{r}_m(t_k)|^2}\vec{n}_{sm}(t_k) + \frac{1}{|\vec{r}_m(t_k)|^2}\vec{n}_m(t_k) \right). \tag{7.83}$$

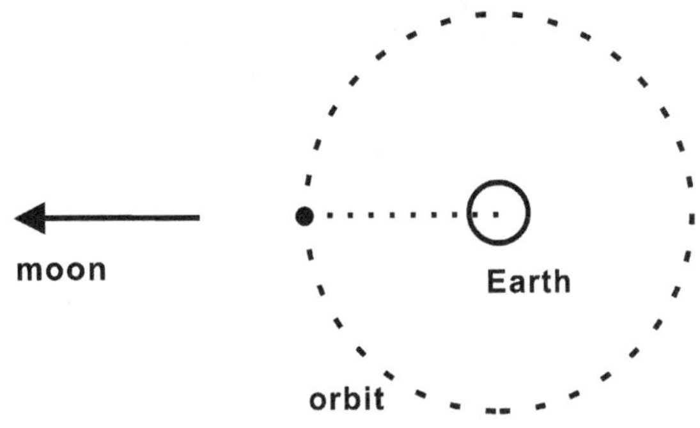

Fig. 7.3 Attracting force disturbance of the moon

The disturbed velocity caused by the moon is then

$$\vec{v}_m(t) = -\sum_{i=1}^{k} \mu_m \left(\frac{1}{|\vec{r}(t_i) - \vec{r}_m(t_i)|^2} \vec{n}_{sm}(t_i) + \frac{1}{|\vec{r}_m(t_i)|^2} \vec{n}_m(t_i) \right) \Delta t. \quad (7.84)$$

The disturbed position caused by the moon is then

$$\vec{\rho}_{\mathrm{m}}(t) = \sum_{j=1}^{k} \vec{v}_m(t_j)\Delta t. \quad (7.85)$$

Equation (7.85) is the discrete solution of the disturbance of the moon on the orbit of the satellite.

Solutions via Gaussian perturbed equations of the moon are very similar to that of the sun. Therefore, the discussions are omitted here.

7.6 Solutions of Disturbance of Planets

The disturbance acceleration of a planet is (see (7.61))

$$\vec{f}_p = -m\mu_p \left(\frac{1}{\left|\vec{r} - \vec{r}_p\right|^2} \vec{n}_{sp} + \frac{1}{\left|\vec{r}_p\right|^2} \vec{n}_p \right), \quad \vec{n}_{sp} = \frac{\vec{r} - \vec{r}_p}{\left|\vec{r} - \vec{r}_p\right|}, \quad \vec{n}_p = \frac{\vec{r}_p}{\left|\vec{r}_p\right|}. \quad (7.86)$$

The identity vectors $\vec{n}_{sp}$ and $\vec{n}_p$ represent the vectors from the satellite to the planet and the geocentric vector of the planet, respectively. The geocentric distance of the planet is far greater than that of the moon. The maximum difference of the two identity vectors is very small for satellite (except the sign difference). Therefore, the difference of the two identity vectors can be neglected. Then (7.86) can be approximated by

$$\vec{f}_p = -m\mu_p \left(\frac{-1}{\left|\vec{r} - \vec{r}_p\right|^2} + \frac{1}{\left|\vec{r}_p\right|^2} \right) \vec{n}_p. \quad (7.87)$$

Because (see (7.3))

$$\frac{1}{(r_p+r)^2} \le \frac{1}{\left|\vec{r}-\vec{r}_{\mathrm{p}}\right|^2} \le \frac{1}{(r_p-r)^2}, \tag{7.88}$$

there are (for Earth's satellite)

$$\frac{1}{(r_p \pm r)^2} = \frac{1}{r_p^2}\left(1 \mp \frac{2r}{r_p} \pm \frac{3r^2}{r_p^2} \mp \cdots\right) \approx \frac{1}{r_p^2}\left(1 \mp \frac{2r}{r_p} \pm \cdots\right). \tag{7.89}$$

Then (7.87) turns out to be

$$\vec{f}_p = m\mu_p \frac{2r}{r_p^3}\vec{n}_p. \tag{7.90}$$

Neglecting the change of the geocentric distance of the satellite r, the potential function of the disturbing force of the planet is

$$V_p = -m\mu_p \frac{r}{r_p^2}. \tag{7.91}$$

The only difference between (7.67) and (7.91) is the index; instead of "s" for the sun, "p" is used for planet. Therefore the solutions of the disturbance of the planet are similar to that of (7.72); one just needs to change the index "s" to "p". For more planets the solutions are still valid; one just needs to change the related parameters to the related planets.

7.7 Summary

Solutions of the extraterrestrial disturbances of the attracting forces of the sun, and the moon, as well as planets, the drag force of the atmosphere, and solar radiation pressure are derived in this chapter.

The solar radiation is a non-conservative disturbing force; of course, the disturbances of the orbit are also non-conservative ones. They are generally non-periodic effects.

The disturbance of the sun has no influence on the orbital plane; however, there are long term effects on the orientation of the ellipse and the position of the satellite as well as short periodic effects on the semi-axis of the satellite and the shape of the ellipse. The effects of the moon and planets are similar to that of the sun.

7.8 Ephemeris of the Moon, the Sun and Planets

The ephemeris of the sun and the moon as well as planets are needed for the computation of shadow functions of the sun and moon (solar radiation pressure), and the disturbance forces of the sun, the moon and planets. The computation of the ephemeris of the sun and the moon can be simplified by considering the orbit of the sun (indeed it is the Earth!) and the moon as Keplerian motion. Consider the orbital right-handed coordinate system, the origin in the geocentre, the xy-plane as the orbital plane, the x-axis pointing to the perigee, and the z-axis pointing in the direction of $\vec{q} \times \dot{\vec{q}}$ where $\vec{q}$ and $\dot{\vec{q}}$ are the position and velocity vectors of the sun or the moon. The two vectors are (see (3.41), (3.42))

$$\vec{q} = \begin{pmatrix} a(\cos E - e) \\ a\sqrt{1-e^2}\sin E \\ 0 \end{pmatrix} = \begin{pmatrix} q\cos f \\ q\sin f \\ 0 \end{pmatrix}, \qquad \dot{\vec{q}} = \begin{pmatrix} -\sin f \\ e+\cos f \\ 0 \end{pmatrix} \frac{na}{\sqrt{1-e^2}}, \quad (7.92)$$

where

$$q = \frac{a(1-e^2)}{1+e\cos f}. \quad (7.93)$$

The position and velocity vectors of the sun or the moon in the ECEI and ECSF coordinate systems are then (see Sect. 2.5 and (3.43))

$$\begin{aligned} \begin{pmatrix} \vec{p} \\ \dot{\vec{p}} \end{pmatrix} &= R_3(-\Omega)R_1(-i)R_3(-\omega) \begin{pmatrix} \vec{q} \\ \dot{\vec{q}} \end{pmatrix}, \\ \begin{pmatrix} \vec{r} \\ \dot{\vec{r}} \end{pmatrix} &= R_1(-\varepsilon) \begin{pmatrix} \vec{p} \\ \dot{\vec{p}} \end{pmatrix}, \end{aligned} \quad (7.94)$$

where a and i are the semi-major axis of the orbit and the inclination angle of the orbital plane of the moon or the sun in the ecliptic coordinate system (ECEI). Ω is the ecliptic right ascension of the ascending node, e is the eccentricity of the ellipse, ω is the argument of perigee, f is the true anomaly of the moon or the sun, and ε is the mean obliquity (the formula is given in Sect. 2.4). Because the sun moves along the ecliptic and the ascending node is defined as the equinox, parameters i and Ω are zero. True anomaly f, eccentric anomaly E and mean anomaly M are given by the Keplerian equation and by the following formulas:

$$\begin{aligned} E - e\sin E &= M, \\ q\cos f &= a\cos E - ae, \\ q\sin f &= b\sin E = a\sqrt{1-e^2}\sin E. \end{aligned} \quad (7.95)$$

For the moon, eccentricity $e_{\rm m} = 0.05490$, inclination $i_{\rm m} = 5.^\circ 145396$ and semi-major axis $a_{\rm m} = 384401\,\text{km}$. For the sun, eccentricity $e_{\rm s} = 0.016709114 -0.000042052T - 0.000000126T^2$ and semi-major axis $a_{\rm s} = 1.0000002\,\text{AU}$. AU signifies the astronomical units ($\text{AU} = 1.49597870691 \times 10^8\,\text{km}$). The fundamental arguments are given in the IERS Conventions (see McCarthy, 1996) as follows:

$$\begin{aligned} l = {} & 134.^\circ 96340251 + 1717915923.''2178T + 31.''8792T^2 + 0.''051635T^3 \\ & -0.''00024470T^4, \end{aligned}$$

$$
\begin{aligned}
l' &= 357.^\circ 52910918 + 129596581.''0481T - 0.''5532T^2 + 0.''000136T^3 \\
&\quad -0.''00001149T^4, \\
F &= 93.^\circ 27209062 + 1739527262''8478T - 12.''7512T^2 - 0.''001037T^3 \\
&\quad +0.''00000417T^4, \\
D &= 297.^\circ 85019547 + 1602961601''2090T - 6.''3706T^2 + 0.''006593T^3 \\
&\quad -0.''00003169T^4, \\
\Omega &= 125.^\circ 04455501 - 6962890.''2665T + 7.''4722T^2 + 0.''007702T^3 \\
&\quad -0.''00005939T^4,
\end{aligned}
\tag{7.96}
$$

where l and l' are the mean anomalies of the moon and the sun, respectively. D is the mean elongation of the moon from the sun. Ω is the mean longitude of the ascending node of the moon. $F = L - \Omega$, L is the mean longitude of the moon (or L_{moon}), and T is the Julian centuries measured from epoch J2000.0. Formulas of (7.96) are the arguments used to compute the nutation. Mean angular velocities n of the sun and moon are the coefficients of the linear terms of l and l' (units: second/century), respectively.

For computation of the ephemeris of the sun, l' is set as M in (7.95), so that E and f of the sun can be computed. Using $D = L_{\text{moon}} - L_{\text{sun}} = F + \Omega - L_{\text{sun}}$, the mean longitude L_{sun} can be computed. ω can be computed by the relation $L_{\text{sun}} = \omega + f$.

For computation of the ephemeris of the moon, l is set as M in (7.95), so that E and f of the moon can be computed. ω can be computed by the spherical triangle formula

$$
\tan(\omega + f) = \tan F / \cos i_{\text{m}}, \tag{7.97}
$$

where angles $u (= \omega + f)$ and F are in the same compartment.

Substituting the earlier-mentioned values of the moon and the sun into (7.92)–(7.94) respectively, ephemeris of the moon and the sun are obtained in the ECSF coordinate system. For more precise computation of the ephemeris of the moon, several corrections have to be considered (see Meeus, 1992; Montenbruck, 1989). Equivalently, a correction $\mathrm{d}F$ can be added to F, and the change of $\mathrm{d}u$ in (7.97) can be considered $\mathrm{d}f$ and added to f, where $\mathrm{d}F$ has the form (units: seconds)

$$
\begin{aligned}
\mathrm{d}F &= 22640 \sin l + 769 \sin(2l) + 36 \sin(3l) - 125 \sin D + 2370 \sin(2D) - 668 \sin l' \\
&\quad - 412 \sin(2F) + 212 \sin(2D - 2l) + 4586 \sin(2D - l) + 192 \sin(2D + l) \\
&\quad + 165 \sin(2D - l') + 206 \sin(2D - l - l') - 110 \sin(l + l') + 148 \sin(l - l').
\end{aligned}
$$

The orbits of the planets are given in the sun-centred ecliptic coordinate system by six Keplerian elements – the mean longitude (L) of the planet, the semi-major axis (a, units: AU) of the orbit of the planet, the eccentricity (e) of the orbit, the inclination (i) of the orbit to the ecliptic plane, the argument (ω) of the perihelion, and the longitude (Ω) of the ascending node. The orbital elements are expressed as a polynomial function of the instant of time T (Julian centuries) for Mercury, Venus, Mars, Jupiter, and Saturn as follows (see Meeus, 1992):

$$\begin{pmatrix} L \\ a \\ e \\ i \\ \omega \\ \Omega \end{pmatrix}_{\text{Mercury}} = \begin{pmatrix} 252.250906 & 149474.0722491 & 0.00030397 & -0.00000002 \\ 0.38709831 & 0 & 0 & 0 \\ 0.20563175 & 0.000020406 & -0.0000000284 & -0.0000000002 \\ 7.0049860 & 0.0018215 & -0{,}00001809 & 0.000000053 \\ 29.1252260 & 0.3702885 & 0.00012002 & -0.000000155 \\ 48.3308930 & 1.1861890 & 0.00017587 & 0.000000211 \end{pmatrix} \begin{pmatrix} 1 \\ T \\ T^2 \\ T^3 \end{pmatrix},$$

$$\begin{pmatrix} L \\ a \\ e \\ i \\ \omega \\ \Omega \end{pmatrix}_{\text{Venus}} = \begin{pmatrix} 181.979801 & 58519.2130302 & 0.00031060 & 0.000000015 \\ 0.72332982 & 0 & 0 & 0 \\ 0.00677118 & -0.000047766 & 0.0000000975 & 0.00000000044 \\ 3.3946620 & 0.00100370 & -0{,}00000088 & -0.000000007 \\ 54.883787 & 0.50109980 & -0.00148002 & -0.000005235 \\ 76.6799200 & 0.90111900 & 0.00040665 & -0.00000008 \end{pmatrix} \begin{pmatrix} 1 \\ T \\ T^2 \\ T^3 \end{pmatrix},$$

$$\begin{pmatrix} L \\ a \\ e \\ i \\ \omega \\ \Omega \end{pmatrix}_{\text{Mars}} = \begin{pmatrix} 355.4332750 & 19141.6964746 & 0.00031097 & 0.000000015 \\ 1.523679342 & 0 & 0 & 0 \\ 0.09340062 & 0.000090483 & -0.0000000806 & -0.00000000035 \\ 1.8497260 & -0.0006010 & 0.00012760 & -0.000000006 \\ 286.502141 & 1.0689408 & 0.00011910 & -0.000002007 \\ 49.558093 & 0.7720923 & 0.00001605 & 0.000002325 \end{pmatrix} \begin{pmatrix} 1 \\ T \\ T^2 \\ T^3 \end{pmatrix}$$

$$\begin{pmatrix} L \\ a \\ e \\ i \\ \omega \\ \Omega \end{pmatrix}_{\text{Jupiter}} = \begin{pmatrix} 34.351484 & 3036.3027889 & 0.00022374 & 0.000000025 \\ 5.202603191 & 0.0000001913 & 0 & 0 \\ 0.04849485 & 0.000163244 & -0.0000004719 & -0.00000000197 \\ 1.303270 & -0.00549660 & 0.00000465 & -0.000000004 \\ 273.866868 & 0.5917118 & 0.00063010 & -0.000005138 \\ 100.464441 & 1.0209550 & 0.00040117 & 0.000000569 \end{pmatrix} \begin{pmatrix} 1 \\ T \\ T^2 \\ T^3 \end{pmatrix}$$

and

$$\begin{pmatrix} L \\ a \\ e \\ i \\ \omega \\ \Omega \end{pmatrix}_{\text{Saturn}} = \begin{pmatrix} 50.0774710 & 1223.5110141 & 0.00051952 & -0.000000003 \\ 9.554909596 & -0.0000021389 & 0 & 0 \\ 0.05550862 & -0.000346818 & -0.0000006456 & 0.00000000338 \\ 2.488878 & -0.0037363 & -0.00001516 & 0.000000089 \\ 339.391263 & 1.0866715 & 0.00095824 & 0.000007279 \\ 113.665524 & 0.8770979 & -0.00012067 & -0.00000238 \end{pmatrix} \begin{pmatrix} 1 \\ T \\ T^2 \\ T^3 \end{pmatrix},$$

where except for the semi-major axis a and eccentricity e, all other elements have units of degrees. $F = L - \Omega$, and f and E can be computed by using (7.97) and (7.95). Mean angular velocities n of the planets are the coefficients of the linear term of L (units: degree/century). The coordinate vector of the planet can then be computed using (7.92)–(7.94). The results are in the sun-centred equatorial coordinate system. The results have to be transformed to the ECSF coordinate system by a translation

$$\begin{pmatrix} \vec{r} \\ \dot{\vec{r}} \end{pmatrix}_{\text{ECSF}} = \begin{pmatrix} \vec{r} \\ \dot{\vec{r}} \end{pmatrix}_{\text{sun}} + \begin{pmatrix} \vec{r} \\ \dot{\vec{r}} \end{pmatrix}_{\text{SCES}}, \tag{7.98}$$

Table 7.1 Gravitational constants of the sun, the moon and planets

Planet	Gravitational constant ($m^3 s^{-2}$)
sun	1.3271240000000E+20
moon	4.9027993000000E+12
Earth	3.9860044180000E+14
Mercury	2.2032070000000E+13
Venus	3.2485850000000E+14
Mars	4.2828300000000E+13
Jupiter	1.2671270000000E+17
Saturn	3.7940610000000E+16

where vectors with an index of sun and SCES are geocentric position and velocity vectors of the sun and the planet in the sun-centred equatorial system.

Gravitational constants of the sun, the moon and planets are given in Table 7.1.

Chapter 8
Numerical Orbit Determination

In this chapter, the principle of numerical orbit determination will be outlined. An algebraic solution of the variation equation is derived. Numerical integration and interpolation algorithms as well as the related partial derivatives are given in detail.

8.1 Principle of GPS Precise Orbit Determination

Recalling the discussions made in Sect. 4.1, the perturbed orbit of the satellite is the solution (or integration)

$$\vec{X}(t) = \vec{X}(t_0) + \int_{t_0}^{t} \vec{F}\mathrm{d}t, \tag{8.1}$$

which can be obtained by integrating the differential state equation under the initial condition

$$\begin{cases} \dot{\vec{X}}(t) = \vec{F}, \\ \vec{X}(t_0) = \vec{X}_0, \end{cases} \tag{8.2}$$

where $\vec{X}(t)$ is the instantaneous state vector of the satellite, $\vec{X}(t_0)$ is the initial state vector at time t_0 (denoted by $\vec{X}_0$), $\vec{F}$ is a function of the state vector $\vec{X}(t)$ and time t, and

$$\vec{X} = \begin{pmatrix} \vec{r} \\ \dot{\vec{r}} \end{pmatrix} \quad \text{and} \quad \vec{F} = \begin{pmatrix} \dot{\vec{r}} \\ \vec{f}/m \end{pmatrix},$$

where $\vec{f}$ is the summated force vector of all possible force vectors acting on the satellite, m is the mass of satellite, and $\vec{r}$, $\dot{\vec{r}}$ are the position and velocity vectors of the satellite.

If the initial state vector and the force vectors are precisely known, then the precise orbits can be computed through the integration in (8.1). Expanding the integration time t into the future, the so-called forecasted orbits can be obtained. Therefore, suitable numerical integration algorithms are needed (see next section).

In practice, the precise initiate state vector and force models, which are related to the approximate initial state vector and force models, have to be determined. These

G. Xu, *Orbits*,

can be realised through suitable parameterisation of the models in the GPS observation equations and then the parameters can be solved by adjustment or filtering.

We generally denote both the range and range rate together by ρ; their partial derivatives with respect to the orbit state vector (see Xu, 2003, 2007) have the forms of

$$\frac{\partial \rho}{\partial \vec{r}}, \frac{\partial \rho}{\partial \dot{\vec{r}}}, \quad \text{or} \quad \frac{\partial \rho}{\partial \vec{X}}.$$

Therefore, the orbit parameter related parts in the linearised GPS observation equation are

$$\frac{\partial \rho}{\partial (\vec{r},\dot{\vec{r}})} \frac{\partial (\vec{r},\dot{\vec{r}})}{\partial \vec{y}} \Delta \vec{y}^T, \quad \text{or} \quad \frac{\partial \rho}{\partial \vec{X}} \frac{\partial \vec{X}}{\partial \vec{y}} \Delta \vec{y}^T, \tag{8.3}$$

where

$$\vec{y} = \left(\vec{X}_0, \vec{Y}\right), \qquad \Delta \vec{y}^T = \left(\Delta \vec{X}_0, \Delta \vec{Y}\right)^T, \quad \frac{\partial \vec{X}}{\partial \vec{y}} = \frac{\partial \vec{X}}{\partial (\vec{X}_0, \vec{Y})}.$$

$\vec{X}, \vec{Y}$ are the state vector of satellite and the parameter vector of the force models, and index 0 denotes the related initial vectors of time t_0. $\vec{y}$ is the total unknown vector of the orbit determination problem, the related correction vector is $\Delta \vec{y} = \vec{y} - \vec{y}_0$, and $\Delta \vec{X}_0$ is the correction vector of the initial state vector. The partial derivative of $\vec{X}$ with respect to $\vec{y}$ is called transition matrix which has the dimension of $6 \times (6+n)$, where n is the dimension of vector $\vec{Y}$. The partial derivatives of the equation of motion of the satellite (see (8.2)) with respect to the vector $\vec{y}$ are

$$\frac{\partial \dot{\vec{X}}(t)}{\partial \vec{y}} = \frac{\partial \vec{F}}{\partial \vec{y}} = \frac{\partial \vec{F}}{\partial \vec{X}} \frac{\partial \vec{X}}{\partial \vec{y}} + \left(\frac{\partial \vec{F}}{\partial \vec{y}}\right)^*, \tag{8.4}$$

where the superscript * denotes the partial derivatives of $\vec{F}$ with respect to the explicit parameter vector $\vec{y}$ in $\vec{F}$, and

$$\begin{aligned} D(t) &= \left(\frac{\partial \vec{F}}{\partial \vec{X}}\right) = \begin{pmatrix} 0_{3\times 3} & E_{3\times 3} \\ \dfrac{1}{m}\dfrac{\partial \vec{f}}{\partial \vec{r}} & \dfrac{1}{m}\dfrac{\partial \vec{f}}{\partial \dot{\vec{r}}} \end{pmatrix} = \begin{pmatrix} 0_{3\times 3} & E_{3\times 3} \\ A(t) & B(t) \end{pmatrix}, \\ C(t) &= \left(\frac{\partial \vec{F}}{\partial \vec{y}}\right)^* = \begin{pmatrix} 0_{3\times 6} & 0_{3\times n} \\ 0_{3\times 6} & \dfrac{1}{m}\dfrac{\partial \vec{f}}{\partial \vec{Y}} \end{pmatrix} = \begin{pmatrix} 0_{3\times(6+n)} \\ G(t) \end{pmatrix}, \end{aligned} \tag{8.5}$$

where E is an identity matrix; the partial derivatives will be discussed and derived in a later section in detail. Notable that the force parameters are not functions of t. Therefore the order of the differentiations can be exchanged. Denoting the transition matrix by $\Phi(t,t_0)$, then (8.4) turns out to be

$$\frac{\mathrm{d}\Phi(t,t_0)}{\mathrm{d}t} = D(t)\Phi(t,t_0) + C(t). \tag{8.6}$$

Equation (8.6) is called a differential equation of the transition matrix or variation equation (see, e.g., Montenbruck and Gill, 2000). Denoting

$$\Phi(t,t_0) = \begin{pmatrix} \Psi(t,t_0) \\ \dot{\Psi}(t,t_0) \end{pmatrix}, \tag{8.7}$$

an alternate expression of (8.6) can be obtained by substituting (8.7) and (8.5) into (8.6)

$$\frac{\mathrm{d}^2\Psi(t,t_0)}{\mathrm{d}t^2} = A(t)\Psi(t,t_0) + B(t)\frac{\mathrm{d}\Psi(t,t_0)}{\mathrm{d}t} + G(t). \tag{8.8}$$

The initial value matrix is (initial state vector does not depend on force parameters):

$$\Phi(t_0,t_0) = (E_{6\times 6} \quad 0_{6\times n}). \tag{8.9}$$

That is, in the GPS observation equation, the transition matrix has to be obtained by solving the initial value problem of the variation equation (8.6) or (8.8). The problem is traditionally solved by integration.

8.1.1 Algebraic Solution of the Variation Equation

The variation equation can also be solved by numerical differentiation.

Equation (8.8) is a matrix differential equation system of size $3\times(6+n)$. Because $A(t)$ and $B(t)$ are 3×3 matrices, the differential equations are independent from column to column. That is, we need to discuss just the solution of the equation of a column. For column j, (8.8) and (8.9) are

$$\frac{\mathrm{d}^2\Psi_{ij}(t)}{\mathrm{d}t^2} = \sum_{k=1}^{3}\left(A_{ik}(t)\Psi_{kj}(t) + B_{ik}(t)\frac{\mathrm{d}\Psi_{kj}(t)}{\mathrm{d}t}\right) + G_{ij}(t), \quad i=1,2,3, \tag{8.10}$$

$$\begin{pmatrix} \Psi_{ij}(t_0) \\ \dot{\Psi}_{ij}(t_0) \end{pmatrix} = \begin{pmatrix} \delta_{ij} \\ \delta_{(i+3)j} \end{pmatrix}, \quad i=1,2,3, \qquad \delta_{kj} = \begin{cases} 1, & if\, k=j, \\ 0, & if\, k\neq j, \end{cases}$$

where index *ij* denotes the related element of the matrix. For time interval $[t_0, t]$ and differentiation step $h=(t-t_0)/m$, one has $t_n = t_0 + nh$, $n=1,\ldots,m$ and

$$\left.\frac{\mathrm{d}^2\Psi_{ij}(t)}{\mathrm{d}t^2}\right|_{t=t_n} = \frac{\Psi_{ij}(t_{n+1}) - 2\Psi_{ij}(t_n) + \Psi_{ij}(t_{n-1})}{h^2}, \quad i=1,2,3,$$

$$\left.\frac{\mathrm{d}\Psi_{ij}(t)}{\mathrm{d}t}\right|_{t=t_n} = \frac{\Psi_{ij}(t_{n+1}) - \Psi_{ij}(t_{n-1})}{2h}, \quad \Psi_{ij}(t)\big|_{t=t_n} = \Psi_{ij}(t_n), \qquad i=1,2,3. \tag{8.11}$$

Then (8.10) turns out to be

$$\Psi_{ij}(t_0) = \Psi_{ij}(t_0), \qquad \Psi_{ij}(t_1) = \Psi_{ij}(t_0) + h\dot{\Psi}_{ij}(t_0), \quad i=1,2,3.$$

$$\frac{\Psi_{ij}(t_{n+1}) - 2\Psi_{ij}(t_n) + \Psi_{ij}(t_{n-1})}{h^2} =$$
$$\sum_{k=1}^{3} \left(A_{ik}(t_n)\Psi_{kj}(t_n) + B_{ik}(t_n)\frac{\Psi_{kj}(t_{n+1}) - \Psi_{kj}(t_{n-1})}{2h} \right) + G_{ij}(t_n), \quad i = 1,2,3, \tag{8.12}$$

where $n = 1,2,\ldots, m-1$. For $i = 1,2,3$ and the sequential number n, there are three equations and three unknowns of time $t_{\mathrm{n}+1}$, so that the initial value problem has a set of unique solutions sequentially. Equation (8.12) can be rewritten as

$$\left(\frac{E}{h^2} - \frac{B(t_n)}{2h} \right) \begin{pmatrix} \Psi_{1j}(t_{n+1}) \\ \Psi_{2j}(t_{n+1}) \\ \Psi_{3j}(t_{n+1}) \end{pmatrix} = \begin{pmatrix} R_1 \\ R_2 \\ R_3 \end{pmatrix}, \tag{8.13}$$

where

$$\begin{pmatrix} R_1 \\ R_2 \\ R_3 \end{pmatrix} = \left(\frac{2E}{h^2} + A(t_n) \right) \begin{pmatrix} \Psi_{1j}(t_n) \\ \Psi_{2j}(t_n) \\ \Psi_{3j}(t_n) \end{pmatrix} - \left(\frac{E}{h^2} + \frac{B(t_n)}{2h} \right) \begin{pmatrix} \Psi_{1j}(t_{n-1}) \\ \Psi_{2j}(t_{n-1}) \\ \Psi_{3j}(t_{n-1}) \end{pmatrix} + \begin{pmatrix} G_{1j}(t_n) \\ G_{2j}(t_n) \\ G_{3j}(t_n) \end{pmatrix}.$$

For $n = 1,\ldots, m-1$, this equation is solvable. Note that the three matrices

$$\left(\frac{E}{h^2} - \frac{B(t_n)}{2h} \right), \quad \left(\frac{2E}{h^2} + A(t_n) \right), \quad \left(\frac{E}{h^2} + \frac{B(t_n)}{2h} \right)$$

are independent from the column number j. The solutions of (8.13) are vectors

$$\begin{pmatrix} \Psi_{1j}(t_{n+1}) \\ \Psi_{2j}(t_{n+1}) \\ \Psi_{3j}(t_{n+1}) \end{pmatrix} \text{ and } \begin{pmatrix} \dot{\Psi}_{1j}(t_{n+1}) \\ \dot{\Psi}_{2j}(t_{n+1}) \\ \dot{\Psi}_{3j}(t_{n+1}) \end{pmatrix}, \quad n = 1,\ldots,m-1, \tag{8.14}$$

where the velocity vector can be computed using the definition of (8.11). Solving the equations of all column j, the solutions of the initial value problem of (8.8) and (8.9) can be obtained. Note that the needed values are the values of t_{n} which can be computed by averaging the values of $t_{\mathrm{n}+1}$ and $t_{\mathrm{n}-1}$.

8.2 Numerical Integration and Interpolation Algorithms

The Runge–Kutta algorithm, Adams algorithm, Cowell algorithm and mixed algorithm as well as interpolation algorithms are discussed in this section (see, e.g., Brouwer and Clemence, 1961; Bate et al., 1971; Herrick, 1972; Xu, 1994; Liu et al., 1996; Press et al., 1992).

8.2.1 Runge-Kutta Algorithms

The Runge-Kutta algorithm is a method that can be used to solve the initial value problem of

$$\begin{aligned} \frac{\mathrm{d}X}{\mathrm{d}t} &= F(t,X), \\ X(t_0) &= X_0, \end{aligned} \tag{8.15}$$

where X_0 is the initial value of variable X at time t_0, and F is the function of t and X. For step size h, the Runge-Kutta algorithm can be used to compute $X(t_0+h)$. By repeating this process, a series of solutions can be obtained as $X(t_0+h)$, $X(t_0+2h)$, $\ldots$, $X(t_0+nh)$, where n is an integer. Denoting $t_n = t_0+nh$, $X(t_n+h)$ can be represented by the Taylor expansion at t_n by

$$X(t_n+h) = X(t_n) + h\left.\frac{\mathrm{d}X}{\mathrm{d}t}\right|_{t=t_n} + \frac{h^2}{2}\left.\frac{\mathrm{d}^2X}{\mathrm{d}t^2}\right|_{t=t_n} + \cdots + \frac{h^n}{n!}\left.\frac{\mathrm{d}^nX}{\mathrm{d}t^n}\right|_{t=t_n} + \cdots, \tag{8.16}$$

where

$$\begin{aligned} \frac{\mathrm{d}X}{\mathrm{d}t} &= F, \\ \frac{\mathrm{d}^2X}{\mathrm{d}t^2} &= \frac{\mathrm{d}F(t,X)}{\mathrm{d}t} = \frac{\partial F}{\partial t} + \frac{\partial F}{\partial X}\frac{\partial X}{\partial t} = \frac{\partial F}{\partial t} + \frac{\partial F}{\partial X}F, \\ \frac{\mathrm{d}^3X}{\mathrm{d}t^3} &= \frac{\partial^2 F}{\partial t^2} + 2\frac{\partial^2 F}{\partial t\partial X}F + \frac{\partial^2 F}{\partial t\partial X} + \frac{\partial^2 F}{\partial X^2}F^2 + \left(\frac{\partial F}{\partial X}\right)^2 F \end{aligned} \tag{8.17}$$

and

$$\begin{aligned} \frac{\mathrm{d}^4X}{\mathrm{d}t^4} &= \frac{\partial^3 F}{\partial t^3} + \frac{\partial^3 F}{\partial t^2\partial X}(3F+1) + \frac{\partial^3 F}{\partial t\partial X^2}(5F^2+2F) + 2\frac{\partial^2 F}{\partial t\partial X}\frac{\partial F}{\partial t} + 4\frac{\partial^3 F}{\partial X^3}F^3 \\ &\quad + 2\frac{\partial^2 F}{\partial X^2}\frac{\partial F}{\partial t}F + 4\frac{\partial F}{\partial X}\frac{\partial^2 F}{\partial t\partial X}F + 6\frac{\partial F}{\partial X}\frac{\partial^2 F}{\partial X^2}F^2 + \left(\frac{\partial F}{\partial X}\right)^2\frac{\partial F}{\partial t} + \left(\frac{\partial F}{\partial X}\right)^2\frac{\partial F}{\partial X}2F \\ &\ \ \vdots \end{aligned}$$

The principle of the Runge-Kutta algorithm is to use a set of combinations of the 1st order partial derivatives around the $(t_n, X(t_n))$ to replace the higher order derivatives in (8.16), that is,

$$X(t_{n+1}) = X(t_n) + \sum_{i=1}^{L} w_i K_i, \tag{8.18}$$

where

$$K_1 = hF(t_n, X(t_n))$$

and

$$K_i = hF\left(t_n + \alpha_i h, X(t_n) + \sum_{j=1}^{i-1} \beta_{ij} K_j\right) \quad (i = 2, 3, \ldots), \tag{8.19}$$

where w_i, α_i, and β_{ij} are constants to be determined, and L is an integer. The Taylor expansions of K_i $(i = 2, 3, \ldots)$ at $(t_n, X(t_n))$ to the 1st order are

$$K_i = hF(t_n, X(t_n)) + h^2 \alpha_i \frac{\partial F}{\partial t} + h \frac{\partial F}{\partial X} \sum_{j=1}^{i-1} \beta_{ij} K_j \tag{8.20}$$

or

$$K_2 = hF(t_n, X(t_n)) + h^2 \left(\alpha_2 \frac{\partial F}{\partial t} + \beta_{21} \frac{\partial F}{\partial X} F\right), \tag{8.21}$$

$$K_3 = hF + h^2 \left(\alpha_3 \frac{\partial F}{\partial t} + (\beta_{31} + \beta_{32}) \frac{\partial F}{\partial X} F\right) + h^3 \beta_{32} \frac{\partial F}{\partial X} \left(\alpha_2 \frac{\partial F}{\partial t} + \beta_{21} \frac{\partial F}{\partial X} F\right),$$

$$\begin{aligned} K_4 = {} & hF + h^2 \left(\alpha_4 \frac{\partial F}{\partial t} + (\beta_{41} + \beta_{42} + \beta_{43}) \frac{\partial F}{\partial X} F\right) \\ & + h^3 \left[(\beta_{42}\alpha_2 + \beta_{43}\alpha_3) \frac{\partial F}{\partial X} \frac{\partial F}{\partial t} + (\beta_{42}\beta_{21} + \beta_{43}(\beta_{31} + \beta_{32})) \frac{\partial F}{\partial X} \frac{\partial F}{\partial X} F\right] \\ & + h^4 \beta_{43}\beta_{32} \frac{\partial F}{\partial X} \frac{\partial F}{\partial X} \left(\alpha_2 \frac{\partial F}{\partial t} + \beta_{21} \frac{\partial F}{\partial X} F\right), \end{aligned}$$

$$\begin{aligned} K_5 = {} & hF + h^2 \left(\alpha_5 \frac{\partial F}{\partial t} + (\beta_{51} + \beta_{52} + \beta_{53} + \beta_{54}) \frac{\partial F}{\partial X} F\right) \\ & + h^3 \frac{\partial F}{\partial X} \left(\begin{array}{l} \alpha_2 \frac{\partial F}{\partial t} + \beta_{21} \frac{\partial F}{\partial X} F + \beta_{53} \left(\alpha_3 \frac{\partial F}{\partial t} + (\beta_{31} + \beta_{32}) \frac{\partial F}{\partial X} F\right) \\ + \beta_{54} \left(\alpha_4 \frac{\partial F}{\partial t} + (\beta_{41} + \beta_{42} + \beta_{43}) \frac{\partial F}{\partial X} F\right) \end{array}\right) \\ & + h^4 \frac{\partial F}{\partial X} \left(\begin{array}{l} (\beta_{53}\beta_{32}\alpha_2 + \beta_{54}(\beta_{42}\alpha_2 + \beta_{43}\alpha_3)) \frac{\partial F}{\partial X} \frac{\partial F}{\partial t} \\ + (\beta_{54}(\beta_{42}\beta_{21} + \beta_{43}(\beta_{31} + \beta_{32})) + \beta_{32}\beta_{21}) \frac{\partial F}{\partial X} \frac{\partial F}{\partial X} F \end{array}\right) \\ & + h^5 \frac{\partial F}{\partial X} \beta_{54} \left(\beta_{43}\beta_{32} \frac{\partial F}{\partial X} \frac{\partial F}{\partial X} \left(\alpha_2 \frac{\partial F}{\partial t} + \beta_{21} \frac{\partial F}{\partial X} F\right)\right) \end{aligned}$$

$$\vdots$$

where F and the related partial derivatives have values at $(t_n, X(t_n))$. Substituting these formulas into (8.18) and comparing the coefficients of $h^n (= 1/n!)$ with (8.16), a group of equations of constants w_i, α_i and β_{ij} can be obtained by separating them through the partial derivative combinations. For example, for $L = 4$, one has

$$\begin{aligned}
&w_1 + w_2 + w_3 + w_4 = 1,\\
&w_2\alpha_2 + w_3\alpha_3 + w_4\alpha_4 = \frac{1}{2},\\
&w_2\beta_{21} + w_3(\beta_{31} + \beta_{32}) + w_4(\beta_{41} + \beta_{42} + \beta_{43}) = \frac{1}{2},\\
&w_3\alpha_2\beta_{32} + w_4(\alpha_2\beta_{42} + \alpha_3\beta_{43}) = \frac{1}{6},\\
&w_3\beta_{21}\beta_{32} + w_4(\beta_{21}\beta_{42} + \beta_{31}\beta_{43} + \beta_{32}\beta_{43}) = \frac{1}{6},\\
&w_4\alpha_2\beta_{43}\beta_{32} = \frac{1}{24}
\end{aligned} \tag{8.22}$$

and

$$w_4\beta_{21}\beta_{43}\beta_{32} = \frac{1}{24}.$$

There are 13 coefficients in the seven equations, so the solution set of (8.22) is not a unique one. Considering w as weight and α as the step factor, one may set, e.g., $w_1 = w_2 = w_3 = w_4 = 1/4$, $\alpha_2 = 1/3$, $\alpha_3 = 2/3$, $\alpha_4 = 1$ and substitute the same into (8.22) and have

$$\begin{aligned}
&\beta_{21} + \beta_{31} + \beta_{32} + \beta_{41} + \beta_{42} + \beta_{43} = 2,\\
&\beta_{32} + \beta_{42} + 2\beta_{43} = 2,\\
&\beta_{21}\beta_{32} + \beta_{21}\beta_{42} + \beta_{31}\beta_{43} + \beta_{32}\beta_{43} = \frac{2}{3},\\
&\beta_{43}\beta_{32} = \frac{1}{2}
\end{aligned}$$

and

$$\beta_{21}\beta_{43}\beta_{32} = \frac{1}{6}.$$

Letting $\beta_{32} = 1$, one has $\beta_{42} = 0$, $\beta_{31} = -1/3$, and $\beta_{41} = 1/2$. Thus, a 4th-order Runge-Kutta formula is

$$X(t_{n+1}) = X(t_n) + \frac{1}{4}\sum_{i=1}^{4} K_i, \tag{8.23}$$

where

$$\begin{aligned}
&K_1 = hF\left(t_n, X(t_n)\right),\\
&K_2 = hF\left(t_n + \frac{1}{3}h, X(t_n) + \frac{1}{3}K_1\right),
\end{aligned} \tag{8.24}$$

$$K_3 = hF\left(t_n + \frac{2}{3}h, X(t_n) - \frac{1}{3}K_1 + K_2\right)$$

and

$$K_4 = hF\left(t_n + h, X(t_n) + \frac{1}{2}K_1 + \frac{1}{2}K_3\right).$$

Similarly, a commonly used 8th order Runge-Kutta formula can be derived. It is quoted as follows (see Xu, 1994; Liu et al., 1996):

$$X(t_{n+1}) = X_n + \frac{1}{840}(41K_1 + 27K_4 + 272K_5 + 27K_6 + 216K_7 + 216K_9 + 41K_{10}), \tag{8.25}$$

where

$$K_1 = hF(t_n, X_n), \quad X_n = X(t_n), \tag{8.26}$$

$$K_2 = hF\left(t_n + \frac{4}{27}h, X_n + \frac{4}{27}K_1\right),$$

$$K_3 = hF\left(t_n + \frac{2}{9}h, X_n + \frac{1}{18}K_1 + \frac{1}{6}K_2\right),$$

$$K_4 = hF\left(t_n + \frac{1}{3}h, X_n + \frac{1}{12}K_1 + \frac{1}{4}K_3\right),$$

$$K_5 = hF\left(t_n + \frac{1}{2}h, X_n + \frac{1}{8}K_1 + \frac{3}{8}K_4\right),$$

$$K_6 = hF\left(t_n + \frac{2}{3}h, X_n + \frac{1}{54}(13K_1 - 27K_3 + 42K_4 + 8K_5)\right),$$

$$K_7 = hF\left(t_n + \frac{1}{6}h, X_n + \frac{1}{4320}(389K_1 - 54K_3 + 966K_4 - 824K_5 + 243K_6)\right),$$

$$K_8 = hF\left(t_n + h, X_n + \frac{1}{20}(-231K_1 + 81K_3 - 1164K_4 + 656K_5 - 122K_6 + 800K_7)\right),$$

$$K_9 = hF\left(\begin{aligned} t_n + \frac{5}{6}h, X_n + \frac{1}{288}(-127K_1 &+ 18K_3 - 678K_4 + 456K_5 \\ &-9K_6 + 576K_7 + 4K_8) \end{aligned}\right)$$

and

$$K_{10} = hF\left(\begin{aligned} t_n + h, X_n + \frac{1}{820}(1481K_1 &- 81K_3 + 7104K_4 - 3376K_5 \\ &+72K_6 - 5040K_7 - 60K_8 + 720K_9) \end{aligned}\right).$$

From the derivation process, it is obvious that the Runge-Kutta algorithm is an approximation of the same order Taylor expansions. For every step of the solution, the function values of F have to be computed several times. The Runge-Kutta algorithm is also called the single step method and is commonly used for computing the start values for other multiple step methods.

Errors of the integration are dependent on the step size and the properties of function F. To ensure the needed accuracy of the orbit integration, a step size adaptive control is also meaningful in computing efficiency (see Press et al., 1992). Because of the periodical motion of the orbit, the step control just needs to be made in a few special cycles of the motion. A step doubling method is suggested by Press et al. (1992). Integration is taken twice for each step, first with a full step, then independently with two half steps. By comparing the results, the step size can be adjusted to fit the accuracy requirement.

To apply the above formulas for solving the initial value problem of the equation of motion (8.2), equation (8.15) shall be rewritten as

$$\frac{\mathrm{d}X_k}{\mathrm{d}t} = \dot{X}_k(t,X), \qquad X_k(t_0) = X_{k0} \quad (k = 1,2,3),$$
$$\frac{\mathrm{d}\dot{X}_k}{\mathrm{d}t} = f_k(t,X)/m, \quad \dot{X}_k(t_0) = \dot{X}_{k0}$$

where $X = (X_1, X_2, X_3, \dot{X}_1, \dot{X}_2, \dot{X}_3)$. Using the Runge-Kutta algorithm to solve the above problem, an additional index k shall be added to all X and K in (8.25):

$$X_k(t_{n+1}) = X_{kn} + \frac{1}{840}(41K_{k1} + 27K_{k4} + 272K_{k5} + 27K_{k6} + 216K_{k7} + 216K_{k9} + 41K_{k10}),$$

and the same index k shall be added to K on the left side and F on the right side of (8.26). For the last three equations, $F_k = f_k/m$, so $\dot{X}_k$ can be computed. For the first three equations, $F_k = \dot{X}_k$, so F_k can be computed through computing $\dot{X}_k$ at the needed coordinates t and X.

8.2.2 Adams Algorithms

For the initial value problem of

$$\begin{aligned} \frac{\mathrm{d}X}{\mathrm{d}t} &= F(t,X), \\ X(t_0) &= X_0, \end{aligned} \tag{8.27}$$

there exists

$$X(t_{n+1}) = X(t_n) + \int_{t_n}^{t_{n+1}} F(t,X)\mathrm{d}t. \tag{8.28}$$

The Adams algorithm uses the Newtonian backward differential interpolation formula to represent the function F by

$$\begin{aligned} F(t,X) &= F_n + \frac{t-t_n}{h}\nabla F_n + \frac{(t-t_n)(t-t_{n-1})}{2!h^2}\nabla^2 F_n \\ &+ \cdots + \frac{(t-t_n)(t-t_{n-1})\cdots(t-t_{n-k+1})}{k!h^k}\nabla^k F_n, \end{aligned} \tag{8.29}$$

where F_n is the value of F at the time t_n, h is the step size, $\nabla^k F$ is the kth order backward numerical difference of F, and

$$\begin{aligned} \nabla F_n &= F_n - F_{n-1} \\ \nabla^2 F_n &= \nabla F_n - \nabla F_{n-1} = F_n - 2F_{n-1} + F_{n-2}, \\ &\vdots \\ \nabla^m F_n &= \sum_{j=0}^{m} (-1)^j C_m^j F_{n-j}, \quad C_m^j = \frac{m!}{j!(m-j)!}, \end{aligned} \tag{8.30}$$

where C_{m}^j is the binomial coefficient. Letting $s = (t-t_n)/h$, then $\mathrm{d}t = h\mathrm{d}s$, $s = 0$ if $t = t_n$, $s = 1$ if $t = t_{n+1}$, so that (8.29) and (8.28) turn out to be

$$F(t,X) = \sum_{m=0}^{k} C_{s+m-1}^m \nabla^m F_n$$

and

$$X(t_{n+1}) = X(t_n) + \int_{t_n}^{t_{n+1}} \sum_{m=0}^{k} C_{s+m-1}^m \sum_{j=0}^{m} (-1)^j C_m^j F_{n-j} h\mathrm{d}s. \tag{8.31}$$

By denoting

$$\begin{aligned} \gamma_m &= \int_0^1 C_{s+m-1}^m \mathrm{d}s, \\ \beta_j &= \sum_{m=j}^{k} (-1)^j C_m^j \gamma_m, \end{aligned} \tag{8.32}$$

one has

$$X(t_{n+1}) = X(t_n) + h\sum_{j=0}^{k} \beta_j F_{n-j}, \tag{8.33}$$

where the sequences of the two sequential summations have been changed. For the first equation of (8.32), there is (see Xu, 1994)

$$\gamma_0 = 1, \quad \gamma_m = 1 - \sum_{j=1}^{m} \frac{1}{j+1}\gamma_{m-j} \qquad (m \geq 1). \tag{8.34}$$

Equation (8.33) is also called the Adams-Bashforth formula. It uses the function values of $\{F_{n-j}, j = 0,\ldots,k\}$ to compute the X_{n+1}. When the order of the algorithm is selected, the coefficients of β_j are constants. This makes the computation using

(8.33) very simple. For every integration step, just one function value of F_n has to be computed. However, the Adams algorithm needs $\{F_{n-j}, j = 0,\ldots,k\}$ as initial values, but to compute those values, the states $\{X_{n-j}, j = 0,\ldots,k\}$ are needed. In other words, the Adams algorithm is not able to start the integration itself. The Runge-Kutta algorithm is usually used for computing the start values.

The Adams-Bashforth formula does not take the function value F_{n+1} into account. Using F_{n+1}, the Adams algorithm is expressed by the Adams-Moulton formula. Similar to the above discussions, function F can be represented by

$$\begin{aligned} F(t,X) &= F_{n+1} + \frac{t-t_{n+1}}{h}\nabla F_{n+1} + \frac{(t-t_{n+1})(t-t_n)}{2!h^2}\nabla^2 F_{n+1} \\ &+ \cdots + \frac{(t-t_{n+1})(t-t_n)\cdots(t-t_{n-k+2})}{k!h^k}\nabla^k F_{n+1}, \end{aligned} \tag{8.35}$$

where

$$\nabla^m F_{n+1} = \sum_{j=0}^{m} (-1)^j C_m^j F_{n+1-j}. \tag{8.36}$$

If one lets $s = (t-t_{n+1})/h$, then $\mathrm{d}t = h\mathrm{d}s$, $s = -1$ if $t = t_n$, and $s = 0$ if $t = t_{n+1}$; similar formulas of (8.33) and (8.32) can be obtained:

$$X(t_{n+1}) = X(t_n) + h\sum_{j=0}^{k} \beta_j^* F_{n+1-j}, \tag{8.37}$$

$$\begin{aligned} \beta_j^* &= \sum_{m=j}^{k} (-1)^j C_m^j \gamma_m^*, \\ \gamma_m^* &= \int_{-1}^{0} C_{s+m-1}^m \mathrm{d}s \end{aligned} \tag{8.38}$$

and (see Xu, 1994)

$$\gamma_0^* = 1, \quad \gamma_m^* = -\sum_{j=1}^{m} \frac{1}{j+1}\gamma_{m-j}^* \qquad (m \geq 1). \tag{8.39}$$

Because of the use of F_{n+1} to approximate F, the Adams-Moulton formula may reach a higher accuracy than that of the Adams-Bashforth formula. However, before X_{n+1} has been computed, F_{n+1} might not have been computed exactly. So an iterative process is needed while using the Adams-Moulton formula. A simple way to use the Adams-Moulton formula is to use the Adams-Bashforth formula to compute X_{n+1} and F_{n+1}, and then to use the Adams-Moulton formula to compute the modified X_{n+1} using F_{n+1}. Experience shows that such a process will be accurate enough for many applications.

8.2.3 Cowell Algorithms

For the initial value problem of

$$\begin{aligned} \frac{d^2X}{dt^2} &= F(t,X), \\ \dot{X}(t_0) &= \dot{X}_0, \\ X(t_0) &= X_0, \end{aligned} \tag{8.40}$$

there is

$$\dot{X}(t) = \dot{X}(t_n) + \int_{t_n}^{t} F(t,X)\,dt. \tag{8.41}$$

Note that here X is the position coordinate of the satellite. In other words, the disturbing force F is not the function of the velocity of the satellite.

By integrating (8.41) in areas of $[t_n, t_{n+1}]$ and $[t_n, t_{n-1}]$ respectively, one has

$$X(t_{n+1}) - X(t_n) - \dot{X}(t_n)(t_{n+1} - t_n) = \int_{t_n}^{t_{n+1}} \int_{t_n}^{t} F(t,X)dtdt \tag{8.42}$$

and

$$X(t_{n-1}) - X(t_n) - \dot{X}(t_n)(t_{n-1} - t_n) = \int_{t_n}^{t_{n-1}} \int_{t_n}^{t} F(t,X)dtdt, \tag{8.43}$$

where $(t_{n+1} - t_n) = h = (t_n - t_{n-1})$. Adding both equations together, one has

$$X(t_{n+1}) - 2X(t_n) + X(t_{n-1}) = \int_{t_n}^{t_{n+1}} \int_{t_n}^{t} + \int_{t_n}^{t_{n-1}} \int_{t_n}^{t} F(t,X)dtdt. \tag{8.44}$$

Similar to the Adams-Bashforth formula, function F can be represented by

$$\begin{aligned} F(t,X) = F_n &+ \frac{t-t_n}{h}\nabla F_n + \frac{(t-t_n)(t-t_{n-1})}{2!h^2}\nabla^2 F_n \\ &+ \cdots + \frac{(t-t_n)(t-t_{n-1})\cdots(t-t_{n-k+1})}{k!h^k}\nabla^k F_n. \end{aligned} \tag{8.45}$$

Substituting (8.45) into (8.44), one has (similar to the derivation of Adams algorithms) (see Xu, 1994)

$$X(t_{n+1}) = 2X(t_n) - X(t_{n-1}) + h^2 \sum_{j=0}^{k} \beta_j F_{n-j}, \tag{8.46}$$

where

$$\beta_j = \sum_{m=j}^{k} (-1)^j C_m^j \sigma_m, \tag{8.47}$$

$$\sigma_0 = 1, \quad \sigma_m = 1 - \sum_{j=1}^{m} \frac{2}{j+2} b_{j+1} \sigma_{m-j} \quad (m \geq 1),$$

$$b_j = \sum_{i=1}^{j} \frac{1}{i}.$$

Equation (8.46) is called the Stormer formula. Similar to the discussions in Adams algorithms, taking F_{n+1} into account, one has

$$F(t,X) = F_{n+1} + \frac{t-t_{n+1}}{h} \nabla F_{n+1} + \frac{(t-t_{n+1})(t-t_n)}{2!h^2} \nabla^2 F_{n+1} + \cdots + \frac{(t-t_{n+1})(t-t_n)\cdots(t-t_{n-k+2})}{k!h^k} \nabla^k F_{n+1} \tag{8.48}$$

and (see Xu, 1994)

$$X(t_{n+1}) = 2X(t_n) - X(t_{n-1}) + h^2 \sum_{j=0}^{k} \beta_j^* F_{n+1-j}, \tag{8.49}$$

where

$$\beta_j^* = \sum_{m=j}^{k} (-1)^j C_m^j \sigma_m^*, \tag{8.50}$$

$$\sigma_0^* = 1, \quad \sigma_m^* = -\sum_{j=1}^{m} \frac{2}{j+2} b_{j+1} \sigma_{m-j}^* \quad (m \geq 1)$$

and

$$b_j = \sum_{i=1}^{j} \frac{1}{i}.$$

Equation (8.49) is called the Cowell formula. Because of the use of F_{n+1} to approximate F, the Cowell formula may reach a higher accuracy than that of the Stormer formula. However, before X_{n+1} has been computed, F_{n+1} may not be computed exactly. So an iterative process is needed while using the Cowell formula. A simple way is to use the Stormer formula first to compute X_{n+1} and F_{n+1}, and then the Cowell formula to compute the modified X_{n+1} using F_{n+1}. Experience shows that such a process will be accurate enough for many applications.

8.2.4 Mixed Algorithms and Discussions

Above we discussed three algorithms for solving the initial value problem of the orbit differential equation. The Runge-Kutta algorithm is a single step method. The formulas of different order Runge-Kutta algorithms do not have simple relationships, and even for a definite order the formulas are not unique. For every step of integration, several function values of F have to be computed. Interestingly, the Runge-Kutta algorithm is a self-starting method. Generally, it is used for providing the starting values for multiple-step algorithms.

Adams algorithms are multiple-step methods. The order of the formulas can be easily raised because of their sequential relationships. However, the Adams algorithms cannot start themselves. For every step of integration, only one function value has to be computed. The disturbing function is considered a function of time and the state of the satellite. So Adams methods can be used in orbit determination without having any problem with the disturbing function. For a higher accuracy requirement, a mixed Adams-Bashforth method and Adams-Moulton methods can be used in an iterative process.

Cowell algorithms are also multiple-step methods. The order can be changed easily. Cowell methods also need starting help from other methods. Analysis shows that Cowell algorithms have a higher accuracy than that of Adams algorithms when the same orders of formulas are used. However, Cowell formulas are only suitable for that kind of disturbing function F, which is the function of the time and the position of the satellite. It is well-known that the atmospheric drag is a disturbing force, which is a function of the velocity of the satellite. Therefore Cowell algorithms can be used only for integrating a part of the disturbing forces. A mixed Cowell method still retains this property.

Obviously, the forces of the equation of motion have to be separated into two parts: one includes the forces that are functions of the velocity of the satellite, and the other includes all remaining forces. The first part can be integrated by Adams methods and the other by Cowell methods. The Runge-Kutta algorithm will be used for providing the needed starting values.

The selections of the order number and step size are dependent on the accuracy requirements and the orbit conditions. Usually the order and the step size are selectable input variables of the software, and can be properly selected after several test runs. Scheinert suggested using 8th-order Runge-Kutta algorithms, as well as 12th-order Adams and Cowell algorithms (see Scheinert, 1996). Note that for the order selection; it is not the higher the order is, the higher the accuracy will be. For the step size selection, it is not the smaller the step size is, the better the results will be.

8.2.5 Interpolation Algorithms

Orbits are given through integration at the step points $t_0 + nh(n = 0, 1, \ldots)$. For GPS satellites, h is usually selected as 300 s. However, GPS observations are made, usually in IGS, every 15 s. For linearisation and formation of the GPS observation equations, the orbit data sometimes have to be interpolated to the needed epochs.

To obtain the ephemeris of any epoch, a Lagrange polynomial is used first to fit the given data and then to interpolate the data in the chosen epoch. The general

Lagrange polynomial is (see Wang et al., 1979):

$$y(t) = \sum_{j=0}^{m} L_j(t) \cdot y(t_j), \tag{8.51}$$

where

$$L_j(t) = \prod_{k=0}^{m} \frac{(t-t_k)}{(t_j-t_k)}, \quad k \neq j, \tag{8.52}$$

where the symbol Π is a multiplying operator from $k = 0$ to $k = m$, m is the order of the polynomial, $y(t_j)$ are given data at the time t_j, $L_j(t)$ is called the base function of order m, and t is the time at which data will be interpolated. Generally speaking, t should be placed around the middle of the time duration (t_0, t_m) if possible. Therefore, m is usually selected as an odd number. For IGS orbit interpolation, a standard m is selected as 7 or 9.

For the equal distance Lagrange interpolation there is

$$\begin{aligned} t_k &= t_0 + k\Delta t, \\ t - t_k &= t - t_0 - k\Delta t, \\ t_j - t_k &= (j-k)\Delta t, \end{aligned}$$

then

$$L_j(t) = \prod_{k=0}^{m} \frac{(t-t_0-k\Delta t)}{(j-k)\Delta t}, \quad k \neq j, \tag{8.53}$$

where Δt is the data interval.

Usually the ephemeris of the sun and the moon are computed or forecasted every half day (12 h). The ephemeris of the sun and the moon at a required epoch are interpolated from the data of the two adjacent epochs (t_1, t_2) by using a 5th-order polynomial:

$$f(t) = a + b(t-t_1) + c(t-t_1)^2 + d(t-t_1)^3 + e(t-t_1)^4 + f(t-t_1)^5.$$

For data at two epochs, e.g.,

$$t_1 : x_1, y_1, z_1, \dot{x}_1, \dot{y}_1, \dot{z}_1, \ddot{x}_1, \ddot{y}_1, \ddot{z}_1$$

and

$$t_2 : x_2, y_2, z_2, \dot{x}_2, \dot{y}_2, \dot{z}_2, \ddot{x}_2, \ddot{y}_2, \ddot{z}_2,$$

where $\dot{x}$ and $\ddot{x}$ are the velocity and acceleration components related to x. Considering the formulas of $f(t)$, $\mathrm{d}f(t)/\mathrm{d}t$, $\mathrm{d}^2 f(t)/\mathrm{d}t^2$ and letting $t = t_1$, one gets $a = x_1$, $b = \dot{x}_1$ and $c = \ddot{x}_1/2$. Letting $t = t_2$, coefficients of d, e, f can be derived theoretically, e.g., in the case of $t_2 - t_1 = 0.5$:

$$\begin{aligned} d &= 80(x_2 - x_1) - 16\dot{x}_2 - 24\dot{x}_1 + \ddot{x}_2 - 3\ddot{x}_1, \\ e &= -240(x_2 - x_1) + 56\dot{x}_2 + 64\dot{x}_1 - 4\ddot{x}_2 + 6\ddot{x}_1, \end{aligned}$$

$$f = 192(x_2 - x_1) - 48\dot{x}_2 - 48\dot{x}_1 + 4\ddot{x}_2 - 4\ddot{x}_1.$$

For components y and z, the formulas are similar. Such an interpolating algorithm is accurate enough for using the given half day ephemeris of the sun and moon to get the data at the required epoch. The computation of the ephemeris of the sun and moon are discussed in Sect. 7.8.

By deriving the Adams and Cowell algorithms, the Newtonian backward differentiation formula has been used to represent the disturbing function F. By simply considering F a function of t (t is any variable), one has

$$\begin{aligned} F(t) = F(t_n) + \frac{t-t_n}{h}\nabla F_n + \frac{(t-t_n)(t-t_{n-1})}{2!h^2}\nabla^2 F_n \\ + \cdots + \frac{(t-t_n)(t-t_{n-1})\cdots(t-t_{n-k+1})}{k!h^k}\nabla^k F_n. \end{aligned} \quad (8.54)$$

This is an interpolating formula of $F(t)$ using a set of function values of $\{F_{n-j}, j = 0,\ldots,k\}$.

8.3 Orbit-Related Partial Derivatives

As mentioned in Sect. 8.1 the partial derivatives of

$$\frac{\partial \vec{f}}{\partial \vec{r}}, \quad \frac{\partial \vec{f}}{\partial \dot{\vec{r}}} \quad \text{and} \quad \frac{\partial \vec{f}}{\partial \vec{Y}} \quad (8.55)$$

will be derived in this section in detail, where the force vector is a summated vector of all disturbing forces in the ECSF coordinate system. If the force vector is given in the ECEF coordinate system, there is

$$\left(\frac{\partial \vec{f}}{\partial \vec{r}}, \frac{\partial \vec{f}}{\partial \dot{\vec{r}}}\right) = R_P^{-1}R_N^{-1}R_S^{-1}R_M^{-1}\left(\frac{\partial \vec{f}_{\mathrm{ECEF}}}{\partial \vec{r}}, \frac{\partial \vec{f}_{\mathrm{ECEF}}}{\partial \dot{\vec{r}}}\right). \quad (8.56)$$

Because

$$\vec{r} = R \cdot \vec{r}_{\mathrm{ECEF}},$$
$$\vec{f} = R \cdot \vec{f}_{\mathrm{ECEF}},$$

one may have the velocity transformation formula

$$\frac{d\vec{r}}{dt} = \frac{dR}{dt} \cdot \vec{r}_{ECEF} + R \cdot \frac{d\vec{r}_{ECEF}}{dt},$$

where

$$R = R_P^{-1}R_N^{-1}R_S^{-1}R_M^{-1}.$$

Therefore one has

$$\frac{\partial \vec{r}_{ECEF}}{\partial \vec{r}} = R^{-1},$$
$$\frac{\partial \dot{\vec{r}}_{ECEF}}{\partial \dot{\vec{r}}} = R^{-1},$$

and

$$\frac{\partial \vec{f}_{ECEF}}{\partial \vec{r}} = \frac{\partial \vec{f}_{ECEF}}{\partial \vec{r}_{ECEF}} \frac{\partial \vec{r}_{ECEF}}{\partial \vec{r}} = \frac{\partial \vec{f}_{ECEF}}{\partial \vec{r}_{ECEF}} R^{-1},$$
$$\frac{\partial \vec{f}_{ECEF}}{\partial \dot{\vec{r}}} = \frac{\partial \vec{f}_{ECEF}}{\partial \dot{\vec{r}}_{ECEF}} \frac{\partial \dot{\vec{r}}_{ECEF}}{\partial \dot{\vec{r}}} = \frac{\partial \vec{f}_{ECEF}}{\partial \dot{\vec{r}}_{ECEF}} R^{-1}.$$

1. Geopotential Disturbing Force

The geopotential disturbing force vector (see Sect. 4.2) has the form

$$\vec{f}_{\text{ECEF}} = \begin{pmatrix} f_{x'} \\ f_{y'} \\ f_{z'} \end{pmatrix} = \begin{pmatrix} b_{11}\frac{\partial V}{\partial r} + b_{21}\frac{\partial V}{\partial \varphi} + b_{31}\frac{\partial V}{\partial \lambda} \\ b_{12}\frac{\partial V}{\partial r} + b_{22}\frac{\partial V}{\partial \varphi} + b_{32}\frac{\partial V}{\partial \lambda} \\ b_{13}\frac{\partial V}{\partial r} + b_{23}\frac{\partial V}{\partial \varphi} \end{pmatrix}, \tag{8.57}$$

where

$$\frac{\partial(r,\varphi,\lambda)}{\partial(x',y',z')} = \begin{pmatrix} b_{11} & b_{12} & b_{13} \\ b_{21} & b_{22} & b_{23} \\ b_{31} & b_{32} & b_{33} \end{pmatrix} = \begin{pmatrix} \cos\varphi\cos\lambda & \cos\varphi\sin\lambda & \sin\varphi \\ -\frac{1}{r}\sin\varphi\cos\lambda & -\frac{1}{r}\sin\varphi\sin\lambda & \frac{1}{r}\cos\varphi \\ -\frac{1}{r\cos\varphi}\sin\lambda & \frac{1}{r\cos\varphi}\cos\lambda & 0 \end{pmatrix},$$

and (x', y', z') are the three orthogonal Cartesian coordinates in the ECEF system. Thus,

$$\frac{\partial \vec{f}_{\text{ECEF}}}{\partial \vec{r}} = \begin{pmatrix} \frac{\partial f_{x'}}{\partial(x',y',z')} \\ \frac{\partial f_{y'}}{\partial(x',y',z')} \\ \frac{\partial f_{z'}}{\partial(x',y',z')} \end{pmatrix} = \left(\frac{\partial(f_{x'},f_{y'},f_{z'})}{\partial(r,\varphi,\lambda)} \frac{\partial(r,\varphi,\lambda)}{\partial(x',y',z')} \right)^T. \tag{8.58}$$

Using index $j(=1,2,3)$ to denote index (x', y', z'), one has

$$\frac{\partial f_j}{\partial(r,\varphi,\lambda)} = \begin{pmatrix} \frac{\partial b_{1j}}{\partial r}\frac{\partial V}{\partial r} + \frac{\partial b_{2j}}{\partial r}\frac{\partial V}{\partial \varphi} + \frac{\partial b_{3j}}{\partial r}\frac{\partial V}{\partial \lambda} + b_{1j}\frac{\partial^2 V}{\partial r^2} + b_{2j}\frac{\partial^2 V}{\partial r\partial\varphi} + b_{3j}\frac{\partial^2 V}{\partial r\partial\lambda} \\ \frac{\partial b_{1j}}{\partial \varphi}\frac{\partial V}{\partial r} + \frac{\partial b_{2j}}{\partial \varphi}\frac{\partial V}{\partial \varphi} + \frac{\partial b_{3j}}{\partial \varphi}\frac{\partial V}{\partial \lambda} + b_{1j}\frac{\partial^2 V}{\partial r\partial\varphi} + b_{2j}\frac{\partial^2 V}{\partial \varphi^2} + b_{3j}\frac{\partial^2 V}{\partial \varphi\partial\lambda} \\ \frac{\partial b_{1j}}{\partial \lambda}\frac{\partial V}{\partial r} + \frac{\partial b_{2j}}{\partial \lambda}\frac{\partial V}{\partial \varphi} + \frac{\partial b_{3j}}{\partial \lambda}\frac{\partial V}{\partial \lambda} + b_{1j}\frac{\partial^2 V}{\partial r\partial\lambda} + b_{2j}\frac{\partial^2 V}{\partial \varphi\partial\lambda} + b_{3j}\frac{\partial^2 V}{\partial \lambda^2} \end{pmatrix}^T, \tag{8.59}$$

where

$$\frac{\partial}{\partial r}\begin{pmatrix} b_{11} & b_{12} & b_{13} \\ b_{21} & b_{22} & b_{23} \\ b_{31} & b_{32} & b_{33} \end{pmatrix} = \begin{pmatrix} 0 & 0 & 0 \\ \frac{1}{r^2}\sin\varphi\cos\lambda & \frac{1}{r^2}\sin\varphi\sin\lambda & \frac{-1}{r^2}\cos\varphi \\ \frac{1}{r^2\cos\varphi}\sin\lambda & \frac{-1}{r^2\cos\varphi}\cos\lambda & 0 \end{pmatrix},$$

$$\frac{\partial}{\partial \varphi}\begin{pmatrix} b_{11} & b_{12} & b_{13} \\ b_{21} & b_{22} & b_{23} \\ b_{31} & b_{32} & b_{33} \end{pmatrix} = \begin{pmatrix} -\sin\varphi\cos\lambda & -\sin\varphi\sin\lambda & \cos\varphi \\ -\frac{1}{r}\cos\varphi\cos\lambda & -\frac{1}{r}\cos\varphi\sin\lambda & -\frac{1}{r}\sin\varphi \\ -\frac{\sin\varphi}{r\cos^2\varphi}\sin\lambda & \frac{\sin\varphi}{r\cos^2\varphi}\cos\lambda & 0 \end{pmatrix}$$

and

$$\frac{\partial}{\partial \lambda}\begin{pmatrix} b_{11} & b_{12} & b_{13} \\ b_{21} & b_{22} & b_{23} \\ b_{31} & b_{32} & b_{33} \end{pmatrix} = \begin{pmatrix} -\cos\varphi\sin\lambda & \cos\varphi\cos\lambda & 0 \\ \frac{1}{r}\sin\varphi\sin\lambda & -\frac{1}{r}\sin\varphi\cos\lambda & 0 \\ -\frac{1}{r\cos\varphi}\cos\lambda & -\frac{1}{r\cos\varphi}\sin\lambda & 0 \end{pmatrix} \tag{8.60}$$

and

$$\frac{\partial^2 V}{\partial r^2} = \frac{\mu}{r^3}\left[2 + \sum_{l=2}^{\infty}\sum_{m=0}^{l}(l+1)(l+2)\left(\frac{a}{r}\right)^l \overline{P}_{lm}(\sin\varphi)[\overline{C}_{lm}\cos m\lambda + \overline{S}_{lm}\sin m\lambda]\right],$$

$$\frac{\partial^2 V}{\partial r\partial\varphi} = -\frac{\mu}{r^2}\sum_{l=2}^{\infty}\sum_{m=0}^{l}(l+1)\left(\frac{a}{r}\right)^l \frac{\mathrm{d}\overline{P}_{lm}(\sin\varphi)}{\mathrm{d}\varphi}[\overline{C}_{lm}\cos m\lambda + \overline{S}_{lm}\sin m\lambda],$$

$$\frac{\partial^2 V}{\partial r\partial\lambda} = -\frac{\mu}{r^2}\left[\sum_{l=2}^{\infty}\sum_{m=0}^{l}(l+1)\left(\frac{a}{r}\right)^l \overline{P}_{lm}(\sin\varphi)m[-\overline{C}_{lm}\sin m\lambda + \overline{S}_{lm}\cos m\lambda]\right],$$

$$\frac{\partial^2 V}{\partial \varphi^2} = \frac{\mu}{r}\sum_{l=2}^{\infty}\sum_{m=0}^{l}\left(\frac{a}{r}\right)^l \frac{\mathrm{d}^2\overline{P}_{lm}(\sin\varphi)}{\mathrm{d}\varphi^2}[\overline{C}_{lm}\cos m\lambda + \overline{S}_{lm}\sin m\lambda],$$

$$\frac{\partial^2 V}{\partial \varphi \partial \lambda} = \frac{\mu}{r} \sum_{l=2}^{\infty} \sum_{m=0}^{l} \left(\frac{a}{r}\right)^l \frac{\mathrm{d}\overline{P}_{lm}(\sin\varphi)}{\mathrm{d}\varphi} m[-\overline{C}_{lm} \sin m\lambda + \overline{S}_{lm} \cos m\lambda]$$

and

$$\frac{\partial^2 V}{\partial \lambda^2} = -\frac{\mu}{r} \sum_{l=2}^{\infty} \sum_{m=0}^{l} m^2 \left(\frac{a}{r}\right)^l \overline{P}_{lm}(\sin\varphi)[\overline{C}_{lm} \cos m\lambda + \overline{S}_{lm} \sin m\lambda], \tag{8.61}$$

where

$$\frac{\mathrm{d}\overline{P}_{lm}(\sin\varphi)}{\mathrm{d}\varphi} = \beta(m)\overline{P}_{l(m+1)}(\sin\varphi) - m\tan\varphi\overline{P}_{lm}(\sin\varphi),$$

$$\frac{\mathrm{d}^2\overline{P}_{lm}(\sin\varphi)}{\mathrm{d}\varphi^2} = \beta(m)\frac{\mathrm{d}\overline{P}_{l(m+1)}(\sin\varphi)}{\mathrm{d}\varphi} - m\frac{1}{\cos^2\varphi}\overline{P}_{lm}(\sin\varphi) - m\tan\varphi\frac{\mathrm{d}\overline{P}_{lm}(\sin\varphi)}{\mathrm{d}\varphi}$$

$$= \beta(m)\beta(m+1)\overline{P}_{l(m+2)}(\sin\varphi) - \beta(m)\tan\varphi(2m+1)\overline{P}_{l(m+1)}(\sin\varphi)$$
$$+ \left(m^2\tan^2\varphi - m\frac{1}{\cos^2\varphi}\right)\overline{P}_{lm}(\sin\varphi),$$

$$\beta(m) = \left[\frac{1}{2}(2-\delta_{0m})(l-m)(l+m+1)\right]^{1/2}$$

and

$$\beta(m+1) = \left[\frac{1}{2}(l-m-1)(l+m+2)\right]^{1/2}. \tag{8.62}$$

Other needed functions are already given in Sect. 4.2. Because the force is not a function of velocity, it is obvious that

$$\frac{\partial \vec{f}_{\mathrm{ECEF}}}{\partial \dot{\vec{r}}} = [0]_{3\times 3}. \tag{8.63}$$

Only non-zero partial derivatives will be given in the text that follows.

Supposing the geopotential parameters $\overline{C}_{lm}^N$, $\overline{S}_{lm}^N$ are known (as initial values), $\overline{C}_{lm}$, $\overline{S}_{lm}$ are true values, and $\Delta\overline{C}_{lm}$, $\Delta\overline{S}_{lm}$ are searched corrections (unknowns), then the geopotential force is

$$\begin{aligned}\vec{f}_{\mathrm{ECEF}}(\overline{C}_{lm}, \overline{S}_{lm}) &= \vec{f}_{\mathrm{ECEF}}(\overline{C}_{lm}^N, \overline{S}_{lm}^N) + \vec{f}_{\mathrm{ECEF}}(\overline{C}_{lm}, \overline{S}_{lm}) - \vec{f}_{\mathrm{ECEF}}(\overline{C}_{lm}^N, \overline{S}_{lm}^N)\\ &= \vec{f}_{\mathrm{ECEF}}(\overline{C}_{lm}^N, \overline{S}_{lm}^N) + \vec{f}_{\mathrm{ECEF}}(\Delta\overline{C}_{lm}, \Delta\overline{S}_{lm}),\end{aligned} \tag{8.64}$$

and

$$\frac{\partial \vec{f}_{\text{ECEF}}}{\partial(\Delta\overline{C}_{lm},\Delta\overline{S}_{lm})} = \begin{pmatrix} b_{11} & b_{12} & b_{13} \\ b_{21} & b_{22} & b_{23} \\ b_{31} & b_{32} & b_{33} \end{pmatrix}^T \frac{\partial}{\partial(\Delta\overline{C}_{lm},\Delta\overline{S}_{lm})} \begin{pmatrix} \frac{\partial V}{\partial r} \\ \frac{\partial V}{\partial \varphi} \\ \frac{\partial V}{\partial \lambda} \end{pmatrix},$$

$$\frac{\partial}{\partial(\Delta\overline{C}_{lm},\Delta\overline{S}_{lm})}\left(\frac{\partial V}{\partial r}\right) = -\frac{\mu}{r^2}(l+1)\left(\frac{a}{r}\right)^l \overline{P}_{lm}(\sin\varphi)\left(\cos m\lambda \;\; \sin m\lambda\right),$$

$$\frac{\partial}{\partial(\Delta\overline{C}_{lm},\Delta\overline{S}_{lm})}\left(\frac{\partial V}{\partial \varphi}\right) = \frac{\mu}{r}\left(\frac{a}{r}\right)^l \frac{\mathrm{d}\overline{P}_{lm}(\sin\varphi)}{\mathrm{d}\varphi}\left(\cos m\lambda \;\; \sin m\lambda\right)$$

and

$$\frac{\partial}{\partial(\Delta\overline{C}_{lm},\Delta\overline{S}_{lm})}\left(\frac{\partial V}{\partial \lambda}\right) = \frac{\mu}{r} m \left(\frac{a}{r}\right)^l \overline{P}_{lm}(\sin\varphi)\left(-\sin m\lambda \;\; \cos m\lambda\right). \tag{8.65}$$

2. Perturbation Forces of the Sun and the Moon as well as Planets

The perturbation forces of the sun, the moon and the planets are given in Sect. 4.2 (see (4.50)) as

$$\vec{f}_m = -m\sum_j \mathrm{Gm}(j)\left[\frac{\vec{r}-\vec{r}_{m(j)}}{\left|\vec{r}-\vec{r}_{m(j)}\right|^3} + \frac{\vec{r}_{m(j)}}{r^3_{m(j)}}\right], \tag{8.66}$$

where Gm(j) are the gravitational constants of the sun and the moon as well as the planets, and the vectors with index $m(j)$ are the geocentric vectors of the sun, the moon and the planets. The partial derivatives of the perturbation force with respect to the satellite vector are then

$$\frac{\partial \vec{f}_m}{\partial \vec{r}} = -m\sum_j \frac{\mathrm{Gm}(j)}{\left|\vec{r}-\vec{r}_{m(j)}\right|^3}\left(E + \frac{3}{\left|\vec{r}-\vec{r}_{m(j)}\right|^2}\begin{pmatrix} x-x_{m(j)} \\ y-y_{m(j)} \\ z-z_{m(j)} \end{pmatrix}\begin{pmatrix} x-x_{m(j)} \\ y-y_{m(j)} \\ z-z_{m(j)} \end{pmatrix}^T\right), \tag{8.67}$$

where E is an identity matrix of size 3×3. The partial derivatives of the force vector with respect to the velocity vector of the satellite are zero. The disturbances of the sun, moon and planets are considered well-modelled; therefore, no parameters will be adjusted. In other words, the partial derivatives of the force vector with respect to the model parameters do not exist.

3. Tidal Disturbing Forces

Similar to the geopotential attracting force, the tidal force (see Sect. 4.2.3) has the form

$$\vec{f}_{\mathrm{ECEF}} = \begin{pmatrix} f_{x'} \\ f_{y'} \\ f_{z'} \end{pmatrix} = \begin{pmatrix} b_{11}\dfrac{\partial V}{\partial r} + b_{21}\dfrac{\partial V}{\partial \varphi} + b_{31}\dfrac{\partial V}{\partial \lambda} \\ b_{12}\dfrac{\partial V}{\partial r} + b_{22}\dfrac{\partial V}{\partial \varphi} + b_{32}\dfrac{\partial V}{\partial \lambda} \\ b_{13}\dfrac{\partial V}{\partial r} + b_{23}\dfrac{\partial V}{\partial \varphi} \end{pmatrix}, \tag{8.68}$$

where $V = \delta V + \delta V_1 + \delta V_2$, which is a summation of the Earth tide potential and the two parts of ocean loading tide potentials. Equation (8.59) is still valid for this case. Other higher order partial derivatives can be derived as follows:

$$\frac{\partial^2 \delta V}{\partial r^2} = \sum_{j=1}^{2} \mu_j \sum_{n=2}^{N} k_n \frac{(n+1)(n+2)a_e^{2n+1}}{r^{n+3} r_j^{n+1}} \begin{bmatrix} P_n(\sin\varphi)P_n(\sin\delta_j) \\ +2\sum\limits_{k=1}^{n} \dfrac{(n-k)!}{(n+k)!} P_{nk}(\sin\varphi)P_{nk}(\sin\delta_j)\cos kh_j \end{bmatrix},$$

$$\frac{\partial^2 \delta V}{\partial r \partial \varphi} = \sum_{j=1}^{2} \mu_j \sum_{n=2}^{N} -k_n \frac{(n+1)a_e^{2n+1}}{r^{n+2} r_j^{n+1}} \begin{bmatrix} \dfrac{\mathrm{d}P_n(\sin\varphi)}{\mathrm{d}\varphi} P_n(\sin\delta_j) \\ +2\sum\limits_{k=1}^{n} \dfrac{(n-k)!}{(n+k)!} \dfrac{\mathrm{d}P_{nk}(\sin\varphi)}{\mathrm{d}\varphi} P_{nk}(\sin\delta_j)\cos kh_j \end{bmatrix},$$

$$\frac{\partial^2 \delta V}{\partial r \partial \lambda} = \sum_{j=1}^{2} \mu_j \sum_{n=2}^{N} -k_n \frac{(n+1)a_e^{2n+1}}{r^{n+2} r_j^{n+1}} \left[2\sum_{k=1}^{n} \frac{(n-k)!}{(n+k)!} P_{nk}(\sin\varphi)P_{nk}(\sin\delta_j) k \sin kh_j \right],$$

$$\frac{\partial^2 \delta V}{\partial \varphi^2} = \sum_{j=1}^{2} \mu_j \sum_{n=2}^{N} k_n \frac{a_e^{2n+1}}{r^{n+1} r_j^{n+1}} \begin{bmatrix} \dfrac{\mathrm{d}^2 P_n(\sin\varphi)}{\mathrm{d}\varphi^2} P_n(\sin\delta_j) \\ +2\sum\limits_{k=1}^{n} \dfrac{(n-k)!}{(n+k)!} \dfrac{\mathrm{d}^2 P_{nk}(\sin\varphi)}{\mathrm{d}\varphi^2} P_{nk}(\sin\delta_j)\cos kh_j \end{bmatrix},$$

$$\frac{\partial^2 \delta V}{\partial \varphi \partial \lambda} = \sum_{j=1}^{2} \mu_j \sum_{n=2}^{N} k_n \frac{a_e^{2n+1}}{r^{n+1} r_j^{n+1}} \left[2\sum_{k=1}^{n} \frac{(n-k)!}{(n+k)!} \frac{\mathrm{d}P_{nk}(\sin\varphi)}{\mathrm{d}\varphi} P_{nk}(\sin\delta_j) k \sin kh_j \right],$$

$$\frac{\partial^2 \delta V}{\partial \lambda^2} = \sum_{j=1}^{2} \mu_j \sum_{n=2}^{N} -k_n \frac{a_e^{2n+1}}{r^{n+1} r_j^{n+1}} \left[2\sum_{k=1}^{n} \frac{(n-k)!}{(n+k)!} k^2 P_{nk}(\sin\varphi)P_{nk}(\sin\delta_j)\cos kh_j \right],$$

$$\frac{\partial^2 \delta V_1}{\partial r^2} = \oint\!\!\!\int_O G\sigma H \sum_{n=0}^{\infty} (1+k'_n) \frac{(n+1)(n+2)a_e^n}{r^{n+3}} \begin{bmatrix} P_n(\sin\varphi)P_n(\sin\varphi_s) + (2-\delta_{0n}) \cdot \\ \sum\limits_{k=0}^{n} \dfrac{(n-k)!}{(n+k)!} P_{nk}(\sin\varphi)P_{nk}(\sin\varphi_s)\cos k(\lambda_s - \lambda) \end{bmatrix} \mathrm{d}s,$$

$$\frac{\partial^2 \delta V_1}{\partial r \partial \varphi} = \oint\!\!\!\int_O G\sigma H \sum_{n=0}^{\infty} (1+k'_n) \frac{-(n+1)a_e^n}{r^{n+2}} \begin{bmatrix} \dfrac{\mathrm{d}P_n(\sin\varphi)}{\mathrm{d}\varphi} P_n(\sin\varphi_s) + (2-\delta_{0n}) \cdot \\ \sum\limits_{k=0}^{n} \dfrac{(n-k)!}{(n+k)!} \dfrac{\mathrm{d}P_{nk}(\sin\varphi)}{\mathrm{d}\varphi} P_{nk}(\sin\varphi_s)\cos k(\lambda_s - \lambda) \end{bmatrix} \mathrm{d}s,$$

$$\frac{\partial^2 \delta V_1}{\partial r \partial \lambda} = \iint\limits_O G\sigma H \sum_{n=0}^{\infty} (1+k'_n) \frac{-(n+1)a_e^n}{r^{n+2}} \begin{bmatrix} (2-\delta_{0n}) \cdot \\ \sum_{k=0}^{n} \frac{(n-k)!}{(n+k)!} P_{nk}(\sin\varphi) P_{nk}(\sin\varphi_s) k \sin k(\lambda_s - \lambda) \end{bmatrix} \mathrm{d}s,$$

$$\frac{\partial^2 \delta V_1}{\partial \varphi^2} = \iint\limits_O G\sigma H \sum_{n=0}^{\infty} (1+k'_n) \frac{a_e^n}{r^{n+1}} \begin{bmatrix} \frac{\mathrm{d}^2 P_n(\sin\varphi)}{\mathrm{d}\varphi^2} P_n(\sin\varphi_s) + (2-\delta_{0n}) \cdot \\ \sum_{k=0}^{n} \frac{(n-k)!}{(n+k)!} \frac{\mathrm{d}^2 P_{nk}(\sin\varphi)}{\mathrm{d}\varphi^2} P_{nk}(\sin\varphi_s) \cos k(\lambda_s - \lambda) \end{bmatrix} \mathrm{d}s,$$

$$\frac{\partial^2 \delta V_1}{\partial \varphi \partial \lambda} = \iint\limits_O G\sigma H \sum_{n=0}^{\infty} (1+k'_n) \frac{a_e^n}{r^{n+1}} \begin{bmatrix} (2-\delta_{0n}) \cdot \\ \sum_{k=0}^{n} \frac{(n-k)!}{(n+k)!} \frac{\mathrm{d} P_{nk}(\sin\varphi)}{\mathrm{d}\varphi} P_{nk}(\sin\varphi_s) k \sin k(\lambda_s - \lambda) \end{bmatrix} \mathrm{d}s,$$

$$\frac{\partial^2 \delta V_1}{\partial \lambda^2} = \iint\limits_O G\sigma H \sum_{n=0}^{\infty} (1+k'_n) \frac{a_e^n}{r^{n+1}} \begin{bmatrix} -(2-\delta_{0n}) \cdot \\ \sum_{k=0}^{n} \frac{(n-k)!}{(n+k)!} P_{nk}(\sin\varphi) P_{nk}(\sin\varphi_s) k^2 \cos k(\lambda_s - \lambda) \end{bmatrix} \mathrm{d}s,$$

$$\frac{\partial^2 \delta V_2}{\partial r^2} = \iint\limits_C G\sigma_e u_r \sum_{n=0}^{\infty} \frac{(n+1)(n+2)a_e^n}{r^{n+3}} \begin{bmatrix} P_n(\sin\varphi) P_n(\sin\varphi_s) + (2-\delta_{0n}) \cdot \\ \sum_{k=0}^{n} \frac{(n-k)!}{(n+k)!} P_{nk}(\sin\varphi) P_{nk}(\sin\varphi_s) \cos k(\lambda_s - \lambda) \end{bmatrix} \mathrm{d}s,$$

$$\frac{\partial^2 \delta V_2}{\partial r \partial \varphi} = \iint\limits_C G\sigma_e u_r \sum_{n=0}^{\infty} \frac{-(n+1)a_e^n}{r^{n+2}} \begin{bmatrix} \frac{\mathrm{d} P_n(\sin\varphi)}{\mathrm{d}\varphi} P_n(\sin\varphi_s) + (2-\delta_{0n}) \cdot \\ \sum_{k=0}^{n} \frac{(n-k)!}{(n+k)!} \frac{\mathrm{d} P_{nk}(\sin\varphi)}{\mathrm{d}\varphi} P_{nk}(\sin\varphi_s) \cos k(\lambda_s - \lambda) \end{bmatrix} \mathrm{d}s,$$

$$\frac{\partial^2 \delta V_2}{\partial r \partial \lambda} = \iint\limits_C G\sigma_e u_r \sum_{n=0}^{\infty} \frac{-(n+1)a_e^n}{r^{n+2}} \begin{bmatrix} (2-\delta_{0n}) \cdot \\ \sum_{k=0}^{n} \frac{(n-k)!}{(n+k)!} P_{nk}(\sin\varphi) P_{nk}(\sin\varphi_s) k \sin k(\lambda_s - \lambda) \end{bmatrix} \mathrm{d}s,$$

$$\frac{\partial^2 \delta V_2}{\partial \varphi^2} = \iint\limits_C G\sigma_e u_r \sum_{n=0}^{\infty} \frac{a_e^n}{r^{n+1}} \begin{bmatrix} \frac{\mathrm{d}^2 P_n(\sin\varphi)}{\mathrm{d}\varphi^2} P_n(\sin\varphi_s) + (2-\delta_{0n}) \cdot \\ \sum_{k=0}^{n} \frac{(n-k)!}{(n+k)!} \frac{\mathrm{d}^2 P_{nk}(\sin\varphi)}{\mathrm{d}\varphi^2} P_{nk}(\sin\varphi_s) \cos k(\lambda_s - \lambda) \end{bmatrix} \mathrm{d}s,$$

$$\frac{\partial^2 \delta V_2}{\partial \varphi \partial \lambda} = \iint\limits_C G\sigma_e u_r \sum_{n=0}^{\infty} \frac{a_e^n}{r^{n+1}} \begin{bmatrix} (2-\delta_{0n}) \cdot \\ \sum_{k=0}^{n} \frac{(n-k)!}{(n+k)!} \frac{\mathrm{d} P_{nk}(\sin\varphi)}{\mathrm{d}\varphi} P_{nk}(\sin\varphi_s) k \sin k(\lambda_s - \lambda) \end{bmatrix} \mathrm{d}s$$

and

$$\frac{\partial^2 \delta V_2}{\partial \lambda^2} = \iint\limits_C G\sigma_e u_r \sum_{n=0}^{\infty} \frac{a_e^n}{r^{n+1}} \begin{bmatrix} -(2-\delta_{0n}) \cdot \\ \sum_{k=0}^{n} \frac{(n-k)!}{(n+k)!} P_{nk}(\sin\varphi) P_{nk}(\sin\varphi_s) k^2 \cos k(\lambda_s - \lambda) \end{bmatrix} \mathrm{d}s, \tag{8.69}$$

where

$$\frac{\mathrm{d}P_n(\sin\varphi)}{\mathrm{d}\varphi} = \frac{n}{\cos\varphi}(P_{n-1}(\sin\varphi) - \sin\varphi P_n(\sin\varphi))$$

and

$$\frac{\mathrm{d}P_{nk}(\sin\varphi)}{\mathrm{d}\varphi} = P_{n(k+1)}(\sin\varphi) - k\tan\varphi P_{nk}(\sin\varphi). \tag{8.70}$$

4. Solar Radiation Pressure

If solar radiation force acting on the satellite's surface (see Sect. 4.2.4) is

$$\vec{f}_{\text{solar}} = m\gamma P_s C_r r_{\text{sun}}^2 \frac{S}{m} \frac{\vec{r} - \vec{r}_{\text{sun}}}{|\vec{r} - \vec{r}_{\text{sun}}|^3}, \tag{8.71}$$

then the partial derivatives of the perturbation force with respect to the satellite vector are

$$\frac{\partial \vec{f}_{\text{solar}}}{\partial \vec{r}} = m\gamma P_s C_r r_{\text{sun}}^2 \frac{S}{m} \frac{1}{|\vec{r} - \vec{r}_{\text{sun}}|^3} \left(E - \frac{3}{|\vec{r} - \vec{r}_{\text{sun}}|^2} \begin{pmatrix} x - x_{\text{sun}} \\ y - y_{\text{sun}} \\ z - z_{\text{sun}} \end{pmatrix} \begin{pmatrix} x - x_{\text{sun}} \\ y - y_{\text{sun}} \\ z - z_{\text{sun}} \end{pmatrix}^T \right), \tag{8.72}$$

where E is an identity matrix of size 3×3. The partial derivatives of the force vector with respect to the velocity vector of the satellite are zero. The disturbance of the solar radiation is considered not well-modelled; therefore, unknown parameters will also be adjusted. The total model (see Sect. 4.2.4) is

$$\vec{f}_{\text{solar-force}} = \vec{f}_{\text{solar}} + \begin{pmatrix} a_{11} & a_{12} & a_{13} \\ a_{21} & a_{22} & a_{23} \\ a_{31} & a_{32} & a_{33} \end{pmatrix} \begin{pmatrix} 1 \\ \cos u \\ \sin u \end{pmatrix}. \tag{8.73}$$

Thus,

$$\frac{\partial \vec{f}_{\text{solar-force}}}{\partial \vec{r}} = \frac{\partial \vec{f}_{\text{solar}}}{\partial \vec{r}} + \begin{pmatrix} a_{11} & a_{12} & a_{13} \\ a_{21} & a_{22} & a_{23} \\ a_{31} & a_{32} & a_{33} \end{pmatrix} \begin{pmatrix} 0 \\ -\sin u \\ \cos u \end{pmatrix} \frac{\partial u}{\partial \vec{r}}, \tag{8.74}$$

where

$$\frac{\partial u}{\partial \vec{r}} = \frac{\partial u}{\partial(\Omega, i, \omega, a, e, M)} \frac{\partial(\Omega, i, \omega, a, e, M)}{\partial(\vec{r}, \dot{\vec{r}})} \frac{\partial(\vec{r}, \dot{\vec{r}})}{\partial \vec{r}}. \tag{8.75}$$

On the right-hand side of above equation there are three matrices, the first one is a 1×6 matrix (vector) and is given in Sect. 4.1.2 (see (4.24)), the second one is given as its inverse in Sect. 5.4 (see (5.59) and (5.61)), and the third one is a 6×3 matrix, or

$$\frac{\partial u}{\partial(\Omega, i, \omega, a, e, M)} = \left(0,\ 0,\ 1,\ 0,\ \frac{2 + e\cos f}{1 - e^2} \sin f,\ \left(\frac{a}{r}\right)^2 \sqrt{1 - e^2}\right),$$

$$\frac{\partial(\Omega,i,\omega,a,e,M)}{\partial(\vec{r},\dot{\vec{r}})} = \left(\frac{\partial(\vec{r},\dot{\vec{r}})}{\partial(\Omega,i,\omega,a,e,M)}\right)^{-1} = \begin{pmatrix} \frac{\partial R}{\partial(\Omega,i,\omega)}\vec{q} & R\frac{\partial\vec{q}}{\partial(a,e,M)} \\ \frac{\partial R}{\partial(\Omega,i,\omega)}\dot{\vec{q}} & R\frac{\partial\dot{\vec{q}}}{\partial(a,e,M)} \end{pmatrix}^{-1}$$

and

$$\frac{\partial(\vec{r},\dot{\vec{r}})}{\partial\vec{r}} = \begin{pmatrix} E_{3\times3} \\ 0_{3\times3} \end{pmatrix}. \tag{8.76}$$

$$\frac{\partial u}{\partial\dot{\vec{r}}} = \frac{\partial u}{\partial(\Omega,i,\omega,a,e,M)} \frac{\partial(\Omega,i,\omega,a,e,M)}{\partial(\vec{r},\dot{\vec{r}})} \frac{\partial(\vec{r},\dot{\vec{r}})}{\partial\dot{\vec{r}}}$$

and

$$\frac{\partial(\vec{r},\dot{\vec{r}})}{\partial\dot{\vec{r}}} = \begin{pmatrix} 0_{3\times3} \\ E_{3\times3} \end{pmatrix}. \tag{8.77}$$

The partial derivatives of the force vector with respect to the model parameters are (for $i = 1,2,3$)

$$\frac{\partial\vec{f}_{\text{solar-force}}}{\partial a_{ij}} = \begin{cases} 1, & \text{if} j = 1, \\ \cos u, & \text{if} j = 2, \\ \sin u, & \text{if} j = 3. \end{cases} \tag{8.78}$$

If the model (4.74)

$$\alpha\vec{f}_{\text{solar}} = \begin{pmatrix} a_1 & b_1 \\ a_2 & b_2 \\ a_3 & b_3 \end{pmatrix} \begin{pmatrix} 1 \\ t \end{pmatrix} \tag{8.79}$$

is used, then one has

$$\frac{\partial\vec{f}_{\text{solar-force}}}{\partial(a_i,b_i)} = (1,t), \quad i = 1,2,3. \tag{8.80}$$

5. Atmospheric Drag

Atmospheric drag force has the form (see Sect. 4.2.5)

$$\vec{f}_{\text{drag}} = -m\frac{1}{2}\left(\frac{C_d S}{m}\right)\sigma\left|\dot{\vec{r}}-\dot{\vec{r}}_{\text{air}}\right|\left(\dot{\vec{r}}-\dot{\vec{r}}_{\text{air}}\right), \tag{8.81}$$

and the air drag force model is

$$\vec{f}_{\text{air-drag}} = \vec{f}_{\text{drag}} + (1+q)\Delta\vec{f}_{\text{drag}}, \tag{8.82}$$

where (see (4.84) and (4.85))

$$\Delta\vec{f}_{\text{drag}} = [a + b\varphi(2\omega)\cos(2f) + c\varphi(3\omega)\cos(3f) + d\varphi(\omega)\cos f]\,\vec{p}, \tag{8.83}$$

$$\varphi(k\omega) = \begin{cases} \sin k\omega, & \text{if } \cos k\omega = 0 \\ \dfrac{1}{\cos k\omega}, & \text{if } \cos k\omega \neq 0 \end{cases} \quad (k = 1, 2, 3). \tag{8.84}$$

It is obvious that the partial derivatives of the air drag force with respect to the satellite position vector are zero, and

$$\frac{\partial \vec{f}_{\text{drag}}}{\partial \dot{\vec{r}}} = -m\frac{1}{2}\left(\frac{C_{\text{d}}S}{m}\right)\sigma\left(\left|\dot{\vec{r}} - \dot{\vec{r}}_{\text{air}}\right|E + \frac{1}{\left|\dot{\vec{r}} - \dot{\vec{r}}_{\text{air}}\right|}\begin{pmatrix} \dot{x} - \dot{x}_{\text{air}} \\ \dot{y} - \dot{y}_{\text{air}} \\ \dot{z} - \dot{z}_{\text{air}} \end{pmatrix}\begin{pmatrix} \dot{x} - \dot{x}_{\text{air}} \\ \dot{y} - \dot{y}_{\text{air}} \\ \dot{z} - \dot{z}_{\text{air}} \end{pmatrix}^T\right), \tag{8.85}$$

$$\frac{\partial \Delta \vec{f}_{\text{drag}}}{\partial f} = [-2b\varphi(2\omega)\sin(2f) - c\varphi(3\omega)\sin(3f) - d\varphi(\omega)\sin f]\,\vec{p}, \tag{8.86}$$

$$\frac{\partial \Delta \vec{f}_{\text{drag}}}{\partial \omega} = \left[b\cos(2f)\frac{\partial \varphi(2\omega)}{\partial \omega} + c\cos(3f)\frac{\partial \varphi(3\omega)}{\partial \omega} + d\cos f\frac{\partial \varphi(\omega)}{\partial \omega}\right]\vec{p}, \tag{8.87}$$

$$\frac{\partial \varphi(k\omega)}{\partial \omega} = \begin{cases} k\cos k\omega, & \text{if } \cos k\omega = 0 \\ \dfrac{k\tan k\omega}{\cos k\omega}, & \text{if } \cos k\omega \neq 0 \end{cases} \quad (k = 1, 2, 3), \tag{8.88}$$

$$\frac{\partial \Delta \vec{f}_{\text{drag}}}{\partial (\vec{r}, \dot{\vec{r}})} = \frac{\partial \Delta \vec{f}_{\text{drag}}}{\partial (\omega, f)}\frac{\partial (\omega, f)}{\partial (\Omega, i, \omega, a, e, M)}\frac{\partial (\Omega, i, \omega, a, e, M)}{\partial (\vec{r}, \dot{\vec{r}})}\frac{\partial (\vec{r}, \dot{\vec{r}})}{\partial (\vec{r}, \dot{\vec{r}})}, \tag{8.89}$$

where

$$\frac{\partial \omega}{\partial (\Omega, i, \omega, a, e, M)} = (0, 0, 1, 0, 0, 0),$$

$$\frac{\partial f}{\partial (\Omega, i, \omega, a, e, M)} = \left(0, 0, 0, 0, \frac{2 + e\cos f}{1 - e^2}\sin f, \left(\frac{a}{r}\right)^2\sqrt{1 - e^2}\right)$$

Some of the formulas in this subsection have been already derived. The partial derivatives of the force vector with respect to the model parameters can be obtained from (8.82) and (8.83).

Chapter 9
Analytic Orbit Determination

Chapters 5, 6 and 7 covered the most important contents of analytic solutions of the disturbed equations of satellite motion. In this chapter, emphasis will be on the applications of the analytic orbit theory.

9.1 Principle of Analytic Orbit Determination

Orbit determination aims to determine the initial orbital elements (i.e., the initial state vector of the satellite) and the unknown model parameters. The technique of numerical orbit determination is developed in a situation that, on one hand, one needs the technique; however on the other hand, one does not have analytic solutions of the disturbed equations of satellite motion. The key difference between the numerical and the analytic orbit determination is that the orbits are represented in the former algorithm by differential equations and in the latter algorithm by analytic formulas.

Recalling the discussions in Chaps. 5, 6 and 7, the perturbed orbit of the satellite is the solution (or integration)

$$\sigma_j(t) = \sigma_j(t_0) + (G_j(t) - G_j(t_0)) \quad \text{where } G_j(t) - G_j(t_0) = \int_{t_0}^{t} F_j \mathrm{d}t. \tag{9.1}$$

$G_j(t)$ are the infinite integrations of the right functions of the equations of motion and are given explicitly by analytic formulas. Equations (9.1) have been obtained by integrating the disturbed equations of motion

$$\begin{cases} \dot{\sigma}_j(t) = F_j, \\ \sigma_j(t_0) = \sigma_{j0}, \end{cases} \tag{9.2}$$

where $\sigma_j(t)$ is the jth Keplerian element, $\sigma_j(t_0)$ is the related initial value at time t_0, F_j is the related right function of the differential equation and is a function of disturbing forces.

If the initial Keplerian elements and the force functions are precisely known, then the precise orbits can be computed by using (9.1). Computing for time t of future, the so-called forecasted orbits can be obtained. That is, for orbit determination using analytic solutions, the traditional numerical integration algorithms are not necessary any more (because the differential equations are theoretically integrated by deriving the solutions).

In practice, the precise initial Keplerian elements are not exactly known and the parameters of the force models have to be co-determined. These can be realised through suitable parameterisation of the models in the GPS observation equations and then solved by adjustment or filtering.

We generally denote both the range and range rate together by ρ; their partial derivatives with respect to the orbit state vector (see Xu, 2003/2007) have the form

$$\frac{\partial\rho}{\partial\vec{r}}, \frac{\partial\rho}{\partial\dot{\vec{r}}}. \tag{9.3}$$

Therefore, the orbit parameter related parts in the linearised GPS observation equation are then

$$\frac{\partial\rho}{\partial(\vec{r},\dot{\vec{r}})}\frac{\partial(\vec{r},\dot{\vec{r}})}{\partial(\sigma_j, j=1,\ldots,6)}\frac{\partial(\sigma_j, j=1,\ldots,6)}{\partial\vec{y}}\Delta\vec{y}^T, \tag{9.4}$$

where

$$\vec{y} = \left(\vec{\sigma}_0, \vec{Y}\right), \quad \Delta\vec{y}^T = \left(\Delta\vec{\sigma}_0, \Delta\vec{Y}\right)^T. \tag{9.5}$$

$\vec{\sigma}, \vec{Y}$ are the Keplerian element vector and the parameter vector of the force models, and index 0 denotes the related initial vectors of time t_0. $\vec{y}$ is the total unknown vector of the orbit determination problem, the related correction vector is $\Delta\vec{y} = \vec{y} - \vec{y}_0$, and $\Delta\vec{\sigma}_0$ is the correction vector of the initial Keplerian element vector. The partial derivatives of the satellite state vector with respect to the Keplerian element vector are known and can be found in Sect. 5.4. The partial derivative of the Keplerian element vector with respect to $\vec{y}$ is called transition matrix which has the dimension of $6\times(6+n)$, where n is the dimension of vector $\vec{Y}$. Because of the analytic solutions of the disturbed equations of motion, the partial derivatives of the Keplerian elements with respect to the vector $\vec{y}$ are almost given by the solutions explicitly. That is to say, by analytic orbit determination, the transition matrix is represented by analytic formulas instead of the so-called variation equations in the numerical algorithm. The variation equation has disappeared from the orbit determination process; so the numerical integration algorithms traditionally used to solve the variation equation are not necessary any more.

Note that the orbit disturbances are mostly linear functions of the parameters of the force models. Therefore, the partial derivatives of Keplerian element vector with respect to parameter vector $\vec{y}$ of the force models are directly the coefficients of the related force parameters. No special derivations of the partial derivatives are needed.

Compared to numerical orbit determination (Chap. 8), in analytic orbit determination, no variation equations need to be solved; no numerical integration algorithms are necessary; no special orbit-related partial derivatives have to be derived. These significant advantages should lead to more efficient algorithms and more accurate orbit determination.

9.2 Real Time Ability

Limitations of the Numerical Orbit Determination

Real time ability of the numerical orbit determination is limited first by the adjustment or filtering algorithms used.

If the classic least squares adjustment algorithm is used to solve the parameters of the orbit determination problem, it is not possible to obtain the solution in real time because of the size and dimension of the equations. The equations of IGS orbit determination are formed and solved daily. It takes from less than an hour to several hours to compute the results depending, of course, on the computer used. The so-called rapid IGS orbits are partly computed using 23 hours past data and one hour updated data. In general, the classic least squares adjustment algorithm is not suitable for real time purpose.

Sequential least squares algorithm and Kalman filtering technique are partly developed for real time applications. Sequential least squares algorithm is a special case of the Kalman filtering; therefore, the discussions will be focused on the filtering method. Kalman filtering solves the equations of every epoch or every epoch-block by taking into account the information from the past to obtain the results. In this way the problem can be solved epoch-wise or epoch-block-wise depending on the property of the problem. For equations of orbit determination the problem is not solvable (or singular) for a few epochs because of the dimension of the unknowns. The equations of orbit determination are generally solvable in half an hour (see Xu, 2004) or longer. That is, the filtering technique and the property of the equations of orbit determination make the real time application of the numerical orbit determination very difficult.

Furthermore, in numerical orbit determination, the numerical integration algorithms have to be used to integrate the orbits and to solve the variation equations. Numerical integrator usually has a so-called integrator length. The selection of the integrator length depends on the accuracy requirement and the physical properties of functions that will be integrated and therefore is not free of choice. Usually in IGS orbit determination, the integrator length is selected as five minutes. This also restricts the real time application of the numerical orbit determination.

Because of the adjustment and filtering techniques and the use of the numerical integrator as well as the properties of the physical problem, numerical orbit determination, is difficult to be at real time.

Real Time Ability of Analytic Orbit Determination

Using the analytic orbit theory the observation equation of the orbit determination problem can be formed easily epoch-wise. The equations are solvable for an epoch-block. Taking past information into account, the solvable equations of an epoch-block can be formed and solved in real time. Taking the information before the solved epoch-block into account, Kalman filtering technique can be used to determine the orbit in real time. This is very significant for applications of satellite technology nowadays and should be further studied intensively.

9.3 Properties of Analytic Orbit Determination

Initial time selection

In numerical orbit determination, the initial time is a matter of free choice. For numerical integration, it really does not matter from which time point one starts to integrate. However, in analytic orbit solution, nearly a half of the formulas are functions of initial time point (another half of the formulas are infinite integrations and functions of instantaneous time). In turn, the functions of the initial time point are in terms of sines and cosines. Of course, theoretically the initial time point of orbit determination can be freely selected. However, if the initial time point is selected at that point such that the sines or cosines of mean anomaly are zero, the intensity of the computations can be reduced by 25%. That is, a suitable initial time selection is very important for analytic orbit determination.

Using general models for 2nd order geopotential disturbances

As shown in Chap. 6, the solutions of the 2nd order geopotential disturbances are very long. Theoretically, any order and any degree of the disturbances can be derived; however, to program all the formulas into software will be definitely a problem. For orbit determination the 2nd order geopotential disturbances are small terms and they can be dealt with like corrections to the initial and nominal orbit. For short periodic terms, the solutions are formed by a set of functions of

$$\{\sin nM, \cos nM, n = 1, \ldots, N\}, \tag{9.6}$$

where M is the mean anomaly of the orbit; n is an integer index and has a truncation number N.

Similarly, for the long periodic terms of the 2nd order geopotential disturbances, the solutions can be formed by the following sets of functions

$$\{\sin n\omega, \cos n\omega, n = 1, \ldots, I\},$$
$$\{\sin n\Omega, \cos n\Omega, n = 1, \ldots, J\},$$
$$\{\sin(n\omega + m\Omega), \cos(n\omega + m\Omega), n, m = 1, \ldots, K\}, \tag{9.7}$$

where m is an integer index; I, J and K are truncation numbers.

The general models of the solutions of the 2nd order geopotential disturbances are then

$$\sum_{n=1}^{N}(A_n \cos nM + B_n \sin nM) + \sum_{n=1}^{I}(C_n \cos n\omega + D_n \sin n\omega) + \sum_{n=1}^{J}(E_n \cos n\Omega + F_n \sin n\Omega) + \sum_{n,m=1}^{K}(G_{nm}\cos(n\omega + m\Omega) + H_{nm}\sin(n\omega + m\Omega)), \tag{9.8}$$

where coefficients (A_n, B_n, C_n, D_n, E_n, F_n,G_{nm}, H_{nm}) can be considered as unknown and should be codetermined by orbit determination. The truncation numbers of (I, J, K) are generally much smaller than N because of the long periodic properties and shall be suitably selected through practical experiments.

Chapter 10
Singularity-Free Theory and Discussions

The previous chapters of this book covered the most important contents of satellite orbit theory, including analytic solutions and applications. Especially, the solutions of the geopotential and extraterrestrial disturbances of the second order were derived and the analytic applications of the theory were discussed. In this, the last chapter of this book, emphasis will be on singularity-free theory and discussions.

10.1 Singularity-Free Orbit Theory

The singularity problem of the solutions of the geopotential disturbances is discussed first. Then the singularity-free disturbed equations of motion are given for three cases; i.e., for the circular orbit, equatorial orbit, circular and equatorial orbit, respectively. If the singularity-free disturbed equations of motion are used, then the derived orbit solutions are singularity-free.

10.1.1 Problem of Singularity of the Solutions

As already discussed in the properties of the solutions (Sects. 5.3 and 6.5), the derived solutions are singular in the cases of $e = 0$ and/or $\sin i = 0$. In other words, the solutions are not valid for the satellite with a circular or an equatorial orbit. An alternative method to overcome the problem of circular orbit has already been discussed in Sect. 4.1.1 by introducing new variables (see (4.12)). The new variables do not have clear geometric meanings and were used to replace the variables (ω, f), which could not be defined in a circular orbit. In the alternative equation of disturbance (4.15), the e factor in the dividend is then eliminated, i.e., the singularity of $e = 0$ disappears. Using another set of variables (a, $h = \sin i \cos\Omega$, $k = -\sin i \sin\Omega$, $\xi = e\cos(\omega+\Omega)$, $\eta = -e\sin(\omega+\Omega)$, $\lambda = M+\omega+\Omega$), both the singularities caused by $e = 0$ and $\sin i = 0$ may disappear. This means that the singularity is not a real problem of the orbits, but a consequence of poor parameterisation of the orbits. Another method to overcome the singularity problem is the canonical transformation. All these methods overcome the singularity problem on one hand and pay the price of losing the geometric meanings of the orbital variables on the other hand.

In the cases of $e = 0$ and/or $\sin i = 0$, the orbits are becoming simpler in practice. However, the equations used to describe a simpler problem are becoming more complicated. This is in conflict with basic scientific philosophy and common sense. A simpler problem should be able to be described in simpler terms.

Looking into the solutions given in Chaps. 5 and 6 carefully, it is obvious that the singular problem is not created by the partial derivations of the potential function with respect to the Keplerian variable. In other words, the singularity problem exists from the beginning in the Lagrangian perturbed equation system (4.11). This may be verified by derivations of (4.11) (see Kaula, 1966/2001).

10.1.2 Disturbed Equations in the Case of Circular Orbit

In the case of a circular orbit, the perigee of the orbit is arbitrary. Then the ascending node of the orbit can be defined as the perigee and the argument of the perigee ω can be considered a constant of zero. (Of course, the eccentricity e is a constant of zero, too.) In this special case the orbit is simpler than the general one. Note that in such a case the eccentric anomaly f is identical with the mean anomaly M. The disturbed equations of motion (similar to (4.11)) in this case can be similarly derived and have accordingly the following simpler forms

$$\begin{aligned}
\frac{\mathrm{d}a}{\mathrm{d}t} &= \frac{2}{na}\frac{\partial R}{\partial M}, \\
\frac{\mathrm{d}e}{\mathrm{d}t} &= 0, \\
\frac{\mathrm{d}\omega}{\mathrm{d}t} &= -\frac{\cos i}{na^2 \sin i}\frac{\partial R}{\partial i}, \\
\frac{\mathrm{d}i}{\mathrm{d}t} &= -\frac{1}{na^2 \sin i}\frac{\partial R}{\partial \Omega}, \\
\frac{\mathrm{d}\Omega}{\mathrm{d}t} &= \frac{1}{na^2 \sin i}\frac{\partial R}{\partial i}, \\
\frac{\mathrm{d}M}{\mathrm{d}t} &= n - \frac{2}{na}\frac{\partial R}{\partial a}.
\end{aligned} \tag{10.1}$$

10.1.3 Disturbed Equations in the Case of Equatorial Orbit

In the case of an equatorial orbit, the ascending node is arbitrary. Then the vernal equinox can be defined as the ascending node of the orbit and the right ascension of the ascending node Ω can be considered a constant of zero. (Of course, $\sin i$, the sine function of inclination angle i, is a constant of zero, too.) In this special case the orbit is simpler than a general one. Especially, the transformed geopotential function with orbital variable is greatly simplified in such a case. The disturbed equations of motion (similar to (4.11)) in this case can be similarly derived and have accordingly the following simpler forms

$$\begin{aligned}
\frac{\mathrm{d}a}{\mathrm{d}t} &= \frac{2}{na}\frac{\partial R}{\partial M}, \\
\frac{\mathrm{d}e}{\mathrm{d}t} &= \frac{1-e^2}{na^2 e}\frac{\partial R}{\partial M} - \frac{\sqrt{1-e^2}}{na^2 e}\frac{\partial R}{\partial \omega}, \\
\frac{\mathrm{d}\omega}{\mathrm{d}t} &= \frac{\sqrt{1-e^2}}{na^2 e}\frac{\partial R}{\partial e}, \\
\frac{\mathrm{d}i}{\mathrm{d}t} &= 0, \\
\frac{\mathrm{d}\Omega}{\mathrm{d}t} &= 0, \\
\frac{\mathrm{d}M}{\mathrm{d}t} &= n - \frac{2}{na}\frac{\partial R}{\partial a} - \frac{1-e^2}{na^2 e}\frac{\partial R}{\partial e}.
\end{aligned} \tag{10.2}$$

10.1.4 Disturbed Equations in the Case of Circular and Equatorial Orbit

In the case of a circular and an equatorial orbit, both the perigee and ascending node are arbitrary. Then the vernal equinox can be defined as the perigee and the ascending node of the orbit; and the argument of the perigee ω and right ascension of the ascending node Ω can be considered constants of zero. (Of course, ω and $\sin i$are constants of zero, too). In this special case the orbit is the simplest one compared with the others. Note that in such a case the eccentric anomaly f is identical with the mean anomaly $M$and the transformed geopotential function with orbital variable is greatly simplified. The disturbed equations of motion (similar to (4.11)) in this case can be similarly derived and have accordingly the following simpler forms

$$\begin{aligned}
\frac{\mathrm{d}a}{\mathrm{d}t} &= \frac{2}{na}\frac{\partial R}{\partial M}, \\
\frac{\mathrm{d}e}{\mathrm{d}t} &= 0, \\
\frac{\mathrm{d}\omega}{\mathrm{d}t} &= 0, \\
\frac{\mathrm{d}i}{\mathrm{d}t} &= 0,
\end{aligned}$$

$$\begin{aligned} \frac{\mathrm{d}\Omega}{\mathrm{d}t} &= 0, \\ \frac{\mathrm{d}M}{\mathrm{d}t} &= n - \frac{2}{na}\frac{\partial R}{\partial a}, \end{aligned} \tag{10.3}$$

10.1.5 Singularity-Free Disturbed Equations of Motion

Define two delta functions as

$$\delta_e = \begin{cases} 1 & \text{if } e \neq 0 \\ e^2 & \text{if } e = 0 \end{cases} \quad \text{and} \quad \delta_i = \begin{cases} 1 & \text{if } \sin i \neq 0, \\ \sin^2 i & \text{if } \sin i = 0. \end{cases} \tag{10.4}$$

Then one has the singularity-free disturbed equations of motion

$$\begin{aligned} \frac{\mathrm{d}a}{\mathrm{d}t} &= \frac{2}{na}\frac{\partial R}{\partial M}, \\ \frac{\mathrm{d}e}{\mathrm{d}t} &= \frac{1-e^2}{na^2 e}\frac{\partial R}{\partial M}\delta_e - \frac{\sqrt{1-e^2}}{na^2 e}\frac{\partial R}{\partial \omega}\delta_e, \\ \frac{\mathrm{d}\omega}{\mathrm{d}t} &= \frac{\sqrt{1-e^2}}{na^2 e}\frac{\partial R}{\partial e}\delta_e - \frac{\cos i}{na^2\sqrt{1-e^2}\sin i}\frac{\partial R}{\partial i}\delta_i, \\ \frac{\mathrm{d}i}{\mathrm{d}t} &= \frac{1}{na^2\sqrt{1-e^2}\sin i}\left(\cos i\frac{\partial R}{\partial \omega} - \frac{\partial R}{\partial \Omega}\right)\delta_i, \\ \frac{\mathrm{d}\Omega}{\mathrm{d}t} &= \frac{1}{na^2\sqrt{1-e^2}\sin i}\frac{\partial R}{\partial i}\delta_i, \\ \frac{\mathrm{d}M}{\mathrm{d}t} &= n - \frac{2}{na}\frac{\partial R}{\partial a} - \frac{1-e^2}{na^2 e}\frac{\partial R}{\partial e}\delta_e. \end{aligned} \tag{10.5}$$

Equations (10.5) are the singularity-free disturbed equations of motion. The solutions derived from these equations are singularity-free. For some reasons, the solutions given in this book are mostly derived from (4.11). To obtain the singularity-free solutions one has to add the two factors of the delta functions (10.4) into the given solutions and the interested readers may attempt these themselves.

10.1.6 Simplified Singularity-Free Disturbed Equations of Motion

Similar to the simplified Gaussian disturbed equations of satellite motion (see (7.25)), the simplified singularity-free disturbed Lagrange equations of motion can be derived and written as

$$\begin{aligned}
\frac{\mathrm{d}a}{\mathrm{d}t} &= \frac{2}{na}\frac{\partial R}{\partial M},\\
\frac{\mathrm{d}e}{\mathrm{d}t} &= \frac{1-e^2}{2ae}\frac{\mathrm{d}a}{\mathrm{d}t}\delta_e - \frac{\sqrt{1-e^2}}{na^2e}\frac{\partial R}{\partial \omega}\delta_e,\\
\frac{\mathrm{d}\omega}{\mathrm{d}t} &= \frac{\sqrt{1-e^2}}{na^2e}\frac{\partial R}{\partial e}\delta_e - \cos i\frac{\mathrm{d}\Omega}{\mathrm{d}t},\\
\frac{\mathrm{d}i}{\mathrm{d}t} &= \frac{1}{na^2\sqrt{1-e^2}\sin i}\left(\cos i\frac{\partial R}{\partial \omega} - \frac{\partial R}{\partial \Omega}\right)\delta_i,\\
\frac{\mathrm{d}\Omega}{\mathrm{d}t} &= \frac{1}{na^2\sqrt{1-e^2}\sin i}\frac{\partial R}{\partial i}\delta_i,\\
\frac{\mathrm{d}M}{\mathrm{d}t} &= n - \frac{2}{na}\frac{\partial R}{\partial a} - \sqrt{1-e^2}\left(\frac{\mathrm{d}\omega}{\mathrm{d}t} + \cos i\frac{\mathrm{d}\Omega}{\mathrm{d}t}\right).
\end{aligned} \tag{10.6}$$

It is obvious that such equations will lead to a simplified process of solving the problems.

10.2 Equations of Motion in Non-Inertial Frame

It is well known that Newton's second law is valid in the inertial coordinate system. This is also the reason why the orbit problem is usually dealt within the ECI frame. However, the geopotential force (or say, geopotential function) is described in an Earth-Centre-Earth-fixed non-inertial system. Without exception, one has to transform the geodetic coordinates of the potential function into orbital elements. However, for the equation of motion expressed in the inertial frame and for a simplified transformation between the CEI and ECEF

$$\frac{\mathrm{d}^2X_{ECI}}{\mathrm{d}t^2} = F_{ECI} \quad \text{and} \quad X_{ECI} = R_3(\omega t)X_{ECEF}, \quad F_{ECI} = R_3(\omega t)F_{ECEF}, \tag{10.7}$$

the equation of motion expressed in ECEF frame can be derived

$$\begin{cases}
\dfrac{\mathrm{d}^2x}{\mathrm{d}t^2} + 2\omega\dfrac{\mathrm{d}y}{\mathrm{d}t} - \omega^2x = f_x(x,y,z),\\
\dfrac{\mathrm{d}^2y}{\mathrm{d}t^2} - 2\omega\dfrac{\mathrm{d}x}{\mathrm{d}t} - \omega^2y = f_y(x,y,z),\\
\dfrac{\mathrm{d}^2z}{\mathrm{d}t^2} = f_z(x,y,z),
\end{cases} \tag{10.8}$$

where ω is the Earth's angular velocity. The homogenous solution of (10.8) is

$$\begin{cases} x=\frac{1}{2}e^{-i\omega t}\left[c_1(1+E+i\omega t-i\omega Et)-c_2(1-iE-\omega t-\omega Et)+c_4(t+Et)-c_5 i(t-Et)\right], \\ y=\frac{1}{2}e^{-i\omega t}\left[c_1(i-iE-\omega t-\omega Et)+c_2(1+E-i\omega t-i\omega Et)+c_4 i(t-Et)+c_5(t+Et)\right], \\ z=c_3+c_6 t, \end{cases}$$

where c_j is integral constants and $E=e^{2i\omega t}, i=\sqrt{-1}$.

As soon as any one special solution of (10.8) is found, then the general solution of (10.8) is equal to the special solution plus the homogenous solution. It may be worthwhile to consider the problem of geopotential disturbance alternatively.

10.3 Discussions

Simplified Singularity-Free Equations of Motion

As seen above (Sect. 10.1), the singularity problem has been solved by using simplified and singularity-free equations. The simplified orbit problem is described using simplified coordinates. The geometric meanings of the variables are remained the same. The use of the traditional and partly non-geometric sensed variable set of (a, $h=\sin i\cos\Omega$, $k=-\sin i\sin\Omega$, $\xi=e\cos(\omega+\Omega)$, $\eta=-e\sin(\omega+\Omega)$, $\lambda=M+\omega+\Omega$) is obviously not an ideal choice. One of the important reasons for using the canonical transformation to represent the orbit equations is that the canonical equations are also singularity-free. After the disturbed equations of motion (10.6) are singularity-free, the advantages of the use of canonical equations have to be carefully re-evaluated.

Analytic Solution vs. Numeric Solution

Solutions of the extraterrestrial disturbances are some times given both in analytic and numerical form (see, e.g., Sects. 7.2 and 7.4). The formulas of the discrete solutions are very easy to be used for computation; however, they do not have clear geometric explanations for the effects of the disturbances.

Potential Functions of the Sun, Moon and Planets

An approximation has been used in the derivation of the potential function of the disturbing force of the sun. Similar means have been used for the moon and can also be used for the planets. Therefore, the related solutions are derived under a precondition that the approximation is allowed.

Confusion of Non Conservative Force with Conservative Effect

Solar radiation is a non conservative disturbing force. It is said that such a non conservative force has a conservative effect. This is confusion and is shown in Fig. 10.1 with an example of solar disturbance on a GEO satellite. One of the possible reasons of such confusion may come from the adjustment model of the solar radiation used in the numerical orbit determination. The models (4.72) are periodic functions of the orbit. No matter what results are obtained from the adjustment, the results are periodic (or conservative). If the determined models are used to interpret the effects of the solar radiation, the confusion is then the consequence. This shows that the parameterisation is very important and the parameterisation should be physically reasonable.

Long Term Effects in Extraterrestrial Disturbances

There exist long term effects in the extraterrestrial disturbances (see (7.26)). The long term perturbations have to be taken into account in the transformation of integral variables. This shall be especially noticed in the practical applications.

Long Term and Long Periodic Effects in Short Periodic Disturbances

There exist long term and long periodic effects in the short periodic geopotential disturbances (see (5.34)). The long term and long periodic effects derived in Sect. 5.2

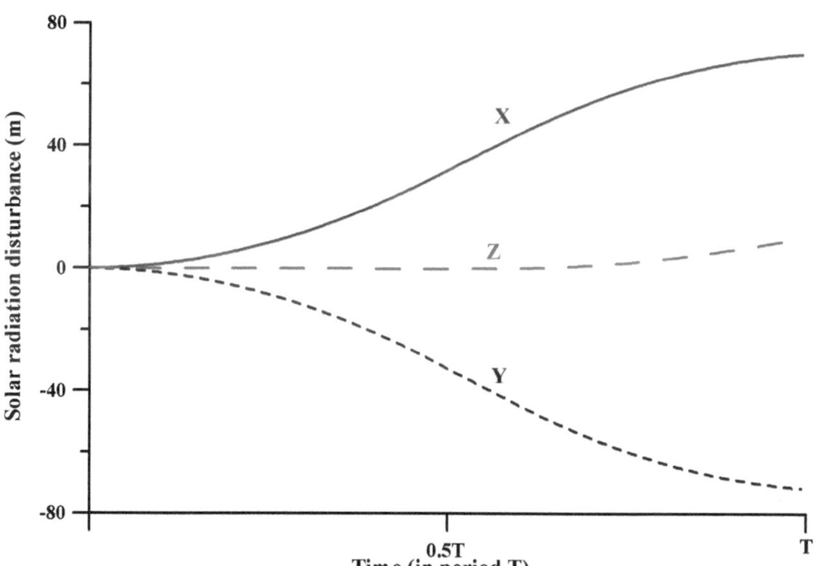

Fig. 10.1 Solar radiation disturbance on a GEO satellite

are not unique and not the complete effects. Note that all the long term and long periodic effects have to be accumulated if the relations will be used.

Further Studies

Further studies have to be carried out on the analytic solutions of the Gaussian equations disturbed by the air drag, on the use of the simplified equations of motion, on the applications of the analytic theory (especially on the analytic orbit determination), on the study of the correlation of the geopotential disturbances on the orbits, on the third order solutions disturbed by the Earth and ocean tides as well as relativity disturbance.

Appendix 1
IAU 1980 Theory of Nutation

Table A.1 The units of A_i and B_i are 0.″0001, units of A_i' and B_i' are 0.″00001 (cf. McCarthy, 1996)

Coefficents of					values of			
l	l′	F	D	Ω	A_i	A_i'	B_i	B_i'
0	0	0	0	1	−171996	−1742	92025	89
0	0	2	−2	2	−13187	16	5736	−31
0	0	2	0	2	−2274	−2	977	−5
0	0	0	0	2	2062	2	−895	5
0	−1	0	0	0	−1426	34	54	−1
1	0	0	0	0	712	1	−7	0
0	1	2	−2	2	−517	12	224	−6
0	0	2	0	1	−386	−4	200	0
1	0	2	0	2	−301	0	129	−1
0	−1	2	−2	2	217	−5	−95	3
−1	0	0	−2	0	158	0	−1	0
0	0	2	−2	1	129	1	−70	0
−1	0	2	0	2	123	0	−53	0
1	0	0	0	1	63	1	−33	0
0	0	0	2	0	63	0	−2	0
−1	0	2	2	2	−59	0	26	0
−1	0	0	0	1	−58	−1	32	0
1	0	2	0	1	−51	0	27	0
−2	0	0	2	0	−48	0	1	0
−2	0	2	0	1	46	0	−24	0
0	0	2	2	2	−38	0	16	0
2	0	2	0	2	−31	0	13	0
1	0	2	−2	2	29	0	−12	0
2	0	0	0	0	29	0	−1	0
0	0	2	0	0	26	0	−1	0
0	0	2	−2	0	−22	0	0	0
−1	0	2	0	1	21	0	−10	0
0	2	0	0	0	17	−1	0	0
−1	0	0	2	1	16	0	−8	0
0	2	2	−2	2	−16	1	7	0
0	1	0	0	1	−15	0	9	0

Table A.1 (continued)

Coefficents of					values of			
l	l′	F	D	Ω	A_i	A_i'	B_i	B_i'
1	0	0	−2	1	−13	0	7	0
0	−1	0	0	1	−12	0	6	0
2	0	−2	0	0	11	0	0	0
−1	0	2	2	1	−10	0	5	0
1	0	2	2	2	−8	0	3	0
0	0	2	2	1	−7	0	3	0
0	−1	2	0	2	−7	0	3	0
0	1	2	0	2	7	0	−3	0
1	1	0	−2	0	−7	0	0	0
1	0	2	−2	1	6	0	−3	0
0	0	0	2	1	−6	0	3	0
2	0	2	−2	2	6	0	−3	0
1	0	0	2	0	6	0	0	0
−2	0	0	2	1	−6	0	3	0
2	0	2	0	1	−5	0	3	0
1	−1	0	0	0	5	0	0	0
0	0	0	−2	1	−5	0	3	0
0	−1	2	−2	1	−5	0	3	0
0	0	0	1	0	−4	0	0	0
1	0	−2	0	0	4	0	0	0
0	1	0	−2	0	−4	0	0	0
1	0	0	−1	0	−4	0	0	0
0	1	2	−2	1	4	0	−2	0
2	0	0	−2	1	4	0	−2	0
0	−1	2	2	2	−3	0	1	0
3	0	2	0	2	−3	0	1	0
−1	−1	2	2	2	−3	0	1	0
1	−1	2	0	2	−3	0	1	0
1	0	2	0	0	3	0	0	0
1	1	0	0	0	−3	0	0	0
1	−1	0	−1	0	−3	0	0	0
−2	0	2	0	2	−3	0	1	0
−1	0	2	4	2	−2	0	1	0
0	0	2	1	2	2	0	−1	0
3	0	0	0	0	2	0	0	0
1	0	0	0	2	−2	0	1	0
2	0	0	0	1	2	0	−1	0
−1	0	2	−2	1	−2	0	1	0
1	1	2	0	2	2	0	−1	0
−2	0	0	0	1	−2	0	1	0
0	−2	2	−2	1	−2	0	1	0
0	1	0	1	0	1	0	0	0
0	0	2	4	2	−1	0	0	0
2	0	0	2	0	1	0	0	0
1	0	−2	2	0	−1	0	0	0
1	1	0	−2	1	−1	0	0	0
0	−1	2	0	1	−1	0	0	0

Table A.1 (continued)

Coefficents of					values of			
l	l′	F	D	Ω	A_i	A_i'	B_i	B_i'
1	0	−2	−2	0	−1	0	0	0
0	1	0	2	0	−1	0	0	0
0	0	2	−1	2	−1	0	0	0
0	0	−2	0	1	−1	0	0	0
−1	−1	0	2	1	1	0	0	0
0	1	2	0	1	1	0	0	0
1	0	2	−2	0	−1	0	0	0
3	0	2	−2	2	1	0	0	0
0	0	4	−2	2	1	0	0	0
1	0	0	2	1	−1	0	0	0
2	0	2	2	2	−1	0	0	0
2	0	2	−2	1	1	0	−1	0
1	−1	0	−2	0	1	0	0	0
−1	0	4	0	2	1	0	0	0
−2	0	2	4	2	−1	0	1	0
1	0	2	2	1	−1	0	1	0
1	1	2	−2	2	1	0	−1	0
2	0	0	−4	0	−1	0	0	0
−2	0	2	2	2	1	0	−1	0
1	0	0	−4	0	−1	0	0	0
−1	0	0	0	2	1	0	−1	0
0	1	2	−2	0	−1	0	0	0
−1	0	0	1	1	1	0	0	0
0	1	0	0	2	1	0	0	0
0	1	−2	2	0	−1	0	0	0
0	0	−2	2	1	1	0	0	0
2	1	0	−2	0	1	0	0	0
2	0	−2	0	1	1	0	0	0

Bibliography

Abramowitz M, Stegun IA (1965) Handbook of mathematical functions. Dover Publications, New York

Ashby N, Spilker JJ (1996) Introduction to relativistic effects on the Global Positioning System. In: Parkinson BW, Spilker JJ (eds) Global Positioning System: Theory and applications, Vol. I, Chapter 18

Axelsson O (1994) Iterative Solution Methods. Cambridge University Press, London/New York

Ayres F (1975) Differential- und Integralrechnung, Schaum's Outline. McGraw-Hill, New York

Babu R (2005) Web-based resources on software GPS receivers. GPS Solutions 9(3):240–242

Balmino G, Schrama E, Sneeuw N (1996) Compatibility of first-order circular orbit perturbations theories: Consequences for cross-track inclination functions. J Geodesy 70(9):554–561

Bar-Sever YE (1996) A new model for GPS yaw attitude. J Geodesy 70:714–723

Bastos L, Osorio J, Hein G (1995) GPS derived displacements in the Azores Triple Junction Region. In: GPS Trends in Precise Terrestrial Airborne, and Spaceborne Applications: 21st IUGG General Assembly, IAG Symposium No. 115, Boulder, USA, July 3–4, 1995, Springer, Berlin, pp 99–104

Bate RR, Mueller DD, White JE (1971) Fundamentals of astrodynamics. Dover Publications, New York

Bauer M (1994) Vermessung und Ortung mit Satelliten. Wichmann Verlag, Karslruhe, Germany

Beutler G (1994) GPS trends in precise terrestrial, airborne, and space borne applications. Springer, Heidelberg

Beutler G (1996) GPS satellite orbits. In Kleusberg A, Teunissen PJG (eds) GPS for geodesy. Springer, Berlin

Beutler G, Brockmann E, Gurtner W, Hugentobler U, Mervart L, Rothacher M, Verdun A (1994) Extended orbit modelling techniques at the CODE Processing Center of the IGS: Theory and initial results. Manuscr Geodaet 19:367–386

Beutler G, Brockmann E, Hugentobler U (1996) Combining consecutive short arcs into long arcs for precise and efficient GPS orbit determination. J Geodesy 70(5):287–299

Beutler G, Schildknecht T, Hugentobler U, Gurtner W (2003) Orbit determination in satellite geodesy. Adv Space Res 31(8):1853–1868

Bian S, Jin J, Fang Z (2005) The Beidou satellite positioning system and its positioning accuracy. Navigation 52(3):123–129

Blewitt G (1998) GPS data processing methodology. In: Teunissen PJG, Kleusberg A (eds) GPS for Geodesy, Springer, Berlin/Heidelberg/New York, pp 231–270.

Bock Y (1996) Reference systems. In: Kleusberg A, Teunissen PJG (eds) GPS for geodesy. Springer-Verlag, Berlin

Boomkamp H, Dow J (2005) Use of double difference observations in combined orbit solutions for LEO and GPS satellites. Adv Space Res 36(3):382–391

Borre K (2003) The GPS Easy Suit-Matlab code for the GPS newcomer. GPS Solutions 7(1):47–51
Borre K, Akos DM, Bertelsen N, Rinder P, Jensen SH (2007) A Software-Defined GPS and Galileo Receiver, A Single-Frequency Approach, Birkhaüser, Boston, M.A. ISBN 978-0-8176-4390-4
Bronstein IN, Semendjajew KA (1987) Taschenbuch der Mathematik. B.G. Teubner Verlagsgesellschaft, Leipzig, Germany. ISBN 3-322-00259-4
Brouwer D, Clemence GM (1961) Methods of celestial mechanics. Academic Press, New York
Brunner FK (1998) Advances in positioning and reference frames. Springer-Verlag, Heidelberg
Cangahuala L, Muellerschoen R, Yuan D-N (1995) TOPEX/Poseidon precision orbit determination with SLR and GPS anti-spoofing data. In: GPS Trends in Precise Terrestrial Airborne, and Spaceborne Applications: 21st IUGG General Assembly, IAG Symposium No. 115, Boulder, CO, USA, July 3–4, 1995, pp 123–127, Springer-Verlag, Berlin
Capitaine N (2002) Comparison of "Old" and "New" Concepts: The Celestial Intermediate Pole and Earth Orientation Parameters. In: Proceedings of the IERS Workshop on the Implementation of the New IAU Resolutions, Paris, France, April 18–19, 2002 (Capitaine N et al., ed.). IERS Technical Note No. 29
Casotto S, Zin A (2000) An assessment of the benefits of including GLONASS data in GPS-based precise orbit determination – I: S/A analysis. Adv Astronaut Sci 105(1):237–256
Chobotov VA (ed) (1991) Orbital mechanics. AIAA, Washington D.C.
Colombo OL (1984a) Altimetry, orbits and tides. NASA Technical Memorandum 86180
Colombo OL (1984b) The global mapping of gravity with two satellites. Netherlands Geodetic Commission, Delft, The Netherlands Publications on Geodesy, Vol. 7, No. 3, 253pp
Cui C (1990) Die Bewegung künstlicher Satelliten im anisotropen Gravitationsfeld einer gleichmässig rotierenden starren Modellerde. Deutsche Geodätische Kommission, Reihe C: Dissertationen, Heft Nr. 357
Cui C (1997) Satellite orbit integration based on canonical transformations with special regard to the resonance and coupling effects. Dtsch Geod Komm bayer Akad Wiss, Reihe A, Nr. 112, 128pp
Cui C, Lelgemann D (1995) Analytical dynamic orbit improvement for the evaluation of geodetic-geodynamic satellite data. J Geodesy 70:83–97
Davis P, Rabinowitz P (1984) Methods of numerical integration, 2nd Edn. Academic Press
Davis PJ (1963) Interpolation and approximation. Dover Publications, New York
Desai SD, Haines BJ (2003) Near-real-time GPS-based orbit determination and sea surface height observations from the Jason-1 mission. Mar. Geod. 26(3/4):383–397
Dick G (1997) Nutzung von GPS zur Bahnbestimmung niedrigfliegender Satelliten. GPS-Anwendungen und Ergebnisse '96: Beiträge zum 41. DVW-Fortbildungsseminar vom 7. bis 8. November 1996 am GeoForschungszentrum Potsdam, pp 241–249
Diggelen F (1998) GPS accuracy: Lies, damm lies, and statistics. GPS World 9(1):41–44
Diggelen F, Martin W (1997) GPS + GLONASS RTK: A quantum leap in RTK performance. Int J Geo-matics 11(11):69–71
Douša J (2004) Precise orbits for ground-based GPS meteorology: Processing strategy and quality assessment of the orbits determined at geodetic observatory Pecny. J Meteor Soc Jpn 82(1B): 371–380
Dow JM (1988) Ocean tides and tectonic plate motions from Lageos. Deutsche Geodätische Kommission, Rheihe C, Dissertation, Heft Nr. 344
Dow JM, Romay-Merino MM, Piriz R, Boomkamp H, Zandbergen R (1993) High precision orbits for ERS-1: 3-day and 35-day repeat cycles. In: Proceedings of the Second ERS-1 symposium – Space at the service of our environment, Hamburg, October 11–14, 1993, pp 1349–1354
Engelhardt G, Mikolaiski H (1996) Concepts and results of the GPS data processing with Bernese and GIPSY Software. In: The Geodetic Antarctic Project GAP95: German Contributions to the SCAR 95 Epoch Campaign, 1996, Muenchen, pp 37–51
Even-Tzur G, Agmon E (2005) Monitoring vertical movements in Mount Carmel by means of GPS and precise leveling. Surv Rev 38(296):146–157
Exertier P, Bonnefond P (1997) Analytical solution of perturbed circular motion: Application to satellite geodesy. J Geodesy 71(3):149–159

Feltens J (1991) Nicht gravitative Störeinflüsse bei der Modellierungen von GPS-Erdumlaufbahnen. Verlag der Bayerischen Akademie der Wissenschaften, Munich, Deutsche Geodätische Kommision, Reihe C, Heft Nr. 371

Fliegel HF, Gallini TE, Swift ER (1992) Global Positioning System radiation force model for geodetic applications. J Geophys Res 97(B1):559–568

Gabor MJ, Nerem RS (2004) Characteristics of satellite-satellite single difference widelane fractional carrier-phase biases. Navigation 51(1):77–92

Goad C, Dorota A, Brzezinska G, Yang M (1996) Determination of high-precision GPS orbits using triple differencing technique. J Geodesy 70:655–662

Gotthardt E (1978) Einführung in die Ausgleichungsrechnung. Herbert Wichmann Verlag, Karlsruhe

Grewal MS, Weill LR, Andrews AP (2001) Global Positioning System, Inertial Navigation, and Integration. Wiley, New York, 392pp.

Groten E (1979) Geodesy and the Earth's gravity field, Vol. I: Principles and conventional methods. Dümmler-Verlag, Bonn

Groten E (1980) Geodesy and the Earth's gravity field, Vol. II: Geodynamics and advanced methods. Dümmler-Verlag, Bonn

Guinn J, Muellerschoen R, Cangahuala L (1995) TOPEX/Poseidon precision orbit determination using combined GPS, SLR and DORIS. In: GPS Trends in Precise Terrestrial, Airborne, and Spaceborne Applications: 21st IUGG General Assembly, IAG Symposium No. 115, Boulder, USA, July 3–4, 1995. pp 128–132, Springer-Verlag, Berlin

Haines B, Bar-Server Y, Bertiger W, Desai S, Willis P (2004) One-centimeter orbit determination for Jason-1: New GPS-based strategies. Mar. Geod. 27(1–2):299–318

Haines BJ, Christensen EJ, Guinn JR (1995) Observations of TOPEX/Poseidon orbit errors due to gravitational and tidal modeling errors using the Global Positioning System. In: GPS Trends in Precise Terrestrial, Airborne, and Spaceborne Applications: 21st IUGG General Assembly, IAG Sym-posium No. 115, Boulder, USA, July 3–4, 1995. Springer-Verlag, Berlin, pp 133–138

Hatch RR (2004) Those scandalous clocks. GPS Solutions 8(2):67–73

Heck B (1995a) Grundlagen der SatellitenGeodaesie. Schriftenreihe des Deutschen Vereins für Vermessungswesen, Bd. 18. Stuttgart, pp 10–31

Heck B (1995b) Grundlagen der erd- und himmelsfesten Referenzsysteme. Schriftenreihe des Deutschen Vereins für Vermessungswesen, Bd. 18, Stuttgart, pp 138–153

Heiskanen WA, Moritz H (1967) Physical geodesy. W.H. Freeman, San Francisco/London

Heitz S (1988) Coordinates in geodesy. Springer-Verlag, Berlin

Herrick S (1972) Astrodynamics, Vol. II. Van Nostrand Reinhold, London

Herring T (2003) MATLAB Tools for viewing GPS velocities and time series. GPS Solutions 7(3):194–199

Hess D, Keller W (1999a) Gradiometrie mit GRACE, Teil I: Fehleranalyse künstlicher Gradiometerdaten. ZfV 5:137–144

Hess D, Keller W (1999b) Gradiometrie mit GRACE, Teil II: Simulationsstudie. ZfV 7:205–211

Hofmann-Wellenhof B, Legat K, Weiser M (2003) Navigation, Principles of Positioning Guidance. Springer-Verlag, New York, 427pp

Hofmann-Wellenhof B, Lichtenegger H, Collins J (1997/2001) GPS theory and practice. Springer-Press, Wien, Austria

Holdridge DB (1967) An alternate expression for light time using general relativity. JPL Space Program Summary 3(37–48):2–4

Hong Y, Ou JK (2006) Algebra solution of the variation equation and its numerical validation (in Chinese)

Hostetter GH (1987) Handbook of digital signal processing: Engineering Applications, Academic Press, San Diego, CA

Hotine M (1991) Differential geodesy. Springer-Verlag, Berlin

Hugentobler U, Ineichen D, Beutler G (2003) GPS satellites: Radiation pressure, attitude and resonance. Adv Space Res 31(8):1917–1926

Hugentobler U, Schaer S, Fridez P (2001) Bernese GPS Software (Version 4.2). Astronomical Institute of University Berne, Switzerland.

The Institute of Navigation (ION). In: Proceedings of ION GPS-87, 88, 89, 90, 91, 92, 93, 94, 95, 96, 97, 98, 99, 00, 01, 02, 03, 04, 05, 06, 07. ION, Fairfax, VA. www.ion.org/publications/toc/gpstoc.html

Jaggi A, Beutler G, Hugentobler U (2005) Reduced-dynamic orbit determination and the use of accelerometer data. Adv Space Res 36(3):438–444

Jerde CL, Visscher DR (2005) GPS measurement error influences on movement model parameterization. Ecol Appl 15(3):806–810

Kang Z (1998) Präzise Bahnbestimmung niedrigfliegender Satelliten mittels GPS und die Nutzung für die globale Schwerefeldmodellierung. Scientific Technical Report STR 98/25, GeoForschungsZentrum (GFZ) Potsdam

Kang Z, Nagel P, Pastor R (2003) Precise orbit determination for GRACE. Adv Space Res 31(8):1875–1881

Karslioglu MO (2005) An interactive program for GPS-based dynamic orbit determination of small satellites. Comput and Geosci 31(3):309–317

Kaula WM (1966/2001) Theory of satellite geodesy. Blaisdell Publishing Company, Dover Publications, New York

Keshin MO (2004) Directional statistics of satellite-satellite single-difference widelane phases biases. Artif Satellites 39(4):305–324

King RW, Masters EG, Rizos C, Stolz A, Collins J (1987) Surveying with Global Positioning System. Dümmler-Verlag, Bonn

Kleusberg A (1995) Mathematics of attitude determination with GPS. GPS World 6(9):72–78

Kleusberg A, Teunissen PJG (eds) (1996) GPS for geodesy. Springer-Verlag, Berlin

Knudsen P, Olsen H, Xu G (1999) GPS-altimetry tests – Measuring GPS signal reflected from the Earth surface. Poster on the 22nd IUGG General Assembly, IAG Symposium, England

Konig R, Reigber C, Zhu SY (2005) Dynamic model orbits and earth system parameters from combined GPS and LEO data. Adv. Space Res. 36(3): 431–437

Konig R, Schwintzer P, Bode A (1996) GFZ-1: A small laser satellite mission for gravity field model improvement. Geophys Res Lett 23(22):3143–3146

Kraus JD (1966) Radio astronomy. McGraw-Hill, New York

Kroes R, Montenbruck O (2004) Spacecraft formation flying: Relative positioning using dual-frequency carrier phase. GPS World 15(7):37–42

Kroes R, Montenbruck O, Bertiger W, Visser P (2005) Precise GRACE baseline determination using GPS. GPS Solutions 9(1):21–31

Kuang D, Rim HJ, Schutz BE, Abusali PAM (1996) Modeling GPS satellite attitude variation for precise orbit determination. J Geodesy 70:572–580

Kwon JH, Grejner-Brzezinska D, Bae TS, Hong CK (2003) A triple difference approach to Low Earth Orbiter precision orbit determination. J Navig 56(3):457–473

Lambeck K (1988) Geophysical geodesy: The slow deformations of the Earth. Oxford Science Publications

Langley RB (2000) GPS, the ionosphere, and the solar maximum. GPS World 11(7):44–49

Leick A (1995/2004) GPS Satellite Surveying, 3rd edn, Wiley, New York, 664pp

Lelgemann D (1983) A linear solution of equation of motion of an Earth-orbiting satellite based on a Lie-series. Celestial Mech 30:309

Lelgemann D (1996) Geodaesie im Weltraumzeitalter. Dtsch Geod Komm 25:59–77

Lelgemann D (2002) Lecture notes of geodesy, Technical University Berlin

Lelgemann D, Xu G (1991) Zur Helmert-Transformation von terrestrischen und GPS-Netzen. ZfV (1)

Lemmens R (2004) Book review: GPS – Theory, Algorithms and Applications, Xu G 2003. Int J Appl Earth Obs Geoinf 5:165–166

Li H, Xu G, Xue H, Zhao H, Chen J,Wang G (1999) Design of GPS application program. Science Press, Peking, 337pp. ISBN 7-03-007204-9/TP.1049 (in Chinese and in C)

Linkwitz K, Hangleiter U (eds) (1995) High precision navigation. Dümmler-Verlag, Bonn

Liu DJ, Shi YM, Guo JJ (1996) Principle of GPS and its data processing. TongJi University Press, Shanghai, China (in Chinese)
Liu L, Zhao D (1979) Orbit theory of the Earth satellite. Nanjing University Press (in Chinese)
Ludwig R (1969) Methoden der Fehler- und Ausgleichsrechnung. Vieweg & Sohn, Braunschweig
MacGougan G, Normark P-L, Ståhlberg C (2005) Satellitenavigation evolution: The software GNSS receiver. GPS World 16(1):48–52
Mackenzie R, Moore P (1997) A geopotential error analysis for a non planar satellite to satellite tracking mission J Geodesy 71(5):262–272
Mackie JB (1985) The elements of astronomy for surveyors. Charles Griffin & Company Ltd., London
Mansfeld W (2004) Satellitenortung und Navigation, 2nd edn, Wiesbanden: Vieweg Verlag, 352pp.
McCarthy DD (1996) International Earth Rotation Service. In: IERS conventions, Paris, 95pp. IERS Technical Note No. 21
McCarthy DD, Capitaine N (2002) Practical Consequences of Resolution B1.6 "IAU2000 Precession-Nutation Model", Resolution B1.7 "Definition of Celestial Intermediate Pole", and Resolution B1.8 "Definition and Use of Celestrial and Terrestrial Ephemeris Origin". In: Proceedings of the IERS Workshop on the Implementation of the New IAU Resolutions, Paris, France, April 18–19, 2002 (Capitaine N et al., ed.) IERS Technical Note No. 29
McCarthy DD, Luzum BJ (1995) Using GPS to determine Earth orientation. In: GPS Trends in Precise Terrestrial, Airborne, and Spaceborne Applications: 21st IUGG General Assembly, IAG Symposium No. 115, Boulder, USA, July 3–4, 1999, pp 52–58. Springer-Verlag, Berlin
McCarthy DD , Petit G (Ed) (2003) International Earth Rotation Service. IERS conventions (2003), IERS Technical Note No. 32
Meeus J (1992) Astronomische Algorithmen. Johann Ambrosius Barth
Melchior P (1978) The tides of the planet Earth. Pergamon Press
Miller KM (2000) A review of GLONASS. Hydrographic J 98:15–21
Mireault Y, Kouba J, Lahaye F (1995) IGS combination of precise GPS satellite ephemerides and clock. In: GPS Trends in Precise Terrestrial, Airborne, and Spaceborne Applications: 21st IUGG General Assembly, IAG Symposium No. 115, Boulder, USA, July 3–4, 1995, pp 14–23. Springer-Verlag, Berlin
Mitchell S, Jackson B, Cubbedge S (1996) Navigation solution accuracy from a spaceborne GPS receiver, GPS World, 7(6):42, 44, 46–48, 50
Montenbruck O (1989) Practical Ephemeris calculations. Springer-Verlag, Heidelberg
Montenbruck O (2003) Kinematic GPS positioning of LEO satellites using ionospheric-free single frequency measurements. Aerosp Sci Technol 7(5):396–405
Montenbruck O, Gill E (2000) Satellite Orbits: Models, Methods and Applications. Springer-Heidelberg
Montenbruck O, Gill E, Kroes R (2005) Rapid orbit determination of LEO satellites using IGS clock and ephemeris products. GPS Solutions 9(3):226–235
Montenbruck O, Kroes R (2003) In flight performance analysis of the CHAMP BlackJack GPS receiver. GPS Solutions 7(2):74–86
Montenbruck O, van Helleputte T, Kroes R, Gill E (2005) Reduced dynamic orbit determination using GPS code and carrier measurements. Aerosp Sci Technol 9(3):261–271
Moritz H (1980) Advanced physical geodesy. Herbert Wichmann Verlag, Karlsruhe
Mueller II (1964) Introduction to satellite geodesy. Frederick Ungar Publishing
Murakami M (1996) Precise determination of the GPS satellite orbits and its new applications: GPS orbit determination at the Geographical Survey Institute. J Geod Soc Jpn 42(1):1–14
Ou J (1995) On atmospheric effects on GPS surveying. In: GPS Trends in Precise Terrestrial, Airborne, and Spaceborne Applications: 21st IUGG General Assembly, IAG Symposium No. 115, Boulder, USA, July 3–4, 1995, pp 243–247. Springer-Verlag, Berlin
Ou JK, Wang ZJ (2004) An improved regularization method to resolve integer ambiguity in rapid positioning using single frequency GPS receivers. Chin Sci Bull 49(2):196–200
Parkinson BW, Spilker JJ (eds) (1996) Global Positioning System: Theory and applications (2 vols). Washington, D.C. Progress in Astronautics and Aeronautics, Vol. 163

Petit G (2002) Comparison of "Old" and "New" Concepts: Coordinate Times and Time Transformations, in Proceedings of the IERS Workshop on the Implementation of the New IAU Resolutions, Paris, France, April 18–19, 2002 (Capitaine N et al, ed), IERS Technical Note No. 29

Press WH, Teukolsky SA, Vetterling WT, Flannery BP (1992) Numerical recipes in C, 2nd edn. Cambridge University Press, New York

Rapp RH (1986) Global geopotential solutions. In: Sunkel H (ed.) Mathematical and numerical techniques in physical geodesy. Springer-Verlag, Heidelberg. Lecture Notes inEarth Sciences, Vol. 7

Reigber C (1997) Geowissenschaftlicher Kleinsatellit CHAMP. GPS-Anwendungen und Ergebnisse '96: Beiträge zum 41. DVW-Fortbildungsseminar vom 7. bis 8. November 1996 am Geo-Forschungs-zentrum Potsdam, pp 266–273

Reigber C, Schwintzer P, Luehr H (1996) CHAMP – a challenging mini-satellite payload for geoscientific research and application. Erste Geodaetische Woche, Stuttgart, Oktober 7–12 1996, 4p

Remondi B (1984) Using the Global Positioning System (GPS) phase observable for relative geodesy: Modelling, processing, and results. PhD Dissertation, Center for Space Research, University of Texas at Austin

Remondi BW (2004) Computing satellite velocity using the broadcast ephemeris. GPS Solutions 8(3):181–183

Rothacher M, Gurtner W, Schaer S (1995) Azimuth- and elevation-dependent phase center corrections for geodetic GPS antennas estimated from calibration campaigns. In: GPS Trends in Precise Terrestrial, Airborne, and Spaceborne Applications: 21st IUGG General Assembly, IAG Symposium No. 115, Boulder, USA, July 3–4, 1995, pp 333–338. Springer-Verlag, Berlin

Rothacher M, Mervart L (1996) Bernese GPS Software (Version 4.0). Astronomical Institute of University of Bern, Switzerland

Rothacher M, Schaer S (1995) GPS-Auswertetechniken. Schriftenreihe des Deutschen Vereins für Vermessungswesen, Bd. 18, Stúttgart, pp 107–121

Roßbach U (2006) Positioning and Navigation Using the Russian Satellite System GLONASS, Universität der Bundeswehr München, URN: de:bvb:707–648

Rummel R, Ilk KH (1995) Height datum connection – the ocean part. Allgemeine Vermessungsnachrichten 102(8/9):321–330

Rummel R, Van Gelderen M (1995) Meissl scheme: Spectral characteristics of physical geodesy. Manuscr Geodaet 20(5):379–385

Rush J (2000) Current issues in the use of the global positioning system aboard satellites. Acta Astronaut 47(2–9):377–387

Rutten JAC (2004) Book review: GPS – Theory, Algorithms and Applications, Xu G 2003, Eur J Navigation 2(1):94

Schaffrin B, Grafarend E (1986) Generating classes of equivalent linear models by nuisance parameter elimination. Manuscr Geodaet 11:262–271

Scheinert M (1996) Zur Bahndynamik niedrigfliegender Satelliten. Verlag der Bayerischen Akademie der Wissenschaften, DGK, Reihe C, Heft 435,

Schneider M (1988) Satellitengeodaesie. Wissenschaftsverlag, Mannheim

Schneider M, Cui CF (2005) Theoreme über Bewegungsintegrale und ihre Anwendung in Bahntheorien, Bayerischen Akad Wiss, Reihe A, Heft Nr. 121, 132pp, München

Schutz BE (2000) Numerical studies in the vicinity of GPS deep resonance. Adv Astronaut Sci 105(1):287–302

Schwintzer P, Kang Z, Reigber C (1995) GPS satellite-to-satellite tracking for TOPEX/Poseidon precise orbit determination and gravity field model improvement. J Geodyn 20(2):155–166

Seeber G (1989, 1993) Satelliten-Geodaesie. Walter de Gruyter

Seeber G (1996) Grundprinzipien zur Vermessung mit GPS. Vermessungsingenieur 47(2):53–64

Seeber G (2003) Satellite Geodesy: Foundations, Methods, and Applications. Walter de Gruyter, Berlin, 589pp

Shen YZ, Chen Y, Zheng DH (2006) A Quaternion-based geodetic datum transformation algorithm, J. Geodesy 80:233–239

Sigl R (1978) Geodätische Astronomie. Wichmann Verlag, Karlsruhe

Sigl R (1989) Einführung in die Potentialtheorie. Wichmann Verlag, Karlsruhe

Smith AJE, Hesper ET, Kuijper DC, Mets GJ, Visser PN, Ambrosius BAC, Wakker KF (1996) TOPEX/ Poseidon orbit error assessment. J Geodesy 70:546–553

Spilker JJ (1996) GPS navigation data. In: Parkinson BW, Spilker JJ (eds) Global Positioning System: Theory and applications, Vol. I, Chapter 4

Springer TA, Beutler G, Rothacher M (1999) Improving the orbit estimates of GPS satellites. J Geodesy 73:147–157

Stoew B, Elgered G (2004) Characterization of atmospheric parameters using a ground based GPS network in north Europe. J Meteor Soc Jpn 82(1B):587–596

Strang G, Borre K (1997) Linear algebra, geodesy, and GPS. Wellesley-Cambridge Press

Sun HP, Ducarme B, Dehant V (1995) Effect of the atmospheric pressure on surface displacements. J Geodesy 70:131–139

Tapley BD, Schutz BE, Eanes RJ, Ries JC, Watkins MM (1993) Lageos laser ranging contributions to geodynamics, geodesy, and orbital dynamics. In: Contributions of Space Geodesy to Geodynamics: Earth Dynamics. Geodyn Ser 24:147–174

Torge W (1989) Gravimetrie. Walter de Gruyter, Berlin

Torge W (1991) Geodesy. Walter de Gruyter, Berlin

Van Kamp PD (1967) Principles of Astronomy. W.H. Freemann and Company, San Francisco, CA/London

Visser PNAM, Van den IJssel J (2000) GPS-based precise orbit determination of the very low Earth-orbiting gravity mission GOCE. J Geodesy 74(7/8):590–602

Wagner JF (2005) GNSS/INS integration: still an attractive candidate for automatic landing systems? GPS Solutions 9(3):179–193

Wang G, Chen Z, Chen W, Xu G (1988) The principle of GPS precise positioning system. Surveying Press, Peking, 345pp. ISBN 7-5030-0141-0/P.58 (in Chinese)

Wang G, Wang H, Xu G (1995) The principle of the satellite altimetry. Science Press, Peking, 390pp. ISBN 7-03-004499-1/P.797 (in Chinese)

Wang LX, Fang ZD, Zhang MY, Lin GB, Gu LK, Zhong TD, Yang XA, She DP, Luo ZH, Xiao BQ, Chai H, Lin DX (1979) Mathematic handbook. Educational Press, Peking, ISBN 13012–0165

Weber R (1996) Monitoring Earth orientation variations at the Center for Orbit Determination in Europe (CODE). Oesterr Z Vermess Geoinf 84(3):269–275

Wells D, Lindlohr W, Schaffrin B, Grafarend E (1987) GPS Design: Undifferenced Carrier Beat Phase Observations and the Fundamental Differencing Theorem, University of New Brunswick, Fredericton, NB, Canada

Wenzel H-G (1985) Hochauflösende Kugelfunktionsmodelle für das Gravitationspotential der Erde. Wissenschaftliche Arbeiten der TU Hannover, Nr. 137, Hannover

Wicki F (1998) Robuste Schätzverfahren für die Parameterschätzung in geodätischen Netzen. Technische Hochschule Zürich, Switzerland

Xia Y, Michel GW, Reigber C, Klotz J, Kaufmann H (2003) Seismic unloading and loading in northern central Chile as observed by differential Synthetic Aperture Radar Interferometry (D-INSAR) and GPS. Int J Remote Sensing 24(22):4375–4391

Xu G (1992) Spectral analysis and geopotential determination (Spektralanalyse und Erdschwerefeldbestimmung). Press of the Bavarian Academy of Sciences, Munich, 100pp. Dissertation, DGK, Reihe C, Heft Nr. 397

Xu G (2003) GPS – Theory, Algorithms and Applications. Springer Heidelberg, 315pp

Xu G (2004) MFGsoft – Multi-Functional GPS/(Galileo) Software: Software User Manual, (Version of 2004), Scientific Technical Report STR04/17 of GeoForschungsZentrum (GFZ) Potsdam, ISSN 1610–0956, 70pp. www.gfz-potsdam.de/bib/pub/str0417/0417.pdf

Xu G (2007) GPS – Theory, Algorithms and Applications. 2ndedn, Springer Heidelberg, 350pp.

Xu G, Guo J, Yeh TK (2006a) Equivalence of the uncombined and combining GPS algorithms. Published as Sect. 6.7 in Xu G (2007) GPS – Theory, Algorithms and Applications, 2nd edn, Springer Heidelberg

Xu G, Schwintzer P, Reigber Ch (1998) KSGSoft (Kinematic/Static GPS Software): Software user manual (Version of 1998). Scientific Technical Report STR98/19 of GeoForschungsZentrum (GFZ) Potsdam

Xu G, Sun H, Shen YZ (2006b) On the parameterisation of the GPS observation models. Published as Sect. 9.1 in Xu G (2007) GPS – Theory, Algorithms and Applications, 2nd edn, Springer Heidelberg

Xu G, Yang YX, Zhang Q (2006c) Equivalence of the GPS data processing algorithms. Published as Sect. 9.2 in Xu G (2007) GPS – Theory, Algorithms and Applications, 2nd edn, Springer Heidelberg

Xu QF (1994) GPS navigation and precise positioning. Army Press, Peking, ISBN 7-5065-0855-9/P.4 (in Chinese)

Yang Y (1997) Robust Kalman filter for dynamic systems. Journal of Zhengzhou Institute of Surveying and Mapping 14:79–84

Yang Y (1999) Robust estimation of geodetic datum transformation. J Geodesy 73:268–274

Yang Y, Cui X (2006) Adaptively Robust Filter with Multi Adaptive Factors. J Surv Eng

Yang Y, Gao W (2005) Comparison of Adaptive Factors on Navigation Results. J Navigation 58:471–478.

Yang Y, Gao W (2006) Optimal Adaptive Kalman Filter with Applications in Navigation. J Geodesy

Yang Y, He H, Xu G (2001) Adaptively robust filtering for kinematic geodetic positioning. J Geodesy 75:109–116

Yang Y, Tang Y, Li Q and Zou Y (2006) Experiments of Adaptive Filters for Kinemetic GPS Positioning Applied in Road Information Updating in GIS. J Surv Eng

Yang YX, Xu TH, Song LJ (2005) Robust estimation of variance components with application in global positioning system network adjustment. J Surv Eng ASCE 131(4):107–112

Yang YX et al. (2005) Combined adjustment project of national astronomical geodetic networks and 2000' national GPS control network. Prog Nat Sci 15(5):435–441

Yeh TK, Chen CS (2006) Clarifying the relationship between the clock errors and positioning precision of GPS receiver. In: VI Hotline-Marussi Symposium of Theoretical and Computational Geodesy Wuhan.

Yeh TK, Chen CS, Lee CW (2004) Sensing of precipitable water vapor in atmosphere using GPS technology. Boll Geod Sci Affini 63(4):251–258

Yoon JC, Lee BS, Choi KH (2000) Spacecraft orbit determination using GPS navigation solutions. Aerosp Sci Technol 4(3):215–221

Yuan YB, Ou JK (1999) The effects of instrumental bias in GPS observations on determining ionospheric delays and the methods of its calibration. Acta Geod Cartogr Sinica 28(2)

Yunck TP, Melbourne WG (1995) Spaceborne GPS for earth science. In: GPS Trends in Precise Terrestrial, Airborne, and Spaceborne Applications: 21st IUGG General Assembly, IAG Symposium No. 115, Boulder, USA, July 3–4, 1995, pp 113–122. Springer-Verlag, Berlin

Zhang Q, Moore P, Hanley J, Martin S (2006) Auto-BAHN: Software for Near Real-time GPS Orbit and Clock Computation. Adv Space Res 39(10):1531–1538

Zhou J (1985) On the Jie factor. Acta Geodaetica et Geophysica 5 (in Chinese)

Zhou J (1989) Classical theory of errors and robust estimation. Acta Geod Cartogr Sinica 18:115–120

Zhou J, Huang Y, Yang Y, Ou J (1997) Robust least squares method. Publishing House of Huazhong University of Science and Technology, Wuhan, China

Zhu J (1996) Robustness and the robust estimate. J Geodesy 70(9):586–590

Zhu S (1997) GPS-Bahnfehler und ihre Auswirkung auf die Positionierung. GPS-Anwendungen und Ergebnisse '96: Beiträge zum 41. DVW-Fortbildungsseminar vom 7. bis 8. November 1996 am GeoForschungszentrum Potsdam, pp 219–226

Zhu S, Reigber C, Koenig R (2004) Integrated adjustment of CHAMP, GRACE, and GPS data. J Geod 78(1–2):103–108

Zhu SY, Reigber Ch, Massmann FH (1996) The German PAF for ERS, ERS standards used at D-PAF. D-PAF/GFZ ERS-D-STD-31101

Index

MIX
Papier | Fördert gute Waldnutzung
FSC® C083411

Zeitfracht Medien GmbH
Ferdinand-Jühlke-Straße 7
99095 Erfurt, Deutschland
produktsicherheit@kolibri360.de